Disarmament and Decommissioning in the Nuclear Domain

Radioactive Risk Set

coordinated by
Jean-Claude Amiard

Volume 4

Disarmament and Decommissioning in the Nuclear Domain

Jean-Claude Amiard

ISTE

WILEY

First published 2021 in Great Britain and the United States by ISTE Ltd and John Wiley & Sons, Inc.

ISTE Ltd
27-37 St George's Road
London SW19 4EU
UK

www.iste.co.uk

John Wiley & Sons, Inc.
111 River Street
Hoboken, NJ 07030
USA

www.wiley.com

Library of Congress Control Number: 2021938220

British Library Cataloguing-in-Publication Data
A CIP record for this book is available from the British Library
ISBN 978-1-78630-721-7

MIX
Paper from responsible sources
FSC® C013604
FSC
www.fsc.org

Contents

Preface

Nuclear energy is used in two main areas: military and civil. In the military field, nine countries are currently manufacturing nuclear weapons of mass destruction, which is a serious threat to humanity and all living beings. Therefore, the UN and many NGOs are working to limit nuclear proliferation, as well as to limit the number of nuclear weapons. The ultimate goal, probably utopian, is a complete eradication of these weapons. Like all human creations, the life span of nuclear weapons is limited. The military must therefore deconstruct obsolete weapons and, unfortunately, manufacture new ones that are more powerful, more precise and therefore more destructive than previous ones. The military must also urgently clean up old nuclear facilities that have become obsolete, as well as highly contaminated.

In the civil field, nuclear energy is mainly used to produce electricity and in many medical applications for diagnostic or therapeutic purposes, with radiotracers (scintigraphy) and ionizing radiation destroying cancer cells (brachytherapy and radiotherapy). Here again, various nuclear installations, nuclear power reactors, plants, particle accelerators, etc. have a limited lifespan and must be decommissioned and dismantled. However, these operations are long, costly and present both environmental and health risks. They must therefore be carefully programmed, seriously supervised and controlled. The ideal is for the site to return to its initial state (often called "greenfield").

The aim of this book, the fourth in the "radioactive risk" series, is to take stock of the state-of-the-art nuclear disarmament and dismantling. As with the previous works in this series, we have tried to use multiple information sources and verify them, in order to get as close as possible to the truth.

The first part of this book is devoted to the international efforts to limit nuclear proliferation, with various international treaties signed and some ratified, as well as agreements to limit the number of nuclear weapons held by the nuclear powers.

The second part deals with the decommissioning and dismantling of various types of nuclear facilities, including power, prototype and research reactors, as well as plants upstream and downstream of the nuclear fuel cycle or nuclear research centers.

May 2021

Acknowledgments

Jean-Claude Zerbib, a former CEA Senior Expert and radiation protection engineer, had the difficult task of reading, annotating and criticizing the original manuscript. He also provided me with the precious documents needed to complete the abundant literature that fed the writing of this book.

Professor Philip Rainbow (former Keeper of Zoology, Natural History Museum, London, UK) has done the same for the English version. I would like to thank them both for their time and effort.

I would also like to thank the members of the ANCCLI Scientific Council who helped me to understand certain subjects. The same goes for all of the members of the *Groupe Radioécologie Nord Cotentin* (GRNC), a pluralist group, for the remarkable work they carried out together courteously.

1

Nuclear Non-Proliferation

1.1. Introduction

Nuclear weapons pose an immense risk to human beings, as well as to all living beings and to the environment. For this reason, many individuals and organizations have been working to limit the number of states capable of manufacturing this type of weapon.

UN efforts to prevent, or at least limit, nuclear proliferation are significant and ongoing. For example, the UN Security Council Resolution 1540 (April 28, 2004) echoes the terms of the UN Security Council's statement from January 31, 1992 by stating that "the proliferation of nuclear, chemical and biological weapons, as well as their means of delivery constitutes a threat to international peace and security". However, in this resolution, it is the risks associated with the acquisition of such weapons by "non-state actors" that are specifically targeted. Some analysts consider that this resolution may justify the use of force to combat proliferation [IRS 19b].

Unfortunately, and in spite of its efforts for 35 years, it is obvious that the UN's policy of nuclear disarmament has been a failure. The traditional vision of security that seems to be shared by the actors involved in the process of arms control and nuclear disarmament limits their capacity and willingness to abolish this type of armament [LAZ 06]. Hence, France and the United States are exerting strong pressure on the French-speaking states and on Japan to not sign the treaty on the Prohibition of Nuclear Weapons (TPNW).

In an interview with the IRIS (*Institut de Relations Internationales et Stratégiques*, the French Institute for International and Strategic Affairs), Jean-Marie Collin [COL 18b], expert and spokesman for ICAN FRANCE, an organization awarded the Nobel Peace Prize in 2017, noted that nuclear weapons constitute a permanent and very real danger for the international community; otherwise, no treaty would have been signed at the UN, and the nuclear deterrent would be null and void. The International Committee of the Red Cross (ICRC) has stressed that no international plan or actor will be able to adequately address the needs of victims in the event of the use of nuclear weapons.

Nuclear weapons operate on the balance of terror. However, the problem with balance is that the risk of falling is always great. This "balance" is undesirable to an overwhelming majority of the world, because their security is directly at stake [COL 18b].

1.2. The first countries to acquire the atomic bomb

Five states have carried out their first atomic tests: the United States on July 16, 1945, the Soviet Union on August 29, 1949, the United Kingdom on October 3, 1952, France on February 13, 1960 and China on October 16, 1964.

In order to prevent the number of states possessing nuclear weapons from becoming too large, the United Nations has been concerned with nuclear disarmament from the outset, and four UN structures are particularly responsible for this task: the Security Council, the Disarmament Commission, the Conference on Disarmament and the Office for Disarmament Affairs [GIL 18].

The Security Council has primary responsibility under the UN Charter for the maintenance of international peace and security. It is composed of five permanent members (China, France, the Russian Federation, the United Kingdom and the United States) and 10 non-permanent members elected by the General Assembly for a period of two years. The general organization of the UN is shown in Figure 1.1.

The United Nations Disarmament Commission (UNDC) was established in 1952, within the framework of the Security Council, by General Assembly Resolution 502 (VI), to deal with disarmament issues. However, it only met occasionally after 1959. In 1978, the first special session of the General Assembly devoted to disarmament established a new Disarmament Commission as a subsidiary organ of the Assembly, composed of all UN member states of the Organization. It was established as a deliberative body to consider various disarmament issues and monitor the implementation of the relevant decisions and recommendations adopted at the special session. It sits in New York for three weeks each year (usually in early spring). It considers a limited number of agenda items on a three-year cycle and submits an annual report to the General Assembly [GIL 18].

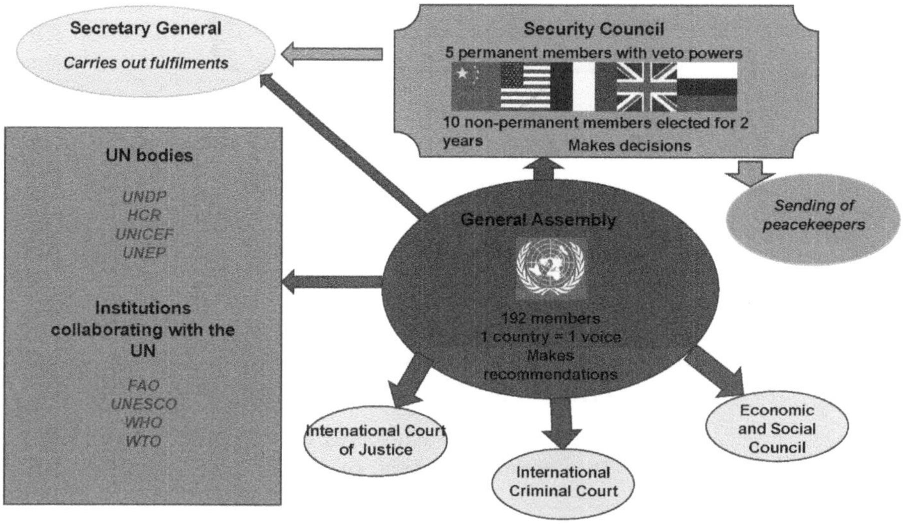

Figure 1.1. *The United Nations organization. For a color version of this figure, see www.iste.co.uk/amiard/disarmament.zip*

The Conference on Disarmament is the only multilateral forum for negotiating disarmament agreements. It is composed of 65 permanent members who hold three sessions a year in Geneva (usually January to March, May to June and August to September). It operates on the basis of consensus to ensure full support for the agreements reached. It has completed negotiations on the Comprehensive Nuclear-Test-Ban Treaty, but has not yet entered into force [GIL 18].

The Office for Disarmament Affairs, established in 1982, works to promote disarmament and non-proliferation and to strengthen disarmament regimes. Part of its work focuses on nuclear weapons [GIL 18] (Figure 1.2).

The first treaty limiting the use of weapons (particularly nuclear weapons), the Antarctic Treaty, signed in Washington, entered into force on June 23, 1961. This treaty stipulates that the Antarctic is an area exclusively reserved for peaceful purposes and prohibits all activities of a military nature, such as military maneuvers and the testing of any kind of weapon.

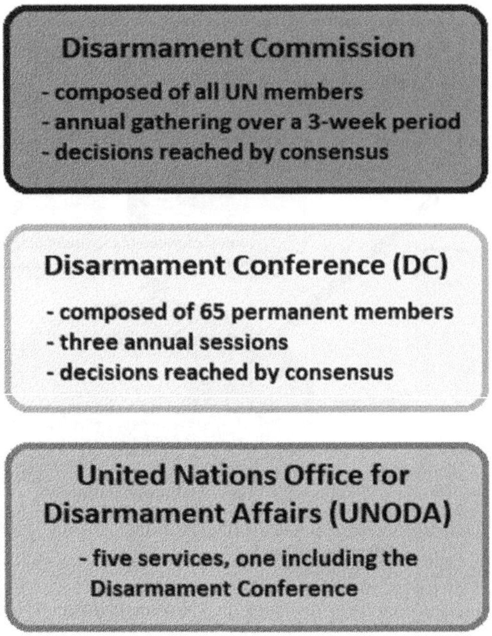

Figure 1.2. *The various UN bodies working on non-proliferation and disarmament. UNODA: United Nations Office for Disarmament Affairs. For a color version of this figure, see www.iste.co.uk/amiard/disarmament.zip*

1.3. The NPT

The International Treaty on the Non-Proliferation of Nuclear Weapons (NPT) was developed within the UN by the Eighteen-Power Committee on Disarmament. Concrete negotiations lasted three years, from 1965 to 1968. The text was mainly drawn up by the United States and the Soviet Union. On July 1, 1968, the treaty was opened for signature in Moscow, Washington and London. It entered into force on March 5, 1970 after ratification by the three depositary Powers and 40 other signatory states [AIE 70].

For the treaty itself, the parties had to decide on its extension 25 years after its entry into force. Despite the reluctance of some non-aligned countries (Mexico and Indonesia) and Arab countries (because of Israel's absence from the treaty), the NPT was extended indefinitely on May 11, 1995 [IRS 19a].

The NPT is now close to universality, with 191 states as members. Only four states have not joined: India, Israel, Pakistan and Southern Sudan. However, in January 2003, North Korea initiated a procedure to withdraw from the treaty [FRA 19b].

1.3.1. *The functioning of the Treaty*

In the operation of the NPT, states are divided into two categories. The first category consists of the nuclear-weapon states (NWS). These are the five official nuclear powers (United States, Russia, United Kingdom, France, China), which according to Article IX "*had manufactured and detonated a nuclear explosive device prior to 1 January 1967*" and are entitled to possess such weapons. The second category includes all non-nuclear weapon states (NNWS), which make up the rest of the NPT membership. For the purpose of control or safeguarding, this last category is divided into two sub-categories depending on whether or not the states have access to nuclear facilities (see section 4.1.1).

For many years, this NPT has not been functioning well. This is mainly due to non-compliance with two clauses. The first clause is the non-implementation of a process to create a zone free of nuclear weapons and weapons of mass destruction in the Middle East. The second clause is related to the hypocrisy of the five weapon states, who do not respect the spirit and letter of the NPT. Indeed, while no one can deny that the nuclear arsenals of these states, with the exception of China, have indeed decreased since the Cold War, it must also be noted that all of these states have undertaken processes of modernizing nuclear weapons and delivery systems.

This treaty seems to have reached its maximum effectiveness with regard to the two pillars of non-proliferation and the peaceful use of nuclear energy, but it seems limited in effectively forcing the NWS to implement the third pillar, that of a real disarmament policy [COL 15].

1.3.2. *Revision of the NPT*

This treaty is subject to a Review Conference (RevCom) every five years: the next one was due to take place in 2020, interspersed with three so-called Preparatory Conferences (PrepCom). The RevCom is an opportunity to take stock of the progress made in the past and the future of the treaty in a roadmap, which is called the Final Document [COL 15].

1.3.3. *Successes of the NPT*

The main success of the treaty is South Africa's renunciation of nuclear weapons. The first, and only South African nuclear research reactor is the SAFARI-1 (South African Fundamental Atomic Research Installation 1), built between 1961 and 1965 on the Pelindaba site. This site also housed a reprocessing plant (Plant Y). The country also had a secret military program for the South African Defence Force, abandoned by the ratification of the Treaty on the Non-Proliferation of Nuclear Weapons before the African National Congress came to power. The six aerial atomic bombs with a power of 15 to 20 kt, built between 1982 and 1989, and a seventh, which was under construction, were dismantled before 1993 [TOU 13].

Historically, the AEC (Atomic Energy Corporation) had two sites – Valindaba, for the production of highly enriched uranium (HEU) and Pelindaba, for research. The ARMSCOR (Armaments Corporation of South Africa) depended on the Ministry of Defense with two sites – Kentron and Naschem. The production of highly enriched uranium by the South African enrichment plant Y in 1994 was 993 kg (enriched on average to 68% in ^{235}U), of which 515 kg were enriched to 85% in ^{235}U [ALB 16].

South Africa's only nuclear test likely occurred in 1979. On the night of September 22, 1979, a double flash was spotted off the Prince Edward Islands by the American satellite Vela 6911. The double flash of light is the signature of an atmospheric atomic explosion. The initial flash burst at the start of the explosion and corresponded to a fireball. Then, the glow was obscured for a fraction of a second by the shock wave, which made the atmosphere opaque. Then, when the cloud dissipated, the second flash appeared. Since South Africa was the only power in the region with the means to enrich uranium, it was immediately suspected, but foreign support, possibly Israeli, was also envisaged [BAT 18].

South Africa's disarmament program was complicated by its division State into two countries (South Africa and Namibia) [DZI 17].

The program to dismantle South Africa's nuclear program began with the dismantling of the six nuclear weapon devices and the seventh, which was in production. The highly enriched uranium was returned to the AEC to be kept under secure surveillance. The second stage was the complete decontamination of the Armscor installations (1990–1991) and the return of highly contaminated equipment, such as melting furnaces, to the AEC. Advena/Circle's facilities were converted into commercial civil or military activities, but with conventional weapons: the destruction and disposal of non-nuclear components, as well as engineering design and manufacturing information provisions. South African nuclear officials informed the de Klerk government of an appropriate timetable for

the state's accession to the NPT, the signing of a comprehensive safeguards agreement with the IAEA (International Atomic Energy Agency), and the submission of a complete national inventory of nuclear materials and facilities, as required by the safeguards agreement. Finally, the closure of plant Y was to take place as soon as possible [ALB 16].

Other countries have developed nuclear programs in the past: Brazil, Algeria, Libya, Ukraine, Sweden, Belarus and Kazakhstan, but have abandoned nuclear deterrence.

1.3.4. *Failures of the NPT*

Despite the NPT's entry into force, four countries have since acquired atomic weapons (Israel, India, Pakistan and North Korea).

Israel has always used deliberately ambiguous language with respect to nuclear weapons and has refused to sign the NPT. The only public revelation of Israel's nuclear capabilities was made by a former Dimona nuclear power plant technician, Mordechai Vanunu, in the Sunday Times on October 5, 1986 [COH 05].

Historically, Israel benefited from French aid, between 1956 and 1961, for scientific advice for the construction of the Dimona reactor, as well as the supply of many nuclear materials by the British between 1950 and 1960. Between 1961 and 1969, American experts visited Israeli installations, but with advanced notice of their arrival. This made it possible to camouflage the real activities of the installations. In 1969, President Richard Nixon agreed to allow Israel to possess nuclear weapons, on the condition that it would not conduct nuclear tests or display its nuclear warheads [COH 98, BIA 09, COH 10, TÉN 10].

No nuclear tests have been officially recognized. However, it is possible that an underground test took place in 1963 in the Negev desert, as well as a 3 kt atmospheric test in 1979 in South African waters, in collaboration with South Africa (double Vela flash). Indeed, on September 22, 1979, the American satellite Vela 6911 detected an optical signal characteristic of an atmospheric nuclear explosion over the southern Indian or Atlantic Ocean. Weiss believes that the flash was an Israeli nuclear test assisted by South Africa [WEI 11]. In a new analysis of radionuclide and hydroacoustic data, it was concluded that the double flash resulted from a low-yield nuclear weapon test [DEG 18]. Among the evidence put forward by these authors, the concentration of iodine-131 found in the thyroids of some Australian sheep could suggest that they had grazed grass in the path of a potential radioactive fallout plume from this test.

In September 1996, when the Comprehensive Test Ban Treaty was signed, it was reasonable to believe that this would become a general rule of behavior accepted by all states. However, in May 1998, India and Pakistan defied the international community by conducting nuclear tests. Faced with the new risk of a regional nuclear conflict involving Kashmir in particular, the Western powers needed to integrate these two states into the nuclear non-proliferation regime. There were only two ways to do this: sanctions or dialogue. However, India and Pakistan were no longer ordinary states that could be dictated to. They had become nuclear powers with all that this implies in terms of prestige and political weight on the international scene [CAP 98].

India developed a nuclear program as soon as it gained independence in 1948, with the adoption of the Atomic Energy Act, aimed at developing nuclear energy for peaceful purposes. Today, India is not a signatory to the Nuclear Non-Proliferation Treaty (NPT), but has more than 20 nuclear reactors, mainly Canadian-designed CANDU pressurized heavy water reactors.

India's Atomic Energy Commission detonated its first underground nuclear weapon, the Smiling Buddha, at a depth of 100 meters in Pokharan on May 18, 1974. This led to the cessation of all forms of international cooperation in favor of India's civilian program and exclusion from the Nuclear Non-Proliferation Treaty (NPT). Subsequent nuclear tests took place underground on May 11 and 13, 1998. India conducted three tests on May 11, testing an A-bomb and an H-bomb, and two more tests on May 13, using small charges [CAP 98]. In 2018, the Indian Nuclear Deterrent Air Force had 48 nuclear warhead-launching aircrafts, 60 ground-based ballistic missiles, and about 18 cruise missiles for a total of 130–140 nuclear bombs [KRI 18a].

Pakistan (Islamic Republic) began its nuclear program in January 1972. This program was originally a response to India's development of nuclear weapons. The military nuclear program began in 1987. Pakistan's first underground nuclear test was conducted on May 28, 1998, codename Chagai-I (Chagai district, Balochistan province). This test was a series of five low-power nuclear tests. The second test (Chagai-II), took place in the Kharan desert two days later. The power was equivalent to 12 kt of TNT. Of the six tests, one used plutonium and the others used enriched uranium. Pakistan, which is not a signatory of the Treaty on the Non-Proliferation of Nuclear Weapons (NPT), thus became the seventh country in the world to develop and test nuclear weapons and is also the only Muslim country to possess nuclear weapons.

Its capacities have not been made public, leading to estimates. Thus, in 2013, one source indicated that the Pakistani armed forces possessed 300 to 350 nuclear warheads (Le Monde, June 3, 2013), compared to around 50 in 2007, while SIPRI estimated Pakistan's arsenal to be 250 to 300 warheads in 2017, much higher than India [GAI 17, SIP 18].

The Indian and Pakistani nuclear programs can be explained by the exacerbated rivalry between the two states. In addition, the Pakistani nuclear physicist A.Q. Khan, has established strong nuclear proliferation ties with Iran, North Korea and Libya because of his personal motivations, religious orientation and anti-Western worldview [ABB 18]. The Pakistani people were overcome with fever dreams for the future and were very proud after the first atomic tests. However, in India, the population was more reserved about the Indian tests and regretted the practice of secrecy in this field [ABR 09].

The third failure of the NPT is that of North Korea, which after signing the treaty withdrew from it. In October 2006, North Korea announced that it had detonated its first nuclear bomb. Other tests were carried out in 2009, 2013, 2016 and 2017. The best estimate of the stocks of highly enriched uranium (HEU) in the Yongbyon nuclear complex at the end of 2015 were between 75 and 320 kg [BRA 18].

In order to stop this North Korean nuclear program, in the early 1990s, successive American presidents tried to negotiate with the North Korean government. American decision-makers were left asking themselves about the usefulness of such a negotiation, the timeframe and the objectives of these diplomatic talks with Pyongyang.

The United States began four major rounds of formal nuclear and missile negotiations with North Korea. These were the bilateral framework (1994–2002), bilateral missile negotiations (1996–2000), multilateral Six-Party Talks (2003–2009) and bilateral Leap Day Deal negotiations (2012). In general, North Korea agreed, ostensibly, to end, or even deactivate its nuclear and/or missile programs in return for economic and diplomatic incentives. But despite some progress in some negotiations, North Korea has continued to advance its nuclear and missile programs [NIK 17].

In April 2018, Pyongyang announced the end of its nuclear tests, and in June 2018, an official meeting between Donald Trump and Kim Jong-un took place in Singapore. On June 30, 2019, another historic meeting between the US and North Korean presidents took place on the border between North Korea and South Korea. However, no specific agreement has been reached.

1.3.5. *Future nuclear-weapon states*

Alongside these four states that have acquired nuclear weapons, nuclear proliferation has been observed in several countries. This is developed in section 4.1.3.

The fact that Israel is the only state in the Middle East that possesses nuclear weapons, and that this state enjoys privileges at the UN and the IAEA, since no control is exercised over its production of fissile material, nor over these nuclear weapons, inevitably leads to strong tension in this region. Also, there are many candidates to acquire nuclear weapons. The first candidate is Iran, which has begun to enrich uranium and has a missile program. Among the other postulants is Saudi Arabia, which is anxious to thwart Iran's influence in the Near East. This state is supported by the United States in this desire. This is obvious and by issuing authorizations for the transfer of American nuclear technology to Saudi Arabia, without any guarantee of non-proliferation, the American president at the time (Donald Trump) was provoking Iran [TIS 19].

1.4. Other nuclear non-proliferation treaties

Three international treaties and several regional treaties restrict the use of nuclear weapons.

1.4.1. *The CTBT Treaty*

In January 1946, the first General Assembly resolution specifically called for the elimination of all nuclear and other similar weapons "capable of being used for purposes of mass destruction", including biological and chemical weapons.

Among the solutions available to the UN, a nuclear test ban can be a very effective way to prevent horizontal proliferation (to states) and vertical proliferation (improvement of nuclear weapons). This solution is not a panacea but could help to limit the number of nuclear weapons.

According to Duarte, a test ban has three advantages. First, it makes it possible to prevent certain forms of "vertical proliferation", in particular, the development of new generations of nuclear weapons or the modernization of existing arsenals. Second, it helps to begin the process of dismantling the complex institutional infrastructures that have been put in place to develop and maintain these arsenals. Third, a ban also contributes to the process of delegitimizing nuclear weapons themselves, in the eyes of the world, and thus makes it more difficult for policy makers to support the need for nuclear deterrence [DUA 19].

Prior to the Comprehensive Nuclear-Test-Ban Treaty (CTBT), five other anti-nuclear test treaties had been signed. The first Partial Test-Ban Treaty, signed on August 5, 1963, prohibited nuclear explosions in the atmosphere, under the seas and in outer space. Its main purpose was to stop radioactive substances from contaminating the environment. The second treaty signed on January 27, 1967 prohibited the placement of nuclear weapons or weapons of mass destruction in space. It entered into force on October 10, 1967. The third treaty, signed on February 11, 1971, prohibited the introduction and placement of nuclear weapons or other weapons of mass destruction on the seabed and ocean floor and in the subsoil. It entered into force on May 18, 1972. The fourth treaty concerned the limitation of underground testing, and was signed on July 3, 1974. The fifth treaty, signed on May 28, 1976, prohibited underground explosions for peaceful purposes. Finally, on September 24, 1996, the CTBT Treaty was signed [COL 19b, IRS 19b].

The Comprehensive Nuclear-Test-Ban Treaty (CTBT), along with the Non-Proliferation Treaty (NPT), is now one of the pillars of the global nuclear non-proliferation structure.

Negotiations took place between 1994 and 1996 and the treaty was opened for signature on September 24, 1996. As of May 1, 2019, only 184 states have signed the CTBT and it has only been ratified by 168 of them. In order for the treaty to enter into force, the negotiators defined, among other things, a list of nuclear-capable countries whose ratification was indispensable (regardless of the number of states that had ratified the treaty elsewhere). These were the 44 so-called Annex 2 states. As of May 1, 2019, eight Annex 2 states had not yet ratified the treaty: Iran, the United States, China, Israel, Egypt, India, Pakistan and North Korea, the last three of which were also expected to sign [RÉP 19].

1.4.2. *The TPNW*

Faced with the non-compliance of one of the pillars of the NPT, effective nuclear disarmament, the non-nuclear weapon states decided to start drafting a new treaty at the level of the UN General Assembly. This made it possible to bypass the undertakings of the five nuclear powers that were hindering this disarmament.

The International Treaty on the Prohibition of Nuclear Weapons (TPNW) was the first multilateral treaty on nuclear disarmament, open for signature in 1996, and is supported by a very large majority of UN member states. It was written by diplomats whose ambition was to both respect the mosaic of treaties that codify nuclear weapons (NPT, CTBT and the future FMCT), in accordance with the safeguarding system currently being formulated by the IAEA, and to move the international community towards a world without nuclear weapons [COL 18a].

In the last decade or so, nuclear weapons have been reclassified, thanks to the humanitarian initiative, as weapons with catastrophic humanitarian consequences. This has led to a change in consciousness. The illegality of nuclear weapons, as well as the illegality of the policy of threatening to use them, appears shocking to the states that possess them because they have not yet made their “moral revolution”. The rest of the world is now aware that the presence of these weapons is no longer acceptable and that they are generating growing insecurity, particularly because of their potential humanitarian, environmental and climatic impact at the global level [COL 18a].

On July 7, 2017, an International Nuclear Weapons Ban Treaty was adopted at the UN. Approved by 122 states and ratified by 50 nations on October 24, 2020, it entered into force on January 22, 2021. Historically, this text will prohibit signatory countries from manufacturing, stockpiling or using nuclear weapons. Will the threat to use these weapons, i.e. the policy of deterrence, also become prohibited [SOR 18]?

It should be noted that many states are revising their policy on nuclear weapons. This is the case for the Vatican, for example. For many years, this state accepted nuclear deterrence as a “tragic illusion” (Paul VI, June 7, 1978), as “a step on the road to progressive disarmament, can still be considered morally acceptable” (John Paul II, June 7, 1982) or as “the prospect [of deterrence based on balance], apart from being fatal, is utterly fallacious” (Benedict XVI, January 1, 2006). Pope Francis, on the contrary, has a much stronger stance on rejecting the use of atomic weapons and balanced nuclear deterrence.

States that possess nuclear weapons, or support a policy of deterrence, can no longer afford to question their understanding of nuclear disarmament. The Nuclear Weapons Treaty offers them this opportunity. This norm of international law exists and is likely to enter into force around the 10th NPT Review Conference (scheduled for May 2020, although it has been postponed). The (non-exhaustive) proposals set out in Annex 1 offer some ideas on how to overcome some of the misunderstandings and demonstrate a common will among states to move forward together and create a world free of nuclear weapons [COL 18a].

1.4.3. *The Fissile Material Cut-Off Treaty (FMCT)*

The treaty banning the production of fissile material or other nuclear explosive devices (FMCT or “ Cut-off ”) is an old draft of the UN General Assembly. It was on December 16, 1993 that this Assembly adopted a resolution (A/RES/48/75L) that called for the opening of negotiations to elaborate such a treaty. The reasoning was simple: stopping the production of fissile materials (military plutonium or ^{239}Pu and highly enriched uranium) constitutes one of the essential locks of nuclear proliferation.

For more than 20 years, the Disarmament Conference (DC) has been working on the text of this treaty. Several drafts have been submitted by France and Canada. However, the impasse in which the DC still finds itself today is due to the abuse of the consensus rule, which allows any state to block the progress of work for any reason whatsoever.

The main causes of divergence on this treaty concern the delimitation of the scope for negotiating this prohibition. Egypt and Pakistan want the stockpiles of nuclear-weapon states (NWS) to be taken into account, which the five strongly reject. China wants to increase its stockpile in the face of India's expanding stockpile. The same is true for Pakistan with regard to India. France has always been favorable and actively supports the process of creating this treaty, but its stockpile (30.6 tons of highly enriched uranium and 6 tons of plutonium 239) would allow it to produce several thousand nuclear weapons [COL 15]. For nearly 15 years, France has been strongly committed to the negotiation of a treaty banning the production of fissile material for nuclear weapons, known as the Cut-Off Treaty. This is one of the French proposals for disarmament, announced by the President of France in his speech in Istres on February 19, 2015 [FRA 15, FRA 19a].

In spite of this lack of progress, there is hope, however, with the shifting of discussions to the level of the General Assembly. In October 2012, the UN General Assembly voted in favor of a Canadian-sponsored resolution that established a Group of Governmental Experts (GGE) on TIPMF. In 2016, the UN General Assembly adopted another resolution introduced by Canada, and co-sponsored by Germany and the Netherlands, establishing a 25-member High-Level Preparatory Group of Experts (Preparatory Group) to build on the work of the GGE and make recommendations on the substantive elements of a treaty being prepared for negotiation [UN 17a, UN 17b].

At its last meeting in June 2018, the Preparatory Group successfully negotiated a consensus report recommending the elements of a future treaty. The UN Secretary General transmitted this report to the DC in the hope that they will begin formal negotiations on a treaty that will help end the global nuclear arms race [GOU 18].

1.4.4. *Regional disarmament treaties*

Nuclear-Weapon-Free Zone Treaties represent a regional approach to strengthening international norms for nuclear non-proliferation and disarmament, coupled with the promotion of international peace and security.

In 2016, there were five nuclear-weapon-free zones established under different treaties: (a) Latin America and the Caribbean, under the Treaty of Tlatelolco; (b) the South Pacific, under the Treaty of Rarotonga; (c) Southeast Asia, under the Treaty of Bangkok; (d) Africa, under the Treaty of Pelindaba; and (e) Central Asia, under the Treaty of Semipalatinsk.

The Treaty of Tlatelolco was the first international instrument establishing a nuclear-weapon-free zone in a densely populated part of the world.

In addition to these regional zones, the UN recognized Mongolia's self-proclaimed nuclear-weapon-free status by adopting the biennial General Assembly resolution entitled "Mongolia's international security and nuclear-weapon-free status".

In 2018, the treaties of Tlateloko and Ranotonga were signed and ratified by the five nuclear powers. The treaties of Pelindaba and Central Asia have been signed by all five powers, but the United States (alone) has not ratified them. The Treaty of Bangkok is in the process of being signed. The text of these treaties and the list of countries that have acceded to them are available at: http://disarmament.un.org/treaties/ [ONU 18]. The geographic scope of these regional treaties is shown in Figure 1.3.

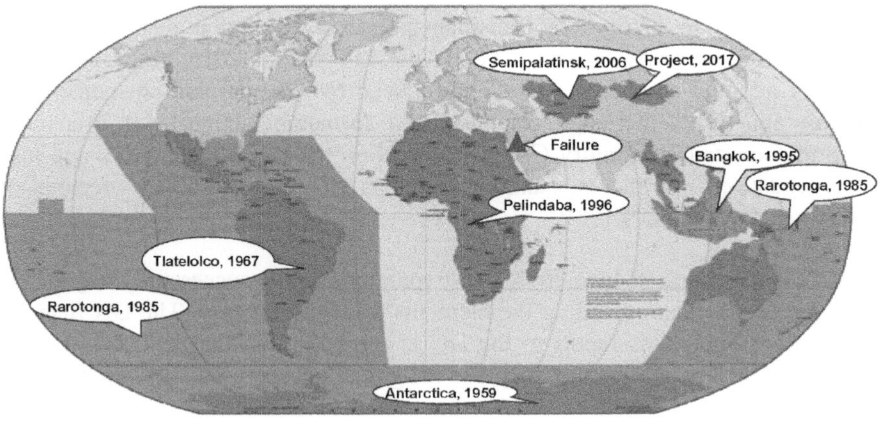

Figure 1.3. *The various regional treaties banning nuclear weapons. For a color version of this figure, see www.iste.co.uk/amiard/disarmament.zip*

1.5. Disarmament controls

The best treaty in the world is worthless if it is not respected and if all parties cannot verify its application. All confidence lies in the effective verification of nuclear disarmament. The main role of a verification system is the same for all arms control treaties, namely, to deter cheating [GOL 07]. Depending on the treaties, verification can be carried out at the international level or bilaterally. International controls mainly concern the NPT and CTBT.

1.5.1. *Principle and practice of disarmament controls*

Disarmament controls are not easy because a large number of factors interfere with them. However, the technologies that serve these controls are increasingly effective.

1.5.1.1. *General*

The verification of nuclear disarmament is influenced by a multitude of factors. First, international politics and inter-state relations play an important role in the design of verification systems, determining which verification techniques can be used and how and when they can be applied. This leads to a compromise between different positions that may not necessarily represent the best technical solution to the various verification problems.

Even within a particular state, different national interest entities may disagree on interpretations and practical aspects of compliance. Some will view violations, or even ambiguities, as clear evidence of deliberate fraud, while others will interpret them as minor and unavoidable oversights or unintended consequences.

In the construction of nuclear verification systems, there is a necessary tension between the levels of intrusiveness and the need to prevent the disclosure of militarily or commercially sensitive information. In the early days of verification, this balance was oriented towards the protection of information, either through the use of national technical means alone, or through a limited number of on-site inspections. Subsequently, verification systems became more intrusive and on-site inspections have become an essential element of nuclear verification regimes, even at sensitive military facilities.

In the context of bilateral disarmament efforts undertaken by the United States and Russia, the potential benefit of fraud through misrepresentation is derisory since both sides have retained large arsenals of nuclear weapons. On the contrary, withholding weapons in small numbers during dismantlement could be a significant strategic advantage for states with limited nuclear arsenals. The case of South Africa's total nuclear disarmament is a hopeful one [BOW 18].

The main issues that need to be considered when verifying the dismantlement of nuclear warheads, are counting and avoiding theft. To do this, it is necessary to ensure the chain of traceability and monitor the access portal to the storage site. Several technical solutions have been implemented (managed access, use of labels and seals, information barrier). During exercises between the United Kingdom and Norway (UKNI, United Kingdom–Norway Initiative), the unexpected result was that human factors could also have a decisive influence on the verification process and its results. This is likely due to the fact that the verification process is characterized by a constant tension between trust, perceived as an evidence-based judgment by informed actors, and confidence, perceived as the perception of intentions in a situation where evidence is lacking [BOW 18].

Some experimental simulations on disarmament verification have been carried out. The design of this simulation was the key element of the preparatory work that formed the basis of the research and required a considerable investment of time and resources. However, this approach offers a flexible way to recreate real-life challenges and problems in an artificial environment that can be, in part, controlled and targeted.

In these simulations, everything from the attitude of the host team, to the manner and speed with which information was provided, was a potential indicator of intentions and was incorporated into the research participants' interpretation scheme. This intellectually demanding cognitive process encompassed the more objective and scientific, evidence-based approach that all participants aspired to experience. Since trust is indeed a crucial influence on the results of verification, it would be simplistic, even dangerous, to only view verification in terms of a purely technical nature, based only on scientific evidence [BOW 18].

1.5.1.2. *Technologies at the service of disarmament*

Future nuclear arms reduction efforts will require technologies that can verify that the warheads to be dismantled are genuine, without revealing any sensitive warhead design information to international inspectors.

How can a claim be proved about an extremely sensitive object, a nuclear weapon, for example, without revealing information about it? This paradox has been challenging the control of nuclear weapons for more than five decades. A mechanism in the form of an interactive evidence system has been proposed, that can validate the structure and composition of an object, such as a nuclear warhead, with arbitrary precision, without revealing its structure and composition [KEM 16].

Confirming the authenticity of nuclear warheads and their components is at the heart of the challenge of verifying future reductions in nuclear arsenals. An overview of the development of verification systems and the challenges and opportunities highlighted for future research in this area are provided in [YAN 15].

For a long time, low-resolution gamma spectrometry (sodium iodide crystal probes) has been used in verification systems. Thus, the programs CIVET (Controlled Intrusiveness Verification Technology) from 1998 to 1991, TRIS (Trusted Radiation Identification System) from 1999 to 2001 and NG-TRIS (Next Generation Trusted Radiation Identification System) since 2007 have been successively used. Since 1984, the FNMIS (Fieldable, Nuclear Material Identification System) has used neutron activation. The parameters most frequently sought are the presence of plutonium, its various isotopes, its mass or its age, and sometimes the presence of uranium 235, its enrichment or its mass. Recently, some authors have proposed the use of intensities of nuclear resonance fluorescence (NRF) close to 2.2 MeV to distinguish between real computer objects and hoax objects, with great confidence and realistic measurement times [VAV 18].

A concept that exploits isotope-specific nuclear resonance phenomena in order to authenticate the fissile components of a warhead by comparing them to a previously authenticated model, has been submitted [HEC 18].

Most of the plutonium in the world resides inside the rods of spent fuel from nuclear reactors. Only the United States, France and Japan have isolated and stored plutonium. This high-activity radioactive waste is generally stored long term in large, heavily shielded casks. The measurements of the diffusion angles of cosmic muons that pass through a storage cask can be used to determine whether the radioactive waste in the casks is in the form of intact spent fuel assemblies, or whether the plutonium has been recovered by reprocessing, without opening the cask [DUR 18].

1.5.2. *NPT controls*

The NPT has been the subject of an impressive number of agreements between states and the IAEA to improve disarmament controls. While there have been some successes, there have unfortunately been some resounding failures.

1.5.2.1. *NPT guarantees*

The UN has entrusted the IAEA with the task of monitoring the application of the NPT, known as safeguards. States are divided into three groups, the five nuclear-weapon states (NWS), the non-nuclear weapon states (NNWS) and the three states (India, Israel and Pakistan) that have not signed the NPT Treaty.

There are four main types of guarantee agreements: General Guarantee Agreements (GGAs), Limited Guarantee Agreements, Voluntary Submission Agreements (VSAs) and emergency measures.

The Generalized Guarantee Agreements concern all the NNWS and the text is the result of compromises, in order to encourage the NNWS to adhere to this new control regime. In order to provide an appropriate and identical legal framework for all signatories, a model Comprehensive Safeguards Agreement, INFCIRC/153 (corrected 1983), was developed in 1971 [AIE 75] by a committee of experts from IAEA Member States. This GGA is currently in progress with 174 NNWS. However, because of Iraq's clandestine military nuclear program, in 1997 it was decided to reinforce the verification standard by an Additional Protocol (AP, i.e. INFCIRC/540) [IEA 97]. This AP – which remains optional – ensures faster access to suspect sites for IAEA inspectors and better control of nuclear materials. As of December 2015, the IAEA had concluded five VSAs and APs, 173 GGAs, 121 APs and three Elemental Agreements. As of December 21, 2018, 134 states have adhered to the AP standard and 14 are in the process of implementation [COL 18].

Limited Warranty Agreements are reserved for certain states (India, Israel, Pakistan), which have not signed the NPT. They have concluded more limited safeguards agreements with the Agency (of the INFCIRC/66 type), which only apply to nuclear material, equipment, non-nuclear material and facilities specified by the agreement (thus designated by the state).

Voluntary Submission Agreements (VSAs) are reserved for NWS that are not required to enter into a safeguards agreement with the Agency. All five states have done so. France, for example, has signed a safeguards agreement modeled on the Comprehensive Safeguards Agreements concluded between the IAEA and the NNWS. Under this agreement (INFCIRC/290), signed on July 27, 1978, France subjects the nuclear material it designates in selected facilities, or parts of facilities, to IAEA safeguards. To take account of the safeguards exercised by the EURATOM, the safeguards agreement concluded by France is trilateral in nature, with EURATOM being a party to the agreement.

The emergency measures decreed by the IAEA concerned three points (declaring a new nuclear installation within 180 days; declaring international transfers and the production of uranium and thorium concentrates; effectively resorting to special inspections) [IRS 19a].

IAEA controls under the NPT Treaty are called "purpose" controls. Their objectives are to ensure that nuclear material in the agency's custody is not used for non-peaceful purposes. Since the Additional Protocols' entry into force, the agency's controls are no longer limited to nuclear material, but extend to the provision of information concerning the nuclear industry.

1.5.2.2. *The various types of collaboration between states to verify disarmament*

Article VI of the NPT states, among other things, that each party to the NPT must undertake effective measures in the field of arms control and disarmament of nuclear weapons.

As a result of the 1996–2002 joint project of the Russian Federation, the United States and the IAEA, called the "Trilateral Initiative", the partners proposed an innovative method for IAEA verification of fissile material in support of nuclear disarmament to the IAEA [SHE 15].

In early 2007, representatives from the UK Ministry of Defence, the Atomic Weapons Establishment, several Norwegian laboratories and the non-governmental organization VERTIC (Verification Research, Training and Information Centre) began work on the best techniques for the verification and control of the effective disarmament of nuclear weapons. This was the first time that a nuclear-weapon state, a non-nuclear weapon state and an independent NGO collaborated in this area of research. This work was presented at the 2009 NPT PrepCom [UKN 10b].

The results of three years of collaboration between experts from Norway and the United Kingdom, to study the technical and procedural problems associated with a possible nuclear disarmament verification regime, were presented in a report. The report encourages the international community as a whole to contribute to the ultimate goal of an effective verification regime for the dismantlement of nuclear weapons [UKN 10a].

In 2010 and 2011, the United States and the United Kingdom undertook a warhead monitored dismantlement (WMD) exercise to investigate the methods for two nuclear-weapon states to verify compliance with an arms reduction treaty. The two fictitious countries negotiated the use of NDA (non-destructive analysis) and CoC (chain-of-custody) measuring equipment [HAU 16].

1.5.2.3. *Practical implementation of NPT safeguards*

By having early detection of the theft of nuclear materials or the misuse of nuclear technology, the IAEA can alert the world to a proliferation risk [AMA 16]. Through safeguards, the IAEA provides credible assurance that states only use nuclear material and technology for peaceful purposes in accordance with their international obligations.

In 2015, the IAEA conducted 2,118 field inspections, collected 967 environmental and nuclear material samples, observed 1,416 surveillance cameras and 407 satellite images through a network of 20 certified laboratories and 883 staff. This enabled it to provide safeguards at 1,286 nuclear facilities.

The implementation of the safeguards is carried out in four steps. The first stage is the collection and evaluation of information about a state. The second step is the development of a monitoring methodology specific to that state. The third step is the planning of safeguards to be carried out in both the field and the Agency's headquarters. The final step is the establishment of a safeguards conclusion for each state [AIE 16a].

A safeguards inspector must perform three types of verification at each nuclear facility. These are to verify the integrity of the reactor seals, to control the correct storage of new and spent fuel and the correct operation of surveillance cameras [HEN 16].

Technicians mainly have five handheld instruments at their disposal to carry out their inspection. An HM-5 (detector of multiple alpha, beta, gamma and neutron radiation) can detect the presence of radioactive materials and the enrichment of uranium. The second is a force sensor that operates in two ranges (up to 5,000 kg and up to 20,000 kg). The third instrument is an electrically cooled germanium gamma radiation detection system; the fourth is an ultrasonic thickness gauge. Finally, there is an attribute tester (gamma ray detector) of irradiated elements for underwater fuel, or a digital device for the observation of the Chernenkov effect, which remains dry and protected from water, associated with a camera that detects ultraviolet radiation emitted by spent fuel assemblies.

In the case of additional protocols, inspections require a special toolkit. The toolkit contains all kinds of instruments to collect information and verify reports: a camera, laser meter, GPS, voice recorder, flashlight, universal radiation meter, such as the HM-5 and an environmental sampling kit [FOU 16a].

The IAEA monitors more than one million sets of encrypted data that are collected by more than 1,400 surveillance cameras and 400 radiation sensors around the world. The containment of materials and equipment is ensured by more than 23,000 seals installed at nuclear facilities.

The IAEA's Next-Generation Surveillance System (NGSS) uses cameras that are protected by casings designed to reveal any attempt at tampering, and are equipped with long-life batteries that ensure their autonomy in the event of an absence of an external power supply for extended periods of time. Approximately 160 systems, with a total of 700 detectors and sensors, are installed in more than 40 countries.

Different types of automatic systems are used, depending on whether it is an enrichment facility, a reactor, a spent fuel storage site or a reprocessing plant.

The IAEA seals are the best known and most widely used safeguarding equipment. Despite their simplicity, these anti-fraud devices are highly effective in preventing unauthorized access to IAEA safeguarding materials and equipment. The IAEA uses several types of seals depending on the case. Some are designed to be installed underwater, others can withstand extreme conditions. The most common seals are metal seals, as well as Cobra seals, which incorporate a multi-core fiber optic cable whose ends are protected by the seal. There are also electro-optical seal systems and laser systems for containment verification [FOU 16b].

The IAEA's scientific safeguards depend on the laboratories located in Seibersdorf, Austria. These laboratories include two modern facilities. The Nuclear Materials Laboratory (NML) handles the sampling of nuclear materials and the Environmental Samples Laboratory receives and analyzes all environmental smear samples for traces of nuclear materials [JAW 16].

The analysis of environmental samples taken by smear tests allows the determination of all elements present in a facility, which have been used in the past or recently. This technique was first used in Iraq in 1990 in a nuclear facility, the Osirak or Tammouz reactor, which had been bombed on June 7, 1981.

Once collected, the samples are anonymously labeled and analyzed at the IAEA laboratories near Vienna, as well as in the IAEA's network of 19 accredited laboratories in many IAEA Member States (Australia, Brazil, France, Germany, Japan, Republic of Korea, Russia, the United Kingdom and the United States). The shipment of samples is also accompanied by "blind samples" for quality control purposes [DIX 16].

Satellite images are used to establish site maps. These provide valuable and accurate information about buildings and structures. For example, satellite imagery helps the IAEA monitor developments in the nuclear program of the Democratic People's Republic of Korea (DPRK), since the IAEA is not in a position to conduct physical verification activities in that country. In particular, the IAEA is monitoring activities at the Yongbyon site [QUE 16].

1.5.2.4. *Minor concealments under the NPT*

Some states voluntarily conceal a number of activities in the nuclear field that could lead to military applications. This was first observed in Iraq in 1981. This NPT signatory country was developing a nuclear program with a military vocation. It was demonstrated that a country bound to the IAEA by a comprehensive safeguard agreement could carry out activities contrary to the treaty without the IAEA's knowledge.

By declaring its renunciation of weapons of mass destruction in December 2003, Libya revealed the existence of a nuclear program for military purposes, unsuspected by the IAEA. While taking advantage, officially, of its adherence to the NPT and of a safeguards agreement with the IAEA, Libya had been able to import certain nuclear materials, carry out conversion activities, set up a centrifuge park intended to enrich uranium and obtain documentation relating to the design and manufacture of nuclear weapons [CHE 10]. This action was the work of A.Q. Khan, who sold the plans for a bomb and first-generation centrifuges to the Libyan President Gaddafi.

Finally, very serious doubts were expressed about Syria's nuclear activities. The American intelligence services transmitted information to the IAEA, that the installation destroyed by the Israeli air force in September 2007 at the Dair Alzour site, was a North Korean-designed nuclear reactor in the process of being completed. Similar to the one at Yongbyon, it was intended for the manufacture of weapon-grade plutonium that could be used to make a nuclear weapon. In 2009, two years later, the IAEA had still not obtained elements that would allow the Syrian authorities to deny the military vocation of this site [CHE 10].

1.5.2.5. *The case of North Korea*

Similarly, North Korea, which joined the NPT in 1985, has delayed concluding a comprehensive safeguards agreement with the IAEA for several years. This agreement was concluded a few days after the January 20, 1992 declaration on the denuclearization of the Korean Peninsula, in which both Koreas "*agree not to test, manufacture, produce, receive, possess, stockpile, deploy or use nuclear weapons*". At that time, suspicions about the conduct of military nuclear activities were already widely shared in the international community. In fact, the North Korean authorities immediately obstructed the verifications that the IAEA intended to carry out as part of the implementation of the safeguards agreement, going so far as to announce, in March 1993, their withdrawal from the NPT, a decision they would go back on three months later.

In early 1994, the Americans discovered that the North Koreans had begun to produce plutonium. President Bill Clinton wanted to make a pre-emptive strike on the reactor, but President Jimmy Carter went to Pyongyang and negotiated with Kim Il Sung, and then, following the latter's death in July 1994, with his son Kim Jong-il. At the end of the agreement, concluded in October 1994 with the United States, North Korea froze its military-scale activities, notably the Yongbyon reactor and reprocessing facilities intended to produce plutonium, in exchange for the construction of two light-water reactors by an international consortium (KEDO) and the delivery, until their entry into service, of 500,000 tons of oil per year. The United States did not respect this agreement. Oil deliveries carried on despite North Korean complaints in 1997, which explained the North Korean position [MOR 18]. Moreover, this agreement did not improve the IAEA's ability to control the non-diversion of nuclear materials. Also, no guarantee could be obtained on the existence of plutonium coming from the Yongbyon reactor, and whether there was a sufficient quantity to build one or more nuclear devices, before the suspension of its operation. On the contrary, from 1994 to 2002, North Korea pursued, without the knowledge of the international community, a program of uranium enrichment. It recognized this fact in 2002, which led to its withdrawal from the NPT in 2003 [CHE 10].

1.5.2.6. *The case of Iran*

In Iran, it was opposition movements that revealed, in the summer of 2002, the existence of the uranium conversion facility in Isfahan, the uranium enrichment facility in Natanz, the heavy water plant and the reactor under construction in Arak. The IAEA was not informed of their existence and could therefore exercise no control over them. As the IAEA later stressed:

> [...] many aspects of Iran's nuclear fuel cycle activities and experiments, particularly in the fields of uranium enrichment and conversion, as well as research concerning plutonium, have not been declared to the Agency, contrary to Iran's obligations under its Safeguards Agreement, since 'Iran's policy of concealment' has led to numerous breaches of its obligation to comply with this agreement.

The exclusively civilian purpose of these activities has still not been attested [CHE 10].

Yet, Iran signed the General Guarantee Agreement (GGA) in 1974 and the Additional Protocol (AP) in 2003. Following these revelations, negotiations began in 2003 and 12 years later, led to the signing of a milestone agreement at the beginning of April 2015 between Iran and the "5+1" (Germany, China, United States, France, United Kingdom, Russia) [CON 15]. This did not prevent Iran negotiations from reducing its cooperation with the IAEA and resuming its activities related to uranium enrichment in February 2006 during these negotiations.

The Vienna Agreement on the establishment of control over Iran's nuclear program in return for the lifting of economic sanctions was signed on July 14, 2015. This agreement presupposed transparency and verifications because "An agreement that cannot be verified is an agreement that is not applied" (Laurent Fabius, July 14, 2015). On January 16, 2016, the IAEA Director General, Yukiya Amano, announced that Iran had completed the preparatory measures for the implementation of the Joint Comprehensive Plan of Action (JCPoA). The JCPoA was concluded in July 2015 between Iran and China, France, Germany, Russia, the United Kingdom, the United States and the European Union, a group known as the "E3/EU+3".

Iran's main commitments under the JCPoA are to refrain from using advanced centrifuges for 8–10 years, to conduct limited centrifuge research and to comply with a limited number of centrifuges at the Natanz Fuel Enrichment Facility. For a period of 15 years, Iran has agreed not to enrich uranium to more than 3.67%, to have its stock of this low-enriched uranium limited to 300 kg, not to practice uranium enrichment in the fuel enrichment facility in Fordou, not to build new enrichment facilities or new heavy water reactors and to export heavy water in excess [AIE 16b].

In addition, Iran has also agreed to provisionally apply the Additional Protocol, a legal agreement giving the IAEA broader access to information and locations, not limited to declared nuclear facilities and materials only. Under the JCPoA, Iran is required to implement voluntary nuclear-related commitments, known as "transparency measures", which include facilitating access by IAEA inspectors to uranium mines and processing facilities, and allowing the ongoing monitoring of centrifuge manufacturing and storage facilities. The number of IAEA staff working on verification and control in Iran doubled in 2016 from what it was before 2014. This leads to a significant increase in additional expenditure (9.2 million euros).

The JCPoA has significantly reduced the proliferation risks associated with Iran. For example, Iran now only has one enrichment site instead of two, and the number of centrifuges available to Iran has been reduced from 21,000 to 5,060. The stock of uranium available to Iran is now limited, and the time needed for Iran to acquire the fissile material necessary for a weapon has been considerably increased. With the restrictions imposed by the Vienna Agreement, it would take Iran at least a year, if it decided to embark on the arms race, to accumulate the material necessary for a bomb by the uranium route. Obtaining plutonium is made even more difficult with the absence of a means of conversion in Arak and of a reprocessing plant.

The US withdrawal from the Iranian nuclear deal on May 8, 2018, led to the reinstatement of US sanctions against Iran, which the US subsequently decided to strengthen. Despite this American withdrawal from the agreement, which France expressed its deep regret about, the agreement remains in place. It involves Iran and the so-called E3/EU+2 states, i.e. France, the United Kingdom, Germany, Russia and China [FRA 19c]. The potential remains for the US to re-enter the agreement with its change of President in January 2021.

As of July 1, 2019, the quantity of uranium enriched to 3.67% ^{235}U in Iran was 205.0 kg. Taking into account the JCPoA and the decisions of the Joint Commission, Iran's total stock of enriched uranium thus exceeded 300 kg of UF_6 enriched to 3.67% ^{235}U (or the equivalent in other chemical forms), since the quantity of 300 kg of UF_6 corresponds to 202.8 kg of uranium [CEA 18; IEA 19a; IEA 19b].

On July 7, 2019, Iran announced that it would resume the enrichment of its uranium at the rate of 4.5% (prohibited by the 2015 agreement), but which remained lower than the necessary rate for the manufacture of a bomb. This quickly became a reality. Every day the stock of low-enriched uranium increases by 5 kg and would have been about 551 kg in November 2019, according to the IAEA. Iran resumed enrichment at Fordo and restarted 1,000 centrifuges in addition to the 5,044 authorized, which represents a qualitative leap. Iran assures that it will not be released from its obligations, in 60-day phases, only to induce the other signatories of the July 2015 agreement to take steps to circumvent US sanctions. The measures announced by Tehran were reversible, but they have become irreversible, and in two months, Iran could decide to move to an enrichment rate of 20%. Beyond that, it is very easy to reach an enrichment rate of more than 90%. In this case, the July 2015 agreement would be broken, with an automatic return of UN Security Council sanctions (snap-back) and European economic sanctions would be added to those of the United States.

Recently, France, the United Kingdom and Germany criticized Iran in a letter addressed to UN Secretary General Antonio Guterres for developing "*nuclear-capable ballistic missiles*". Indeed, an Iranian long-range Shahab-3 missile was tested in April 2019 [MAL 19].

1.6. Actions of NGOs

The group of NGOs working on disarmament is difficult to quantify. Initiatives abound, statutes, funding and doctrinal orientation vary. They are mostly American and European, among democratic states, and on the contrary, there are very few in countries with authoritarian regimes. The size of NGOs varies greatly, from two-member NGOs to coalition NGOs, such as Abolition 2000, which brings together 2030 organizations from 93 countries. Some have an impressive number of activists, such as the International Physicians for the Prevention of Nuclear War

(IPPNW). Organizations can be very hierarchical, like Greenpeace, or very loose, like the Parliamentarians for Global Action [MAR 02b].

The British example for NGOs is significant. 173 British disarmament NGOs are grouped together in the Abolition 2000 network. Many have a local or regional dimension, some with a religious or professional notation, and many use the words "nuclear" and "peace" in their titles. The most active British NGOs in the field of disarmament are the Acronym Institute, created in 1995, BASIC (British American Security Information Council), ISIS (International Security Information Service), VERTIC, PPNN (Programme for Promoting Nuclear Non-proliferation), Pugwash, born of the Cold War and awarded the Nobel Peace Prize in 1995, CND (Campaign for Nuclear Disarmament) and Labour Action for Peace [MAR 02b].

1.6.1. *The main actions of NGOs for disarmament*

Historically, after the 1948 People's Congress for Peace in Wroclaw, the World Movement of Peace Supporters was formed and encouraged the creation of national committees. In France, the "*Combattants de la Liberté*", created by Charles Tillon at the end of 1947, became the "*Conseil national français du movement de la paix*" in 1951, commonly called the "*Mouvement de la paix*".

The *Mouvement de la paix* became the first NGO concerned with disarmament. Its first major action was when the World Peace Council, meeting in Stockholm, issued the Stockholm Appeal [AMI 19a] on March 19, 1950.

Thereafter, a multitude of NGOs was created and had common and coordinated actions on a cyclical basis. The first UN Special Session on Disarmament (SSD1) in 1978 was an opportunity for NGOs to organize their relations and cooperation. For example, with the creation of the International Mobilization for Survivals (ImfS) in October 1977, a coalition of national and international peace and disarmament NGOs. This coalition sent messages to the UN and above all organized an International NGO Conference on Disarmament in the spring of 1978 in Geneva, before the SSD1. This Conference brought together 500 representatives of 85 transnational NGOs and 200 national NGOs from 46 countries. At the SSD1, about 800 people from 236 NGOs were present, plus 500 Japanese delegates from 200 Japanese NGOs. However, this initiative was a failure because the positions of the states did not change at all. The roles of NGOs were therefore limited to raising awareness and educating the public.

The second UN session on disarmament in 1982 (SSD2) was the occasion for an anti-nuclear campaign, attracting one million people to Central Park, New York. The American NWFC (Nuclear Weapons Freeze Campaign) mainly mobilized on

US-Soviet nuclear weapons negotiations and testing, while European organizations (the British Campaign for Nuclear Disarmament, the Dutch Interchurch Peace Council and the German Greens) were interested in intermediate-range nuclear forces (Pershing 2 and SS-20).

The next action opposed the National Missile Defense (NMD) and mobilized several hundred NGOs. NMD had two drawbacks in the eyes of NGOs: it promoted nuclear proliferation and it was very expensive. However, anti-NMD NGOs were confronted by pro-NMD NGOs, such as High Frontier, or the Centre for Policy Studies [MAR 02b].

While non-governmental organizations have become, in recent years, the driving force behind disarmament processes, it is largely as a result of the paralysis that has hit the United Nations, particularly the Geneva-based Conference on Disarmament, which "constitutes the single multilateral disarmament negotiating forum" [BOU 14].

The Luxembourg International Forum on Nuclear Disaster Reduction, held in early June 2019 in Rome, brought together experts from around the world to consolidate efforts in nuclear safety. The leaders of five major international organizations committed to the control of nuclear weapons took part in the discussions: the Nuclear Threat Initiative, the Pugwash Movement, the Global Zero Movement, the Russian Council on International Affairs and the James Martin Center for Nonproliferation Studies (United States) [SKO 19].

Unlike biological and chemical weapons, the atomic bomb is not yet the subject of a treaty prohibiting its development, production, acquisition, financing, stockpiling and transfer. On March 27, 2017, and again on June 15 of the same year, United Nations conferences were held "for the purpose of negotiating a legally binding instrument to prohibit nuclear weapons with a view to their complete elimination". However, the chairs of the five nuclear powers, the United States, Russia, the United Kingdom, France and China, remained empty. This is the reason for the international campaign for the abolition of nuclear weapons that ICAN has begun. The Arms Observatory, the International Network of Parliamentarians for Nuclear Non-Proliferation and Disarmament and Greenpeace France are members of the International Campaign to Abolish Nuclear Weapons (ICAN) [BOU 17]. In 2019, the ICAN campaign, launched in 2007, brought together more than 540 non-governmental organization partners, including about 50 in France, in 103 countries. The ICAN-France collective was created in the spring of 2009, taking over from the French campaign for nuclear disarmament, which already included many associations (see http://icanfrance.org/).

1.6.2. *NGOs and the Nobel Peace Prize*

The Nobel Prize in Physics has been awarded to more than 30 nuclear physicists and the Nobel Prize in Chemistry has been awarded to about nine nuclear chemists, as well as a Nobel Prize in Physiology, for their scientific work in the nuclear field [AMI 13].

The idea that the use of nuclear weapons could lead to catastrophic consequences remains a basis for their deterrent effect. The horror experienced by the victims of the Hiroshima and Nagasaki bombings was tangible proof of this. The Norwegian jury for the Nobel Peace Prize has also honored many personalities and organizations.

The first laureates to be rewarded were American chemist and physicist Linus Carl Pauling (in 1962), for his commitment against nuclear weapons testing; the Japanese Prime Minister Eisaku Sato (in 1974), for his renunciation of the nuclear option for Japan; the physicist Andrei Sakharov (in 1975), notably for his commitment to disarmament; the diplomats Alva Myrdal and Alfonso Garcia Robles (in 1982), for their "central role" in disarmament negotiations at the UN.

In 1985, the International Physicians for the Prevention of Nuclear War (IPPNW) was recognized for "services to humanity" in disseminating information about the catastrophic consequences of nuclear war, and in 1995, the NGO Pugwash and Jozef Rotblat were recognized for "their efforts in reducing the role of nuclear weapons in international politics and, in the long term, to eliminate them". Still on the theme of abolition, the Nobel Committee provided a powerful sounding board for President Obama's Prague speech in 2009, in which he shared his vision of a "world without nuclear weapons" [DEC 17].

In 2005, the Nobel Peace Prize was awarded to the International Atomic Energy Agency (IAEA) and to Mr. El Baradei, its director at the time, "for their efforts in preventing the use of nuclear energy for military purposes".

Finally, in 2017, the anti-nuclear NGO ICAN, which used the theme of the consequences of a nuclear war, saw its action rewarded with the Nobel Peace Prize, a few months after the adoption of the Treaty on the Prohibition of Nuclear Weapons (NPT). However, the concrete benefits of this treaty for international peace and security are much less obvious than the risks it presents of blocking any progress in nuclear disarmament [DEC 17].

In addition, the "Don't bank on the bomb" project reflects the NGOs' search for other ways to delegitimize nuclear weapons. This initiative, implemented by the Netherlands-based NGO PAX, establishes a ranking of companies and investments in the field of nuclear weapons in an annual report, denouncing the financial institutions ("Hall of shame") investing in the 27 companies identified as being involved in the production and maintenance of nuclear weapons [DEC 17].

1.7. The military denuclearization of a state

Several countries have voluntarily stopped their military nuclear programs. Of these, South Africa was the most advanced state in its program as tests had been conducted. However, its nuclear weapons were never really operational because they were too large to be delivered. However, the first bombs, including those from Hiroshima and Nagasaki, were designed for air delivery. They were not vectorizable, but they became so.

1.7.1. *South Africa: the example of the complete denuclearization of a country*

The denuclearization of South Africa is a unique case, since it is the only state that possesses atomic weapons that voluntarily renounces them. Moreover, during this denuclearization, South Africa split into two new states: South Africa and Namibia.

In 1979, South Africa's nuclear weapons manufacturing facilities were transferred from the Pelindaba Nuclear Research Centre to the South African Weapons Corporation (Armscor), which developed the Kentron Circle facility. This facility was built in 1980 and is located 20 km west of Pretoria. It was later renamed Advena.

The organization of South Africa's militarized nuclear program is described in Figure 1.4.

At Pelindaba, pistol-type mini nuclear weapons have been developed. Armscor installed a production line there in 1981 and produces at least one nuclear device each year with an output of 10 to 18 kilotons.

Advena's central laboratories have built platforms for launching intercontinental ballistic missiles (ICBMs). Work to produce warheads with advanced designs has also been developed. On the same site, in collaboration with Israel, a missile with a range of 2,000 km was designed and tested. It was close to the ICBM Jericho II. The construction of Advena was completed at the same time as the end of South Africa's nuclear program, before the 1994 election of Nelson Mandela [GLO 11].

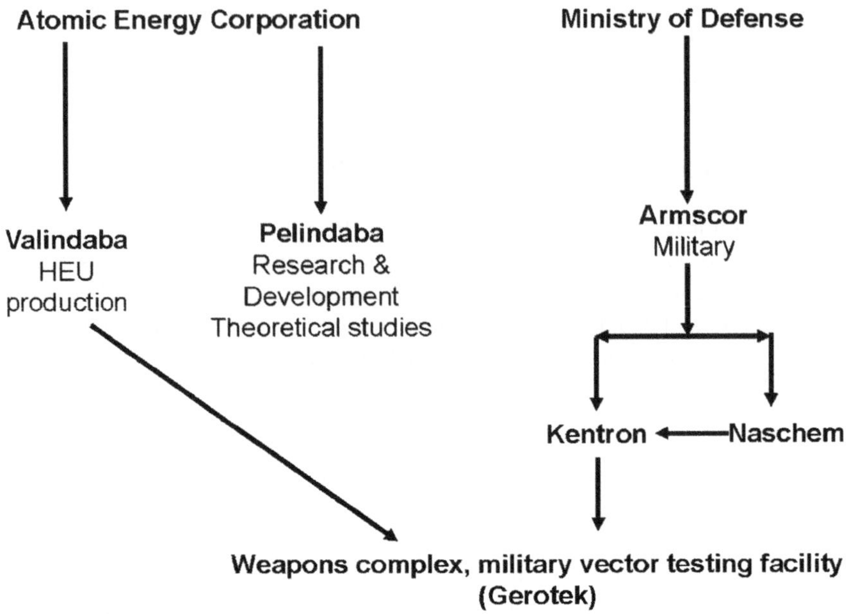

Figure 1.4. *Organizational structure of the South African nuclear program according to an IAEA member in August 1992 (source: [ALB 16])*

The end of South Africa's nuclear program should be seen in the context of major political developments in the region. First, in December 1988, South Africa, Angola and Cuba signed a tripartite agreement for the gradual withdrawal of Cuban troops from Angola. In April 1989, Namibia gained its independence. In September 1989, F. W. de Klerk was elected president. He immediately took steps to bring about fundamental political reforms aimed at ending apartheid and creating a democratic South Africa. Within a short time, the nuclear weapons program had become a liability. In November 1989, the government decided to stop the production of nuclear weapons. On February 26, 1990, de Klerk issued written instructions to end the nuclear weapons program and dismantle all existing weapons. The nuclear materials were to be melted down and returned to the Atomic Energy Corporation (AEC) in preparation for South Africa's accession to the NPT.

The government also decided to not admit the existence of the nuclear weapons program before joining the NPT. As a result, the decommissioning project, like the nuclear weapons project, was classified as top secret. Dismantling began in July 1990. On September 6, 1991, all of the highly enriched uranium (HEU) had been removed from the weapons, melted down, and returned to the AEC for storage. Shortly after the last of the material was sent to the AEC, the Circle building was completely decontaminated, and the equipment that had been used to remelt and sink the HEU was sent to the AEC. The Advena/Circle facility was converted to a non-nuclear commercial operation. De Klerk's revelation that South Africa had developed a nuclear arsenal was finally made in 1993 [ALB 94].

At the time of the dismantling of its military nuclear program, South Africa possessed six atomic weapons, with a seventh in the process of assembly. In 1991, South Africa's enrichment plant Y had produced 993 kg of highly enriched uranium, including 677 kg of the isotope 235 [ALB 16]. The conversion rate of 68.2% was close to the threshold of 90% necessary to make a weapon. The South Africans were thus close to the goal.

The split of South Africa had various consequences on the demobilization of troops and their disarmament [DZI 17].

Namibia is a successful model of an international approach to disarmament and demobilization (DD). Namibia has been assisted by two UN structures, the United Nations Transitional Assistance Group (UNTAG) and the United Nations High Commissioner for Refugees (UNHCR), which have effectively carried out their military and repatriation tasks in Namibia's successful transition to independence. This was facilitated by the fact that Namibia's liberation movement comprised a single armed formation, the People's Liberation Army of Namibia (PLAN). The official law enforcement force, SWATF (South West Africa Territorial Force) and the unofficial Koevoet (the counter-insurgency arm of the South West African Police Force) were eventually disarmed. The government decided to implement the aptly named Peace Project. This plan focused on the placement of ex-combatants in activities that were paid for, mainly in the civil service [DZI 17].

In South Africa, the unstable political and security contexts that followed the splitting of Namibia paralyzed disarmament efforts, contributing to the proliferation of illicit small arms and light weapons, some of which were subsequently used as instruments of armed crime [DZI 17].

1.7.2. Other states that have renounced nuclear weapons

Switzerland committed itself very early on, as early as 1945, to the acquisition of nuclear weapons. This process of nuclearization was halted in 1968, when it signed the NPT, and was then definitively abandoned in 1988.

Swedish research on nuclear weapons began in 1945, conducted by the Swedish National Defence Research Institute (FOA, *Forsvarets Forskningsanstalt*). In 1955,

Sweden was well advanced in its research on atomic weapons. In 1958, the Swedish Parliament had to choose between two programs, L (Loading Program) and S (Protection Program), depending on whether or not it was producing nuclear weapons. In 1968, Sweden was preparing to test an atomic bomb but, for various economic and diplomatic reasons, the government decided to dismantle its military nuclear program to join the negotiation process that would lead to the signing of the Non-Proliferation Treaty [JON 01, JON 10].

Two South American countries, Brazil and Argentina, had secretly developed military nuclear programs [GOL 06]. Changes in political regimes, with the fall of the military and the takeover by civilians in both countries, led to a renunciation of the bomb in the 1980s. This led to the creation of a regional bi-national safeguards agency in 1991, known as the Brazilian-Argentine Agency for Accounting and Control of Nuclear Materials (ABACC) [GOL 18].

In 1991, at the time of the dissolution of the USSR, Belarus, Kazakhstan and Ukraine possessed nuclear weapons on their territory. These three countries became independent and returned them to Russia. Thus, an agreement was signed in 1991 (Budapest Memorandum) between Russia and Ukraine. Under this agreement, Ukraine received sovereignty and territorial safeguards and, in return, it returned its nuclear arsenal to Russia. The annexation of Crimea and continued Russian aggression on Ukraine's eastern border directly defied this agreement [MEY 19].

1.8. Conclusions

The UN policy of non-proliferation has had mixed success as a number of countries (South Africa, Libya) have abandoned their military programs and signed the NPT. However, this policy has also suffered many setbacks since a number of states have acquired control of atomic bombs (India, Pakistan and North Korea), and this number is likely to increase in the short and medium term (Iran and Saudi Arabia).

International treaties are currently undergoing a serious crisis. For example, only one international treaty has been ratified (NPT), but it is on hold. In addition, nuclear deterrence is increasingly being challenged by some non-nuclear weapon states. This is the case, for example, for the Vatican, which no longer seems to accept the concept. In the face of this accumulation of negative facts, non-nuclear-weapon states have recently adopted the TIAN Treaty and 50 states ratified it on October 24, 2020 (mainly states from South America, the Caribbean, Oceania and Africa). It therefore entered into force on January 22, 2021.

The probability of an intentional or accidental explosion is increased today due to the considerable investment in nuclear weapons, their associated systems and their modernization. The increased reliance on automated false alarm detection systems also increases the risks. In addition, technological advances favor terrorists because their needs for specific expertise will be less. Therefore, to completely eliminate nuclear risk, the only solution is the total elimination of nuclear weapons [UNI 17].

2

Disarmament of Atomic Weapons

2.1. Introduction

Recently, for the majority of states, the arms race has resumed. This is shown by an accelerated development of new technologies, especially cybertechnologies. This means that nuclear weapons safety systems are even more uncertain and fragile. This is accentuated by the fact that the established invisibility of submarines is only temporary, as new generations of satellites that are being developed will see them. In addition, states are significantly increasing their military spending. Thus, in France, the budget of the new military programming law (*Loi de Programmation Militaire*, LPM) has been increased to re-equip the French armed forces. This increase goes mainly to nuclear power and no less than 37 billion euros will be devoted to it in the LPM 2019/2025, an increase of 60% compared to the previous LPM (which allocated 23.3 billion euros) [COL 18].

There are two meanings of the term "disarmament". One results from the obsolescence of a weapon, and explains its destruction and replacement by a new, more sophisticated and powerful weapon. The other is due to a weapon's definitive elimination as a result of diplomatic agreements and disarmament treaties.

2.2. Limitations on the number of nuclear weapons

The limitations on the number of nuclear weapons have mainly been achieved through bilateral agreements between the Americans and the Russians. At the international level, a few agreements have been designed to restrict the trade in arms and sensitive supplies.

2.2.1. *Bilateral disarmament agreements and treaties between the Americans and Soviets*

The Americans concluded bilateral disarmament agreements with the Soviets, and then with the Russians. A total of seven agreements were signed between the two superpowers.

2.2.1.1. *The SALT I agreements*

On May 26, 1972, Richard Nixon and Leonid Brezhnev signed the SALT (Strategic Arms Limitation Talks) agreements, limiting strategic weapons. This was the culmination of negotiations that began in November 1969. The text of this first agreement includes an ABM (Anti-Ballistic Missile) Treaty (see below). This 1972 agreement, which lasted five years, decided to freeze the nuclear arsenals of the two powers. The Soviets, not having mirroring[1] technology, obtained a quantitative advantage [IRS 19b]. This treaty also included a provisional convention of freezing the manufacture of land launchers of intercontinental ballistic missiles (ICBMs) and limiting the launchers of submarine ballistic missiles for 20 years [LAR 17a].

2.2.1.2. *The ABM Treaty*

In May 1972, the United States and Russia signed the ABM (Anti-Ballistic Missile) Treaty, which restricts the deployment of anti-missile systems in each country. This treaty is of unlimited duration. ABM systems, including weapons, their location and composition, are limited to two sites. Amended in 1974, the treaty authorizes the deployment of an ABM system at only one site. This treaty was to be terminated by the United States in 2001, with effect from June 2002 [LAR 17a]. In March 1983, President Reagan announced the Strategic Defense Initiative (SDI), known as the "*Star Wars Program*", a major research program aimed at providing the United States with a missile shield. This initiative appears contrary to the ABM Treaty [IRS 19b]. The SDI program was abandoned in 1993 for a more limited program of shielding around deployed troops.

2.2.1.3. *The SALT II agreement*

The United States and the USSR concluded the six-year SALT II agreement in Vienna. In this agreement, the quantitative ceilings for nuclear weapons were identical for both states.

---

1 Multiple Independently Targeted Reentry Vehicle (MIRV) is a technique in the field of military armament that allows a missile to be equipped with several warheads (nuclear or conventional) that each follow an independent trajectory upon entry into the atmosphere.

2.2.1.4. *The INF Treaty*

On December 8, 1987, the United States and the USSR (then led by Ronald Reagan and Mikhail Gorbachev) adopted the Washington Treaty for the elimination of intermediate-range nuclear forces (INF), commonly known as "Euromissiles". This treaty entered into force in June 1988, providing for the elimination of intermediate-range and short-range missiles – i.e. rockets with a range of 500 to 5,500 km – within three years. However, these missiles represented only 3–4% of the arsenal available. Mutual *in situ* verifications were planned. On May 31, 2001, Russia and the United States conducted the last inspection related to the INF Treaty. In October 2018, Trump announced his intention to withdraw from the INF Treaty and on July 3, 2019, Vladimir Putin signed the law on Russia's exit from the INF Treaty.

The now official demise of the INF Treaty, announced by Donald Trump and signed on Friday, August 2, 2019, is another disturbing sign of growing disorder in international relations. The INF allowed, at the time, for the destruction of all nuclear missiles with a range between 500 and 5,500 km. Thus, before 1991, 2,692 missiles were destroyed [STR 19]. However, this 30-year-old treaty did not take into account the rise in power, or even the hubris, of China, which has massively developed this type of weaponry. Moreover, the United States wanted to have its hands free to face the Russian 9M729 mobile missiles with an effective range of 2,500 km, although Moscow, against all evidence, assured the world that their range does not exceed 480 km [ANO 19a].

For their part, in August 2019, the Americans tested their first intermediate-range missile (500 km) since the Cold War [ANO 19b].

2.2.1.5. *The START I agreement*

On July 31, 1991, just after the fall of the Berlin Wall, the Soviet Union and the United States, represented by Presidents George H.W. Bush and Mikhail Gorbachev, signed the START I (Strategic Arms Reduction Treaty) agreement in Moscow. This agreement required nine years of negotiations. The two nuclear powers committed to reducing the number of their strategic nuclear warheads to 6,000 each and their delivery systems to 1,600. This text was a reference point in the field of disarmament, because of the multiple agreements on cooperation, control and monitoring of actions to destroy these weapons, created by both parties. The dissolution of the Soviet Union in December 1991 complicated the adoption and application of START I. In fact, parts of the Soviet arsenals that had to be disarmed were in the territories of Belarus, Kazakhstan and Ukraine [COL 15].

START I was the first text that effectively reduced the strategic arsenals of both powers. On December 5, 2001, the reductions in nuclear arsenals provided for in this agreement were achieved.

2.2.1.6. *The Lisbon Protocol*

Following the dissolution of the USSR, the new states of Ukraine, Belarus and Kazakhstan came to possess nuclear weapons on their territory. The Lisbon Protocol, signed on May 23, 1992, associated these three states with the START I Treaty. These states became NNWS. The repatriation of nuclear weapons to Russia ended in 1996 [IRS 19b].

2.2.1.7. *The START II treaty*

START II was signed by Boris Yeltsin and George H.W. Bush on January 3, 1993, in Moscow. It provided for a two-thirds reduction of arsenals. In 2003, then postponed to 2007, the American arsenal was to go from 9,986 nuclear warheads to 3,500 and the Russian arsenal from 10,237 warheads to 3,027. This treaty also provided for the complete removal of rockets with multiple independent and remote-controlled warheads (mirrored ICBMs). This treaty was ratified in 1996 by the United States and in 2000 by Russia.

2.2.1.8. *The SORT treaty*

After the era of interminable negotiations, a new way of thinking about arms control emerged with the implementation of the Strategic Offensive Reduction Treaty (SORT). On May 24, 2002, George W. Bush and Vladimir Putin signed, in Moscow, what would pass for a historic agreement, since it was a question of reaching, for each party, a maximum remaining arsenal of between 1,700 and 2,200 nuclear warheads, operationally deployed, by December 31, 2012. This represented a two-thirds reduction of the arsenals in 10 years. This is a treaty that only concerned strategic weapons and, once, again, excluded any disarmament measures on tactical weapons[2] [COL 15].

Criticisms of this treaty were numerous. In particular, it did not apply to vectors, the head count was not specified (operational or in reserve), no verification measures were provided for and the reduction measures were not irreversible.

2.2.1.9. *The New START or START III treaty*

Barack Obama and Dmitry Medvedev (the new Russian Prime Minister), signed, in Prague, the New START or START III treaty, the successor to the START I treaty which expired on December 5, 2009. This new treaty also included the SORT

---

2 Strategic weapons, which are mainly nuclear, are embarked on long-range, often intercontinental, delivery systems (bombers, cruise missiles, submarines), whose objectives may be demographic, economic or military. In contrast, tactical weapons are only at the medium range.

treaty. The New START was officially signed on April 8, 2010 and entered into force on February 5, 2011 for a period of 10 years. It provided for the limitation of the number of warheads in each of the two countries to 1,550, the limitation of the number of ICBMs on board submarines and bombers to 800, the on-site verification of nuclear facilities and the exchange of data between the two countries [COL 15]. In addition, there would be joint inspections to verify destruction.

2.2.2. *Delivery of nuclear supplies*

One way to limit nuclear proliferation is to strictly control the exchange of nuclear information, equipment and materials with states that might acquire the atomic bomb. Thus, the Zangger Committee was created in 1971, which brought together states party to the NPT, those countries likely to supply nuclear technology. In 1974, this committee adopted a list of material equipment whose export was to be made conditional on the application of IAEA safeguards by the importing country [IRS 19b].

On September 21, 1977, 15 nuclear technology supplier countries (including France, which was not a party to the NPT at the time) met in the “London Club” (now the NSG, Nuclear Suppliers Group). In 1978, this group adopted the London Guidelines, which included a list of nuclear materials and equipment that could only be exported to countries that agreed to submit them to IAEA control. However, these guidelines were of national application and constituted, for each member, only a political commitment which was, therefore, not legally binding.

On April 3, 1992, following the discovery of Iraq’s secret nuclear program, the NSG guidelines were strengthened by the addition of a list of dual-use goods (i.e. goods used for both civilian and military purposes) and by the adoption of the principle of “full control”, prohibiting any nuclear export to a country that had not submitted control of all its nuclear activities to comprehensive IAEA safeguards, as required by the NPT. These lists of goods are periodically revised.

2.2.3. *Controlling the delivery systems of atomic weapons*

An atomic weapon is of no use if it cannot hit a specific target in the opponent’s territory and this territory is usually far from the starting site of the atomic bomb. Therefore, a variety of delivery systems must be used to propel the weapon. The most commonly used are missiles with various ranges. Controlling the development of missile technology is therefore another solution to limiting nuclear proliferation.

In December 1979, in response to the deployment of Soviet Euromissiles (notably the SS-20) against Western Europe in 1977, NATO member states decided to deploy intermediate-range nuclear missiles (Pershing and cruise missiles) in Europe. The European NATO countries wanted to prevent their security from being inferior to that of the United States.

On April 7, 1987, the G7 countries created a Missile Technology Export Control Regime (MTCR). The MTCR member countries agreed on common criteria for the control of transfers of missile equipment and technology that could carry weapons of mass destruction (WMD).

On September 11, 2003, following a call by the United States on May 31, 2003, 10 other countries joined the Proliferation Security Initiative (PSI) and issued a statement on interception principles. This initiative aims to support efforts to prevent the proliferation of WMD, their means of delivery and related materials, and their transport by land, air and sea to and from countries of concern regarding proliferation. States are encouraged to cooperate and combine their legal, diplomatic, economic and military means to facilitate the interception of such weapons.

2.2.4. *The Hague Code of Conduct (HCoC)*

States are also aware that it is absolutely necessary to limit missiles, particularly ICBMs, which have a long range (more than 3,000 km). This type of missile can carry several warheads that can strike different targets in the same area.

By 2017, the five permanent members of the United Nations Security Council had operational systems in place to launch ICBMs; they all had nuclear-powered ballistic missile submarines. In addition, Russia, the United States and China had land bases capable of launching ICBMs, and China and Russia had mobile land-based systems.

Other states are developing ICBMs. This is the case of India with the Agni-IV missile, which would have a range of 6,000 km, and would also be the case for North Korea.

Ballistic flights can be divided into three phases: first of all, a thrust phase (3–4 minutes at 7 $km.s^{-1}$ at an altitude of 200 km); then the intermediate phase of about 25 minutes for a distance of 12,000 km with an apogee at 1,200 km; finally, the re-entry phase, of about 2 minutes, where the missile releases several warheads and a large number of decoys.

Missiles are frequently divided into six categories. First, there are short-range ballistic missiles (SRBMs); their range is less than 800 km and the main ones are Iskander-E, Pluto, Scud and MGM-31 Pershing. Second, there are medium-range ballistic missiles (MRBMs), which have a range between 1,000 and 3,000 km, such as the Shahab-3, Nodong-1 and Jericho II. Third, there are intermediate-range ballistic missiles (IRBMs), which have a range between 2,400 and 6,400 km, such as the S3 and SS-20. Fourth, there are long-range ballistic missiles, or intercontinental ballistic missiles (ICBMs), which have a range between 6,000 and 13,000 km, such as Topol-M, Peacekeeper and SS-18. Fifth, there are air-launched ballistic missiles (ALBMs), which are launched from aircraft, such as the Russian Kinjal, the Israeli Rampage and the Chinese CH-AS-X-13. Finally, when launched from submarines, the missiles are called submarine-launched ballistic missiles (SLBMs). These are, notably, the M51 missile (France), the Trident 2D5 missile (United States) and the Bulava missile (Russia).

Missiles that have been created since 2000 are China's DF-31 (2000, 8,000 km), Israel's Jericho III (2008, 6,500 km), France's M51 (2010, 9,000 km), India's Agni-III (2012, 3,500–5,000 km), China's Julang 2 (2015, 8,600–14,000 km), India's Agni-V (2016, 5,000 km) and North Korea's Hwasong-14 and Hwasong-15 (2007, 10,000 and 13,000 km, respectively). Various missile projects are underway, such as, for Russia, the RS-28, which was planned for 2020, launched from a silo with a range of 17,000 km and a precision of 10 m! Or indeed, the R-30, which can be launched from a submarine with a range of 8,000 km.

Some states, for strategic reasons, are developing intermediate-range ballistic missiles. These are China, Pakistan, India, Israel and Iran.

The Hague Code of Conduct against Ballistic Missile Proliferation (HCoC) was an initiative born in 2000 that took the form of a draft Code of Conduct in 2002, a non-binding instrument establishing confidence-building and transparency measures to combat ballistic missile proliferation, with the support of the European Union and approximately 30 partner states (in particular, the member states of the MTCR). Today, 139 states are signatories to the HCoC.

The code is a multilateral instrument of trust and transparency and operates on a voluntary basis. To be a signatory state to the code consists of complying with confidence-building and transparency measures in space and ballistic activities, including pre-notification of launches and providing all signatory states with annual declarations on its space and ballistic activities.

The Pakistani crisis, born in 2004 from the discovery of a nuclear proliferation network, carried out by A.Q. Khan, was violent. In recent years, Pakistan has taken a number of measures to increase international confidence in the security of its nuclear arsenal. However, instability in Pakistan has called into question the extent and sustainability of these reforms [KER 16].

The number of atomic bomb delivery vehicles held by the various states in 2018 is provided in Table 2.1.

Country	Bombers or fighter aircraft		ICBM (SLBM) (land or water)	
	Number	Warheads	Number	Warheads
United States	60/107*	880	400 (240)**	800 (1,920)
Russia	50/68	616	318 (11/176)	1,138 (768)
United Kingdom			48	215
France	50	50	48	180
China			131 + 48	180 + 48
India	48	48	60 + 2	60 + 4
Pakistan	36	36	102	102
Israel		30		50

Table 2.1. *Strategic forces bomb vectors in January 2018 [SIP 18]. * The first number corresponds to the bombs assigned to nuclear missions, the second to the total inventory. ** The ocean forces given in brackets*

2.2.5. *The disarmament of France*

France has long been committed to nuclear disarmament. In 1996, it was the first state to decide to close and dismantle its facilities dedicated to the production of fissile material for nuclear weapons in Pierrelatte and Marcoule. As a result, France no longer has facilities producing fissile material for nuclear weapons [FRA 19a]. In 2010, it was the only state to have closed and dismantled all its facilities producing fissile material for nuclear weapons [RÉP 10].

France's record on nuclear proliferation and disarmament is exemplary, and, in some areas, even unparalleled. For example, it is the first state, along with the United Kingdom, to have signed and ratified the Comprehensive Nuclear-Test-Ban Treaty (CTBT). It is also the first state to have decided to close and dismantle its facilities producing fissile material for nuclear weapons, the only nuclear-weapon state to have transparently dismantled its nuclear test site in the Pacific, the only state to have dismantled its ground-to-ground nuclear missiles and the only state to have voluntarily reduced the number of its nuclear-powered ballistic missile submarines (NSBMs) by one-third. It has similarly reduced its nuclear weapons, missiles and airborne aircraft by one-third. One of France's future objectives is the negotiation, at the Conference on Disarmament, of a Fissile Material Cut-off Treaty (FMCT) [FRA 19b].

In addition, France actively participates in the four international control regimes – or "supplier regimes" – created between 1974 and 1996, which define common standards for the export of goods and technologies that could be used in the development of WMD: the Zangger Committee, the NSG, the Australia Group and the MTCR. It also subscribes to the HCoC and is also active in the PSI. To this end, at the national level, in February 2009 France adopted the Governmental Interdiction Proliferation Plan (*Plan Gouvernemental Interception Prolifération*, PIP), which establishes the framework for action to respond to trafficking in goods or technologies related to the proliferation of WMD [DGR 17a].

Finally, *Direction générale des relations internationales et de la stratégie (du ministère de la Défense)* (DGRIS) actively participates in the International Partnership for Nuclear Disarmament Verification (IPNDV), an initiative that brings together more than 20 nuclear-weapon and non-nuclear weapon states to increase understanding of the technical challenges of verifying nuclear disarmament [DGR 17b].

2.3. Nuclear deterrent forces

Nuclear deterrent forces are important in the world. Nine states now possess nuclear weapons: the United States, Russia, France, the United Kingdom, China, India, Pakistan, Israel and North Korea. Iran is also in the process of preparing for them. Deterrent forces are based on three different components: the land, the air and the ocean. Only the United States, Russia and China are developing all three with submarines, aircraft and rockets (Table 2.2). France now has only two components (oceanic and air). The United Kingdom only has the oceanic component and leases nuclear warheads from the United States. India, Pakistan and Israel deploy both the land and air components and continue to develop them intensely. India and Pakistan also have programs that are working on acquiring an oceanic component. North Korea is rapidly developing its land component.

Nuclear deterrent forces rely on both the various types of atomic bombs (A-bombs and H-bombs) (Figure 2.1) and on delivery systems capable of delivering these bombs to the enemy.

Country	Nuclear power component			Number of nuclear warheads
	Terrestrial	Aerial	Oceanic	
Russia	X	X	X	7,000
United States	X	X	X	6,800
France		X	X	300
China	X	X	X	250
United Kingdom			X	165
Pakistan	X	X		130
India	X	X		120
Israel	X	X		80
North Korea	X			10

Table 2.2. *Components of nuclear power and number of nuclear warheads in 2017*

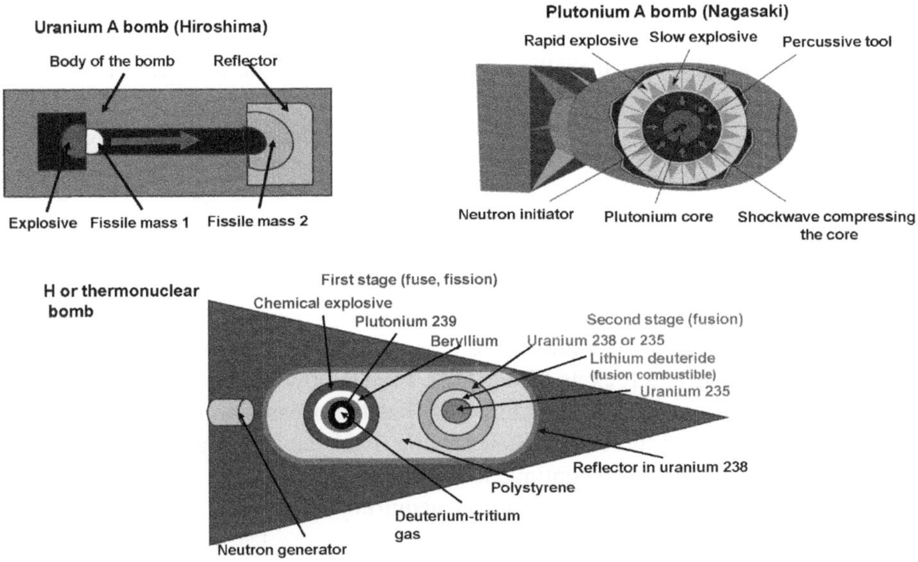

Figure 2.1. *Various types of atomic bombs. For a color version of this figure, see www.iste.co.uk/amiard/disarmament.zip*

2.3.1. *Land forces of nuclear deterrence*

With the exception of France and the United Kingdom, ICBMs are an essential element of nuclear deterrence. Two factors concerning these missiles are important: their range and the weight load they can carry.

The United States has 600 nuclear warheads available for its missiles, which are, essentially, the 200 LGM-30G Minuteman III Mk-12A (13,000 km) and the Mk-21 SERV (13,000 km) with 600 nuclear warheads [KRI 19a]. The 400 Minuteman III intercontinental missiles operated by the US Air Force are based in silos in three areas: the 90th Missile Wing at Francis E. Warren Air Force Base in Colorado, Nebraska and Wyoming, the 91st Missile Wing at Minot Air Force Base in North Dakota and the 341st Missile Wing at Malmstrom Air Force Base in Montana [KRI 20].

The Russian Strategic Land Missile Forces held the majority of the Soviet army's arsenal, were the main pillar of the Soviet nuclear triad and grew rapidly. By the 1970s, they surpassed the land-based component of the US Strategic Air Command in both number of weapons and delivery systems. Currently, Russia possesses 318 ICBMs carrying 1,138 nuclear warheads. These are, mainly, 46 RS-20Vs (11,000–15,000 km) with 460 warheads, 20 RS-18s (10,000 km) with 120 warheads, 72 RS-12M Topols (10,500 km) with 72 warheads, 60 RS-12M2 Topol-Ms (10,500 km) with 60 warheads, 18 RS-12M1 Topol-Ms (10,500 km) with 18 warheads and 90 RS-24 Yars (10,500 km) with 360 warheads. A new generation (RS-26 Yars-M) is being tested [KRI 19b]. The strategic missile troops also have a fleet of planes and helicopters consisting of the Mi-8, An-12, An-24, An-26 and An-72 for transport and measurements, stationed at seven airfields and eight heliports.

Ukraine, a former member of the Soviet Republic, had the third-largest nuclear arsenal in the world with 1,900 Soviet strategic nuclear warheads and between 2,650 and 4,200 Soviet tactical nuclear weapons deployed on its territory at the time of independence in 1991 [NTI 19a]. It returned its strategic nuclear weapons to Russia between 1994 and 1996 to be dismantled, in accordance with the Budapest Memorandum. Similarly, Kazakhstan had the fourth-largest nuclear arsenal in the world (1,410 nuclear warheads), but chose total nuclear disarmament by returning all these nuclear weapons to Russia [NTI 19b].

In China, 131 nuclear warheads are carried by land-based ballistic missiles. These are 5 DF-4s (5,500 km) with 10 warheads, 10 DF-5As (12,000 km) with 10 warheads, 10 DF-5Bs with 30 warheads, less than 50 DF-21s (2,100 km) with 50 warheads, 16 DF-26s (2,100 km) with 16 warheads, 8 DF-31s (>7,000 km) with 8 warheads and 32 DF-31As (>11,200 km) with 32 nuclear warheads [KIL 19c]. The DF-5 Dong Fengs (East Wind) are liquid-fueled missiles that can strike the American continent. Since 2017, they have been equpped with a solid fuel. The

DF-4 can reach Europe and carry a warhead with a mass of 2 tons and a power of 3 megatons. The DF-21s are missiles fired from mobile launchers.

India's nuclear arsenal has been under the supervision of the Nuclear Command Authority since January 2003. India has reaffirmed its doctrine of no first use of nuclear weapons, reserving the right to use them in the event of an atomic, chemical or biological attack. India's land-based missiles are loaded with 60 nuclear warheads. The missiles are 24 Prithvi-IIs (250 km) with 24 warheads, 20 Agni-Is (>700 km) with 20 warheads, 8 Agni-IIs (>2000 km) with 8 warheads and 8 Agni-IIIs (>3200 km) with 8 nuclear warheads [KIL 19d].

In Pakistan, their land-based missiles carry 102 nuclear warheads. The main missiles are 10 Abdalis (200 km), 16 Ghaznavis (290 km), 16 Shaheen-Is (750 km), 12 Shaheen-IIs (1,500 km), 24 Ghauris (1,250 km) and 24 Nasrs (60–70 km). Each missile carries a single nuclear warhead [KIL 19e].

Israel possesses Jericho II land missiles (1,500–1,800 km) with 25 warheads and Jericho III (>4,000 km) with 25 nuclear warheads [KIL 19f].

North Korea has developed many types of land-based missiles such as Hwasong-7 (>1,200 km), Hwasong-9 (1,000 km), Bukkeuseong-2 (1,000 km), Hwasong-10 (>3,000 km), Hwagsong-12 (3,300–4,500 km), Hwasong-13 (>5,500 km), Hwasong-14 (6,700–10,400 km), Hwasong-15 (8,500–13,000 km) and Taepodong-2 (12,000 km) [KIL 19g].

2.3.2. *Air forces of nuclear deterrence*

Strategic air forces use various types of aircraft carrying nuclear warheads. The limiting factor is the range of the aircraft. This can be compensated for either by refueling in mid-air or by using aircraft carriers that bring the carrier closer to its target.

In the United States, strategic air forces are provided by bombers. Currently, 44 B-52Hs (16,000 km) and 16 B-2As (11,000 km) carrying 880 nuclear warheads are available and operational [KRI 19a]. These bombers are organized into nine squadrons spread over five areas and three bases: Minot Air Force Base in North Dakota, Barksdale Air Force Base in Louisiana and Whiteman Air Force Base in Missouri. By early 2018, the Chief of the Air Staff had estimated that 175 B-21s would be needed, and it was announced in May of that year that the bombers would be based at Dyess Air Force Base, in Texas, Ellsworth Air Force Base, in South Dakota and Whiteman Air Force Base in Missouri [KRI 20].

In Russia the strategic air forces also depend on bombers, with 616 nuclear warheads. This state possesses 14 TU-95S6s (6,500–10,500 km) with 84 warheads, 25 Tu-95MS16s (6,500–10,500 km) with 400 warheads and 11 Tu-160s (10,500–15,200 km) with nuclear warheads [KRI 19b].

In France, deterrence missions are carried out by the Strategic Air Force (*Forces Aériennes Stratégiques*, FAS), which has improved medium-range air-to-ground missiles (ASMP-A). They are carried by 20 Mirage-2000Ns (2,750 km) with 20 nuclear warheads, 20 Rafale F3s (2,000 km) with 20 warheads and 10 Rafale MF3s (2,000 km) with 10 nuclear warheads from the French Air Force and Navy [KIL 19b]. Since 2019, all Mirage aircraft have been replaced by Rafales [PIN 17].

The airborne component was completely renewed, starting in 2007 with the Rafale Marine and Air entering into service, the replacement of the ASMP with the upgraded ASMP (ASMP-A) and the replacement of the TN81 with the new airborne warhead (*Tête Nucléaire Aéroportée*, TNA). Since 2015, the four fighter squadrons (Gascogne, La Fayette, Aquitaine and the technical support squadron) have been grouped together at Saint-Dizier Air Force Base (BA 113). The Nuclear Naval Air Force (*Force Aéronavale Nucléaire*, FANU) is made up of the aircraft carrier Charles de Gaulle and the 28 Rafale Ms of the two flotillas on board, while the three naval aeronautics flotillas (11F, 12F and 17F) are located at the Landivisiau Naval Air Force Base and qualified ASMP-A. Delivery of the new fighter aircraft tankers, a fleet of 15 Airbus A330-200 "Phoenix" Multi-Role Tanker Transports (MRTTs), is now scheduled for 2023, and the first Phoenix arrived at Istres, its future home base, on September 27, 2018 [PIN 17].

China's strategic aviation consists of 20 H-6s (3,100 km) with 20 nuclear warheads [KIL 19c]. The Xian H-6 bombers of the Chinese air force are a licensed copy of the Tupolev Tu-16.

Strategic aviation in India has 48 nuclear warheads carried by 32 Mirage 2000Hs (1,850 km) with 32 warheads and 16 Jaguar ISs (1,600 km) with 16 warheads [KIL 19d]. On September 23, 2016, India and France signed an agreement for the delivery of 36 Rafale aircraft. The Rafales would be deployed in two squadrons of equal size, one at Ambala Airbase (located only 220 km from the Pakistani border) and the other at Hasimara Airbase, in West Bengal [KRI 18a].

Pakistani aviation has 36 nuclear warheads equipping 24 F-16As (2,100 km) with 24 warheads and 12 Mirage III/Vs (2,100 km) with 12 warheads [KIL 19e].

Israel has about 30 nuclear warheads with F-16A/B/C/D/I (1,600 km) and Falcon carriers [KIL 19f].

The North Korean Air Force could deploy its atomic arsenal with the 60 or so Ilyushin Il-28s, from the 1950s, and the 15 or so Sukhoi Su-7s at its disposal. However, some South Korean media believe that some of the Chengdu J-7s (Chinese copy of the MIG-21) and the Mikoyan-Gurevitch MiG-23s would be suitable for this type of mission [AMI 17].

2.3.3. *Oceanic nuclear deterrent forces*

Initially only five states (United States, Russia, United Kingdom, France and China) had a strategic nuclear submarine force that was implemented based on SSBNs, soon joined by India, which is currently immobilized due to water entering the SSBN. An instrument of deterrence, the SSBN is a very large nuclear-powered submarine, equipped with nuclear-armed SLBMs, arranged in vertical launch tubes and launched by diving. As of January 1, 2011, 36 nuclear-powered ballistic missile submarines were in service worldwide. The United States and Russia have 26 between them, or more than two-thirds [LEG 11].

The US Strategic Oceanic Force has 14 Ohio-class SSBNs. Each submarine can carry 24 Trident II D5 missiles with an announced range of approximately 8,300 km. Each missile has a payload of 2,800 kg, or 12 W76 warheads with a power of 100 kilotons (kt), which represents, for each SSBN, a theoretical firepower of 28,800 kt, 1,900 times the power of the Hiroshima bomb. The American cruise missiles are the UGM-133A Trident II (D5/DSLE) Mk-4, Mk-4A and Mk-5 (>7,500 km) with 1,920 nuclear warheads, to which are added 200 warheads in non-strategic forces, making a total of 3,800 warheads, as well as 2,650 warheads that have been withdrawn and are in the process of being dismantled [KRI 19a]. Eight missile submarines operate in the Pacific from their base (Kitsap) near Bangor, Washington, and six operate in the Atlantic from their base in Kings Bay, Georgia [KRI 20].

As of January 1, 2011, the Russian Navy had 12 SSBNs of four different types. Four are Delta III class submarines (three assigned to the Pacific Fleet and one to the Northern Fleet), six Delta IV class submarines assigned to the Northern Fleet and one Typhoon class submarine, assigned, since 2005, to test the new SS-N-30 Bulava ballistic missile. The Russians have 11 operational cruise missiles (176 in reserve) with 768 nuclear warheads. These are mainly 2 RSM-50 Volnas (6,500 km) with 96 warheads, 6 RSM-54 Sinevas (9,000 km) with 384 warheads and 3 RSM-56 Bulavas (>8,500 km) with 288 warheads [KRI 19b].

The *Force océanique stratégique* (FOST, Strategic Oceanic Force) is the main component of France's strategic nuclear forces. As of January 1, 2011, the FOST is composed of four new-generation SSBNs (SSBN/NG) of the *Le Triomphant* class, three armed with 16 M-45 strategic missiles and the fourth, *Le Terrible*, being equipped with the MSBS/NG M-51. The other three SSBNs were adapted to the M-51 between 2011 and 2018. France currently has 32 M51-1s (>6,000 km) carrying 160 warheads and 16 M51-2 (>6,000 km) carrying 80 warheads, as well as a reserve of 10 nuclear warheads [KIL 19b]. Starting in 2015, the new "TNO" (*Tête Nucléaire Océanique*, oceanic nuclear warhead) gradually replaced the TN75, which equipped the M45s. Its estimated power is 100 kt, as opposed to 75 kt. It also benefits from a new penetration aid system. The M45 and M51 missiles can carry six nuclear warheads using the technique known as "MIRV".

As a delivery vehicle for its nuclear weapons, the United Kingdom only uses the oceanic component of the three components of the "nuclear triad", equipped with four SSBN nuclear submarines of the Vanguard class (*HMS Vanguard*, *HMS Victorious*, *HMS Vigilant* and *HMS Vengeance*). However, the United Kingdom remains dependent on the United States for its 48 Trident III D5 ballistic missiles (>7,400 km), carrying 215 nuclear warheads, manufactured and maintained by the Americans [KIL 19a]. Since these missiles are rented from the United States, British submarines must regularly visit the Kings Bay naval base in the United States. British nuclear submarines are based in southwestern Scotland at Her Majesty's Naval Base Clyde at Faslane, which has access to the Irish Sea. Non-operational warheads are stored at the Royal Naval Armaments Depot in Coulport, about 3 km west of the base [KRI 11].

The navy of the People's Liberation Army of China has three nuclear-powered ballistic missile submarines: one of the Daqingyu class (Xia, in NATO terminology) and two of the Jin class type 094. According to some sources, only one submarine is operational. It has 48 Ju Lang-2 cruise missiles (>7,000 km) carrying 48 nuclear warheads [KIL 19c]. The Ju Lang 2 has either a 2.5 megaton warhead or three 90 kiloton MIRVs.

On July 26, 2009, Indian Prime Minister Manmohan Singh attended the launch of the Indian Navy's first SSBN, the INS *Arihant*. Launched in 1998, the vessel was developed as part of a project known as the Advanced Technology Vessel. It was implemented at the Visakhapatnam shipyard by the Indian Navy, the Bhabha Atomic Research Center (BARC) and the Defence Research and Development Organisation (DRDO) jointly [KIL 11]. India has 14 ballistic cruise missiles, 2 Dhamush (400 km) with 4 nuclear warheads and 12 K-15 (700 km) with 12 nuclear warheads [KIL 19d]. India's missile development program began in the 1980s (the Integrated Guided Missile Development Program) and was officially closed in 2008. A naval base for Indian nuclear submarines called INS *Varsha* is currently under construction near Rambilli on India's east coast [KRI 18a].

The Pakistani Navy created a Strategic Naval Force Command in 2012. By 2015, it had five submarines: two Agosta 70 (PNS/M *Hasmat* and *Hurmat*, commissioned in 1979) and three Agosta 90B (PNS/M *Khalid*, *Saad* and *Hamza*). In April 2015, the Pakistani government announced its intention to acquire eight S-20 submarines derived from the Chinese type 039A. Pakistan has 12 Babur GLCM cruise missiles (350 km) carrying 12 nuclear warheads and capable of being launched from the Agosta [KIL 19e].

North Korea has developed a cruise missile, the Bukkeuseong-1 [KIL 19g].

2.3.4. *Nuclear weapon manufacturing sites*

Information on the existence and location of nuclear weapon manufacturing plants is not always easy to obtain because it is not always disclosed. Many uncertainties exist in this area.

In the United States, there is a relatively large number of sites used for the design and manufacture of nuclear weapons. They are managed by the National Nuclear Security Administration (NNSA) under the Department of Energy (DoE), whose headquarters are in Washington, DC and the Albuquerque Complex, New Mexico. This institution employs 39,000 employees. The NNSA includes four national laboratories, three production plants, an assembly and dismantling site (Pantex, Texas), and a testing site (Nevada National Security Site, Nevada). The laboratories are Lawrence Livermore National Laboratory (LLNL, California), Los Alamos National Laboratory (LANL, New Mexico) and two Sandia National Laboratories (SNL, California and New Mexico).

The three production sites are the Kansas City National Security Campus (KCNSC, Missouri), the Y-12 National Security Complex (Tennessee) and the Savannah River Site (SRS, South Carolina). The NNSA has identified 30 capabilities that are needed and are to be developed to maintain the nuclear weapons stockpile. The NNSA's roadmap for the next 25 years to maintain the US nuclear deterrent is contained in the Stockpile Stewardship and Management Plan (SSMP) [WIL 18].

Russian nuclear weapons production sites are in Lesny (Sverdiovsk-45/16) and Trekhgorny (Zlalust-36). The Russian experimental laboratory is in Snezhinsk (Chelyabinsk-70) [KRI 14].

Currently, the United Kingdom does not manufacture nuclear weapons since it leases its submarine-launched missiles from the United States. The Atomic Weapons Establishment (AWE) is headquartered on a 300-hectare site in Aldermaston, Berkshire, but the management of weapons is delegated to a consortium of private companies. The warheads were previously assembled at the Royal Ordnance Factory (ROF) in Burghfield (91 ha) and tested on the beaches of Foulness and Orford Ness.

Components for nuclear weapons were also produced at the former ROF Cardiff site. These two sites have been partially closed since 1971 and 1997.

In France, all military nuclear facilities for nuclear weapons are managed by the *Direction des applications militaires* (DAM), which is part of the *Commissariat à l'énergie atomique et aux énergies renouvelables* (CEA), the French Alternative Energies and Atomic Energy Commission. The DAM is responsible for the research, design, manufacturing, operational maintenance and dismantling of nuclear warheads. The DAM comprises of three main sites: Bruyères-le-Châtel, Valduc and CESTAS.

The design and the simulation of the action of nuclear warheads take place at the DAM center in Bruyères-le-Châtel, approximately 30 km south of Paris. The center houses the Tera 1000, a very powerful European supercomputer with a capacity of 25 Petaflops.

The CEA/DAM Valduc center is located approximately 30 km northwest of Dijon. This center is responsible for the production, maintenance and dismantling of nuclear warheads. The Franco-British Teutates Treaty of 2010 extends this center to the respective nuclear weapon stockpiles of both countries. In addition, the Epure facility at Valduc includes three high-power radiographic axes, including the AIRIX X-ray generator, which studies the condition and hydrodynamic behavior of materials under the conditions encountered during the pre-nuclear phase of the weapons' operation. This is an essential function for the maintenance and development of France's nuclear weapons in the absence of real nuclear explosions.

Finally, the CESTA (*Centre d'études scientifiques et techniques d'Aquitaine*) of the CEA near Le Barp is in charge of designing equipment for nuclear weapons and re-entry vehicles, as well as coordinating the development of nuclear warheads. The Megajoule laser, the French equivalent of the US National Ignition Facility, is located on the same site. It aims to validate theoretical models of nuclear weapon detonations and thus plays an important role in the French nuclear simulation program [KRI 19f].

In China, there is little information on the location of factories producing nuclear weapons. Gaseous diffusion plants for the production of uranium are in Heiping (Chin-k'ou-ho/Chinkouko), Chengdu and Lanzhou (Lan-chou, plant 504). Reactors for plutonium production are in Guangyuan (Kuang-yuan, plant 821) and Jiuquan (Yumen/Subei, plant 404). Plutonium processing takes place in Yibin (I-pan, plant 812) and beryllium production takes place in Helan Shan (plant 905). The production of uranium oxide is at Hengyang (plant 414). Weapons are designed in Jiangsu (7th Institute) and in Mianyang/Zitong (Mien-yang/Tzu-t'ung, CAEP) with an experimental reactor in Chongqing (Ch'ung-ch'ing). The plants for the design and manufacture of nuclear weapons are in Haiyan (Hai-yen/Koko Nor, 9th Academy). The assembly and production of nuclear weapons are in Harbin. Weapons testing is carried out at Lop Nor, with the support of Malan. These two entities constitute base 21 [FAS 20].

For Israel, without any certainty it seems that the major site for the manufacture of fissile material and nuclear weapons is in Dimona. The nuclear reactor in Dimona is a heavy water reactor fueled by natural (not enriched) uranium that uses both heavy water and graphite as nuclear moderators. It is therefore particularly effective at producing plutonium because it can be recharged in working order. It is designed on a model similar to Marcoule (G1) and Saclay (EL2) French atomic batteries. It seems that in 1963, the initial power of the reactor was 24–26 MW. It was then secretly increased, first to 70 MW and then to 150 MW.

According to an uncertified source[3], next to the reactor, a 20 m high steel dome constituting Unit 1, there is a nuclear complex organized into nine other units. Unit 2 is a two-story building 60 m high with six additional basement floors for the processing of spent nuclear fuel, the separation of plutonium and the production of tritium. Unit 3 is the plant for manufacturing nuclear fuel rods and lithium rods used for bombs. Unit 4 houses the treatment of radioactive waste. Unit 5 is a workshop for cladding fuel rods with aluminum. Unit 6 is the engine room (generator and steam turbine). Unit 7 has an unknown role. Unit 8 is the plant for enriching uranium by centrifugation. Unit 9 is an experimental laboratory for enriching uranium using a laser process (Israeli patent). Unit 10 is a plant for the production of depleted uranium. Rumors report that, since the commissioning of the Dimona site, radioactive waste has been buried near the plant or in the Golan Heights zone.

For India, the sites for the manufacture of nuclear weapons are unknown. Its main atomic research center is located in Bhabha (BARC). The BARC seems to be linked to its nuclear weapons program. In addition to its operational Dhruva plutonium production reactor near Mumbai, India plans to build at least one other plutonium production reactor. In addition, the prototype fast neutron reactor under construction at the Indira Gandhi Centre for Atomic Research (IGCAR), near Kalpakkam, could significantly increase India's plutonium production capacity in the near future if it achieves criticality as planned. The director of the IGCAR has further stated that six additional fast breeder reactors will come online over the next 15 years. Construction of the first two, which will be located at IGCAR, is expected to begin in 2021 and will be ready for commercial power generation in the early 2030s [KRI 18a].

Pakistan has a well-established and diversified fissile material production complex that is growing. It includes the uranium enrichment plant in Kahuta, east of

3 Available at: http://www.armagedon.org.il/dimona_english.htm.

Islamabad, with, presumably, a second enrichment plant on the same site, as well as a third enrichment plant in Gadwal, north of Islamabad. Four heavy water plutonium production reactors appear to have been completed in what is normally called the Khushab complex, about 33 km south of Khushab in the Punjab province. The Khushab complex is rumored to produce tritium, and the PINSTECH (Pakistan Institute of Nuclear Science and Technology) complex near Nilore may also produce tritium [KRI 18c].

The New Labs spent fuel reprocessing plant in Nilore, east of Islamabad, which extracts uranium and plutonium, has been expanded. Meanwhile, a second reprocessing plant in Chashma in the northwestern part of the Punjab province may have been completed.

Nuclear-capable missiles and their mobile launchers are developed and produced at the National Defense Complex in the Kala Chitta Dahr mountain range west of Islamabad. The complex is divided into two zones. A western zone, located south of Attock, appears to be involved in the development, production and test launching of missiles and rocket engines. The eastern zone, located north of Fateh Jang, is involved in the production and assembly of mobile launchers (TEL, Transporter Erector Launchers). Other production and maintenance facilities related to launchers and missiles may be located near Tarnawa and Taxila. However, the geographical location of nuclear warhead production plants is not known. They could be near Wah, northwest of Islamabad [KRI 18c].

North Korea is very discreet about the location of its nuclear activities. The major North Korean nuclear site is Yongbyon, which hosts one or two gaseous uranium enrichment plants with numerous centrifuges, a 5 MWe plutonium-producing reactor, and, probably, a tritium separation plant and an experimental light-water reactor (ELWR) [ALB 17].

2.3.5. *Nuclear weapon deployment and storage sites*

There is little disclosure on the location of nuclear warhead storage sites. For example, Russia and the United States refuse to disclose how many tactical nuclear weapons they have or where they are stored, and none of these weapons are restricted by arms control agreements.

Kristensen and Norris [KRI 17] estimate that by mid-2017, there were nearly 15,000 nuclear weapons in the world, located at some 107 sites in 14 countries. The majority (about 9,400) of these weapons are in military arsenals, while the remainder (about 5,600) have been withdrawn and are awaiting dismantling. Nearly 4,000 are operational, and 1,800 are on high alert and ready for use at short notice.

In the United States, Kristensen and Norris [KRI 17] identify 12 sites with nuclear weapons. The United States Navy's nuclear weapons are located at the naval bases in Kitsap (near Bangor, Washington), where eight Ohio Class SSBNs are stationed, and Kings Bay, Georgia, where six Ohio Class SSBNs are stationed [NOR 06].

The United States Air Force air bases are in Barksdale (Louisiana) with B-52H strategic bombers, in Minot (North Dakota) with B-52H strategic bombers and 150 ICBM LGM-30G Minuteman III, in Malmstrom (Montana) with 150 ICBM LGM-30G Minuteman III, in Nellis (Nevada) with some arms depots, in Kirtland (New Mexico) with some arms depots, in Francis E. Warren (Wyoming) with 150 ICBM LGM-30G Minuteman IIIs and in Whiteman (Missouri) with 20 Northrop B-2 Spirit strategic bombers [NOR 06]. To these are added the American air bases located in Europe, where about 400 nuclear bombs are deployed, in Germany (Büchel), Belgium (Kleine-Brogel), Italy (Aviano and Ghedi-Torre), the Netherlands (Volkel), Turkey (Incirlik) and the United Kingdom. In 2010, 150–200 US non-strategic nuclear weapons remained based in Europe in five countries (Belgium, Germany, Italy, the Netherlands and Turkey) after the withdrawal of its weapons from Greece in 2001 and the United Kingdom, starting in 2004 [KNO 10]. There is also a training base (Ramstein) in addition to these bases. The Incirlik air base in Turkey stores the largest number, about one-third, of the weapons in Europe [NOR 11]. These atomic weapons are involved in NATO policy [AND 18].

US nuclear weapons are stored in approximately 24 geographic locations, spread across 11 US states and five European countries. By far the largest nuclear weapons location is the large Kirtland Underground Ammunition Storage and Maintenance Complex south of Albuquerque. Most of the weapons at this location are removed weapons awaiting shipment for dismantling at the Pantex plant in Texas. The second largest inventory is in Washington State, which houses the Strategic Weapons Facility Pacific (SWFPAC) and the ballistic missile submarines at Naval Submarine Base Kitsap.

Among the air-launched cruise missiles (ALCMs), 300 are deployed, 200 at Minot Air Force Base and about 100 at Whiteman Air Force Base. The US Air Force operates a force of 400 Minuteman III ICBMs based in silos at three locations: the 90th Missile Wing at Francis E. Warren Air Force Base in Colorado, Nebraska and Wyoming, the 91st Missile Wing at Minot Air Force Base in North Dakota and the 341st Missile Wing at Malmstrom Air Force Base in Montana [KRI 20].

In Russia, Kristensen and Norris [KRI 17] identify 48 sites with nuclear weapons. Intercontinental missiles (ICBMs) of the Russian deterrent forces are located in at least 11 sites (Barnaul, Dombarovsky, Irkutsk, Kozelsk, Novosibirsk, Tagil, Tashishchevo, Teykovo, Uzhur, Vypolsovo and Yoshkar-Ola).

Kristensen and Korda [KRI 19d] estimate that Russia stores its nuclear weapons at about 40 permanent storage sites across the country, including about 10 central storage sites at the national level. Since 2016, Russia has been upgrading its national nuclear warhead storage sites. This started with Kaliningrad in order to support the nuclear-capable forces in this isolated enclave. It was then the turn of the Air Force nuclear weapons storage site near Tver, about 145 km northeast of Moscow. This site includes two nuclear weapons bunkers as well as service and security buildings and is located near Migalovo Air Force Base. Significant upgrades are also underway at the Mozhaysk-10 storage site, approximately 114 km west of Moscow. The latter site includes six underground igloos [KRI 19b].

After the initial 2010–2012 deployment of the first 18 missiles in two regiments with the 54th Guards Missile Division in Teykovo, the deployment of the 54 SS-27 Mod 2 mobile missiles now appears to be almost complete in the Novosibirsk and Tagil divisions, while an upgrade to the first of the garrisons has begun in Irkutsk and Yoshkar-Ola.

The new Russian submarines are of the Borei class. The first submarine, *Yuri Dolgoruki* (K-535), is based in Yagelnaya in the Northern Fleet. The second, *Alexander Nevsky* (K-550), arrived at its home port in Rybachiy near Petropavlovsk in September 2015, where it was joined by the third Borei, *Vladimir Monomakh* (K-551), in September 2016. All Delta-class sub-surface ballistic nuclear (SSBN) missiles are part of the Northern Fleet and are based in Yagelnaya Bay (Gadzhiyevo) on the Kola Peninsula [KRI 19d].

In 2017, the United Kingdom had four sites involved in the storage of nuclear weapons: Aldermaston Atomic Weapons Establishment England, Burghfield Atomic Weapons Establishment, Coulport Royal Navy Ammunition Depot and Faslane Royal Navy Base [KRI 17].

The French ocean force (SSBN) is based at the Île Longue naval base near Brest in Brittany, which includes two dry docks, a nuclear warhead storage facility and 24 vertical silos for storing missiles that are not loaded on submarines. The missiles are assembled about 4 km south of the base at the Guenvénez Missile Storage and Maintenance Area and the nuclear warheads are then stored south of Île Longue in Saint Jean (Finistère) [KRI 17].

Long-term submarine repairs and refueling are carried out at the Brest Naval Base, which has three large dry docks. The SSBN and SNA are built and dismantled at the Cherbourg shipyard. The FOST command center is located at the Île Longue base and the submarine communication facilities use the HWU transmitter in Rosnay [KRI 19f].

The management and storage of airborne nuclear weapons are managed by the *Dépôts-Ateliers de Munitions Spéciales* (DAMS, Special Ammunition Workshop Depots) located at the Saint-Dizier, Istres and Avord air bases, adapted since 2009 for ASMPA storage (K Buildings). In addition, Avord, Istres or both serve as storage sites for ASMPAs assigned to the Charles de Gaulle for the FOST strike mission [KRI 19f].

China currently operates a fleet of four Jin-class (Type 094) nuclear-powered submarines (SSBNs) based at Longposan Naval Base near Yulin on Hainan Island. Two other Jin-class submarines are currently under construction at the Huludao shipyard [KRI 19e].

In 2017, Israel had at least five sites with nuclear weapons: the nuclear sites of Dimona and Soreq Nuclear Research Center, without any certainty as to their actual use. The site of Sdot Micha is the Jericho II missile base, and the sites of Nevatim and Tei Not are used to store bombs [KRI 17].

India has at least five sites with nuclear weapons. These are the factories in Chandighar (Punjab) for various activities and in Jodhpur (Rajasthan) for the Prithvi and Agni missiles. There are also two bomb storage sites (for air and naval forces) and one missile storage site which do not have a geographic location listed [KRI 17].

Pakistan has spread its nuclear weapons over nine sites. For missile activities, presumably but without certainty, the sites involved are Akro Garrison (Sindh), Gujrawada Garrison (Punjab), Khuzdar Garrison (Balochistan), National Development Complex (Farehjang, Punjab), Pano Akill Garrison (Sindh), Masroor Depot (Karachi, Sindh) and Sangodha Depot (Punjab). Two sites have a variety of activities, Tarbala Underground Depot (Khyber Pakhunkhwa) and Wah Ordonance Facility (Punjab) [KRI 17].

For North Korea, information on the location of nuclear weapons is poor [KRI 18b].

2.3.6. *The state of stocks of nuclear weapons and fissile materials*

Several organizations, often private, regularly make an inventory of the number of nuclear weapons present in the world by state. Among these organizations are the Federation of American Scientists (FAS) which publishes the *Bulletin of the Atomic Scientists* or the Stockholm International Peace Research Institute (SIPRI). These inventories are uncertain because they affect the military domains of each state and therefore are often covered by defense secrecy. Moreover, in counting nuclear warheads, several categories should be retained. These include strategic warheads, non-strategic warheads, those that are deployed, those that are in reserve and, finally, those that are in the process of being dismantled.

The temporal evolution of the nuclear weapon stockpiles of the first two nuclear powers decreased significantly between 2009 and 2018 (Table 2.3). However, this decrease concerned only the number of nuclear warheads, because, during the same period, these weapons were renewed and modernized, meaning increased power and better precision, with the appearance of mirrored warheads (see footnote 1), i.e. multiple warheads where each sub-unit aims at a particular target with an "intelligent" trajectory.

Since the end of the Cold War, the stockpiles of tactical nuclear weapons have decreased by an order of magnitude, from 20,000–30,000 to about 2,500 today. The United States, Pakistan and Russia are modernizing their tactical nuclear arsenals, adding new types to the inventory and increasing the role and importance of tactical nuclear weapons in their military strategies. In addition, tactical nuclear weapons are being used to bias and defraud existing arms control agreements. These trends threaten to recreate some of the dynamics that triggered an arms race and dangerous escalation strategies that increased the risk of nuclear war during the Cold War [KRI 19g].

In addition, the United States continues to improve its four major missile defense systems. One is designed for homeland defense and the other three for regional defense. This enhancement will be without "any limitation or constraint". It is likely to be destabilizing, as potential adversaries will attempt to build offensive systems to offset US defensive systems. This dynamic is currently being achieved with Russia and China, both of which are developing missiles specifically designed to counter US missile defenses [KOR 19].

The storage of fissile materials remains a huge safety problem. There are 1,380 tons of highly enriched uranium and 495 tons of plutonium in the world. Knowing that it takes only about 50 kg or 5 kg, respectively, to produce a bomb, these stocks give a theoretical production capacity of 27,600 uranium bombs and 99,000 plutonium bombs [MAM 16a]. The number of bombs that it is possible to produce will be different if we use the values of 25 kg of highly enriched uranium or 8 kg of plutonium necessary, values cited by the AFP [AFP 17].

Country	Number of nuclear warheads			
	2009[a]	2013[b]	2017[c]	2018[d]
United States	9,400	7,700 (including 3,000*)	6,800	6,185
Russia	13,000	8,500 (including 4,000*)	7,000	6,500
France	300	300	300	300
China	400	250	215	290
United Kingdom	<200	225	270	200
Israel	100–200	100–200	80	80–90
Pakistan	60	100–120	130–140	150–160
India	50–60	90–110	120–130	130–140
North Korea	<10	<10	10–20	(20–30)

Table 2.3. *Temporal changes in nuclear weapons stockpiles in 2009, 2013, 2017 and 2018 (sources: a [CHE 10], b [FEI 14]; c [SIP 18], d [SIP 19]). * Awaiting dismantling*

The state of inventories is always difficult to estimate. Thus, for Israel, the number of plutonium bombs estimated varies from 60 to 400. Similarly, the stock of plutonium for this state was estimated to be between 510 and 630 kg at the end of 2003 [COH 10] and that of highly enriched uranium was estimated at 347 kg in 1996 [MAT 16].

A synthesis on fissile materials in nuclear disarmament and non-proliferation was prepared by Feiveson *et al.* [FEI 14].

In Russia, the stocks of highly enriched uranium (HEU) of military quality are very large, and they are also significant in the United States (Table 2.4). Most of the stockpile is of military origin. However, a non-negligible part has been decommissioned.

Since the 1990s, some weapon states have declared significant quantities of HEU in excess, for weapons or military purposes. This includes the landmark 1993 agreement between Russia and the US to reduce surplus Russian weapons-grade HEU to LEU by 500 tonnes by the end of 2013 [FEI 14].

In 2017, global stocks decreased significantly, compared to 2012 (Table 2.5). This was true for the two superpowers, while stocks in India and Pakistan were growing steadily.

	Military	Naval	Civil	Eliminated
Russia	616	30	20	517
United States	260	252	20	141
Other nuclearized states	59	6	6	-
Other non-nuclearized states	-	-	15	-

Table 2.4. *Stocks of highly enriched uranium (in tonnes) at the end of 2012 (source: [FEI 14])*

State	Enriched uranium reserve (tonne)
China	14 ± 3
France	30 ± 6
India	4 ± 14
Israel	0.3
Pakistan	3.4 ± 0.4
Russia	679 ± 120
United Kingdom	21.1
United States	574.5
Others	15
Total	**1,340**

Table 2.5. *World stock of enriched uranium in 2017 (source: [KÜT 19])*

Plutonium stocks are very large in the United States and smaller in Russia (Table 2.6). As a result of post-Cold War reductions in their nuclear arsenals, three countries reported a combined total of more than 90 tons of excess plutonium. They are the United States, Russia and the United Kingdom (4.4 tons was eliminated by the end of 2012). France has reduced its nuclear weapon stockpile by about 50% since the end of the Cold War, but did not declare any excess plutonium for military purposes until 2013 [FEI 14].

Civilian stocks are generally higher than military stocks (Table 2.7). Russian stocks changed little between 2012 and 2017, while US civilian stocks disappeared (though the reason is unknown). On the other hand, the civil stocks of some states (France, Japan and the United Kingdom) tended to increase.

	Military	Excess	Civil	Eliminated
Russia	88	40	50.7	-
United States	38.3	49.3	149	4.4
Other nuclearized states	17.2	-	-	-
Other non-nuclearized states	-	-	61.1	-

Table 2.6. *Stocks of plutonium (in tonnes) at the end of 2012 (source: [FEI 14])*

State	Military stock (t)	Civilian stock (t)
China	2.9 ± 0.6	0.04
France	6 ± 1	65.4
Germany	-	0.6
India	0.58 ± 0.13	6.4 ± 3.5
Israel	0.9 ± 0.13	-
Japan	-	47
North Korea	0.04	-
Pakistan	0.28 ± 0.09	-
Russia	128 ± 8	57.2
United Kingdom	3.2	110.3
United States	87.8	-
Others	-	2.3
Total	**230**	**290**

Table 2.7. *Global stockpile of military and civilian plutonium (source: [KÜT 19])*

North Korea's stocks at the end of 2016 were 33.1 ± 1.56 kg (23.2–37.3 kg) for plutonium and 175 ± 28.5 kg (116–276 kg) for highly enriched uranium. HEU was detected as early as 2006–2007. On the basis of this cumulative estimate, North Korea is currently increasing its nuclear weapons at the rate of about three to five weapons per year [ALB 17].

Nuclear "disarmament" is slowing down, as revealed by the Swedish Research Institute SIPRI in a new report, published in 2018. In January 2017, nine countries shared 14,395 nuclear warheads, of which nearly 4,150 were considered operational. According to SIPRI, the world's nuclear arsenal decreased by 460 warheads in 2016, a number almost equivalent to the previous year (455) but still small compared to the 2,000 warheads neutralized in 2010 [GAI 17, SIP 18].

2.4. Disarmament controls

The best treaty in the world is worthless if it is not respected and if all parties to the treaty cannot verify its application. All confidence lies in the effective verification of nuclear disarmament. The main role of a verification system is the same for all arms control treaties, namely, to deter cheating [GOL 07]. Depending on the treaty, verification can be carried out at the international level or bilaterally. International controls mainly concern the NPT (Chapter 1) and CTBT.

2.4.1. *Controls of the CTBT*

The CTBT has not entered into force. Among the countries with nuclear forces, only the United Kingdom, France and Russia have signed and ratified it. The United States signed it in 1996, but Congress has refused to ratify it since then. France was among the first states to sign, in September 1996, and ratified in April 1998, this instrument that is aimed at banning all nuclear explosions [RÉP 19].

Despite this state of affairs, the UN created the Comprehensive Nuclear-Test-Ban Treaty Organization (CTBTO) within the framework of this treaty. The Provisional Technical Secretariat of the Preparatory Commission of the CTBTO has been established in Vienna since 1997. It is responsible for preparing for the treaty coming into force by installing and developing the technology necessary for monitoring and verifying all forms of nuclear explosions around the world. This secretariat is composed of approximately 300 international civil servants. The regular budget of the CTBTO in 2019 was US$121.27 million [RÉP 19].

The verification regime, based on an International Monitoring System (IMS), has been set up under the aegis of the CTBTO to detect, locate and confirm any nuclear test with a power greater than 1 kiloton. This IMS is, in fact, a network of 321 detection stations (170 seismic stations, 80 radionuclide stations, 60 infrasound stations and 11 hydroacoustic stations) spread around the world, as well as 16 radionuclide analysis laboratories [LAG 19]. The measurements are received at the International Data Centre (IDC) in Vienna, mostly in (near) real time, with the exception of radionuclide measurements, due to the time required for sample collection and counting in the laboratory. The CTBTO makes these data and the associated analytical bulletins available to the signatory states so that they can determine, in full sovereignty, the nature of the events detected. Today, this network is approximately 85% operational. The verification regime also includes a consultation and clarification mechanism and possible on-site inspections, which any state may request in case of persistent doubt. States have the possibility to voluntarily submit pre-event information, which are known as confidence-building measures [RÉP 19].

Overall, the CTBT, even though it is not universally ratified, appears to be relatively well followed. Exceptions exist for North Korea and doubts are expressed for Russia. For example, on May 29, 2019, General Robert Ashley, the head of the Defense Intelligence Agency (DIA), stated that Russia "probably does not respect the moratorium on nuclear testing", i.e. its commitments under the CTBT. The unexpected dispatch to the United Kingdom in February 2017 of one of the two WC-135 "Constant Phoenix" aircraft of the US Air Force, whose mission is to detect the presence of radioactive particles, could support this assertion [LAG 19].

2.4.2. *Bilateral controls*

Early Soviet–US agreements and treaties provided for relatively effective disarmament controls. In contrast, the most recent agreements were very weak on verifying the reality of disarmament.

2.5. Conclusions

Since 1945, states have engaged in numerous discussions within the United Nations, or bilaterally, to achieve control of nuclear arsenals. As a result, more than 10 treaties and bilateral agreements have been signed. Only the United States and the Soviet Union (then Russia) have ratified bilateral treaties on both nuclear test limitation and their strategic arsenals. On the surface, these arms control and limitation treaties discussed during the Cold War were very detailed, particularly on the accounting of nuclear weapons and the process of verifying destruction or stockpile limitation. In reality, they also provide the means to ensure the sustainability of arsenals [COL 15].

The bilateral agreements are currently undergoing a serious crisis. Some bilateral agreements, such as the INF, are being abandoned and the latest agreements are not very binding. A significant slowdown in disarmament has been observed in recent years.

Disarmament and arms control are prevented by lies from the states. The existence of technological innovation allows the sovereign state to conspire to present arms controllers with formidable and often insoluble problems. It is therefore essential to provide an honest diagnosis of the situation in order to achieve a minimum of progress for humanity [CAR 18].

The Cold War nuclear arms race is over. However, the thousands of nuclear weapons and hundreds of tons of weapons-grade fissile material will not disappear overnight. Preventing their spread to rogue states and terrorist groups is of paramount importance to international peace and security. In particular, there remains a serious security problem with Russia's huge stockpile of highly enriched uranium (HEU) and plutonium. The Russian government and the international community have been working for many years to strengthen the security of fissile materials [VON 18].

However, much remains to be done. The dismantling of nuclear warheads, the safe management and disposal of fissile materials and the restoration of the environment must become the main tasks of the American and Russian nuclear military complexes. Fissile material operations must emphasize international cooperation and transparency. The complexes themselves must be downsized and reconfigured to be able to fulfill these post-Cold War tasks. Workers and facilities must be redirected to non-nuclear commercial activities [VON 18].

The return to conventional deterrence seems difficult for nuclear-armed and geopolitically important states. Indeed, for them, nuclear weapons are perceived as a compensation for the supposed conventional superiority of the adversary. This is exemplified in NATO's response to Soviet conventional superiority during the Cold War, or Pakistan's response to India's military superiority.

It should be remembered that nuclear weapons coincided with the beginning of a global strategic rivalry between the United States and the Soviet Union. Thus, nuclear weapons played an important role in the United States' decision to forge alliances in Western Europe and the Asia–Pacific region, as well as its decision to forge these alliances against a higher conventional adversary. It should be noted that the United States is, to date, the only state to have extended nuclear deterrence.

Moreover, the significance of nuclear weapons is increasing, particularly in the Asia–Pacific region, where China has become the main competitor of the regional order formerly led by the United States.

The prospect of general and complete disarmament has never been taken seriously by the five powers. In any case, it is imprudent to think that a world without nuclear weapons will also be a world without missiles. It seems that nuclear weapons will continue to play a crucial role in international security beyond the initial (and short-lived) context of the Cold War [LEA 17].

The United Nations considered that a new disarmament agenda was needed. Indeed, the tensions of the Cold War have returned, but in a much more complex and dangerous environment. Armed conflicts have become more frequent, longer and more devastating for civilians. The costs of the resulting insecurity are enormous. For example, in 2017, more than one-eighth of the world's gross product was spent to contain all forms of violence, and global military spending is at its highest level since the fall of the Berlin Wall [UN 18].

In many recent conflicts, the laws of humanity have been ignored, and prohibited weapons, such as chemical munitions, have returned to the battlefield. In addition, many of the disarmament commitments and promises made at the end of the Cold War have not been fulfilled.

To save humanity, nuclear disarmament is indispensable. The existence of nuclear weapons is a permanent threat to the world. Their total elimination can only be achieved through reinvigorated dialogue, serious negotiations and a return to a common vision leading to nuclear disarmament. All states must work together to put in place concrete and irreversible measures to prepare for a world free of nuclear weapons, including making the nuclear test ban permanent, developing methods to verify nuclear disarmament and ending the production of fissile material for weapons [UN 18].

3

International Recommendations and National Policies in Decommissioning

3.1. Introduction

The terminology used for operations that occur when a Basic Nuclear Installation (BNI) is permanently shut down varies from country to country. The same applies to the various phases in the life of a BNI. In this chapter, the general principles of decommissioning and dismantling are described together with the recommendations of international bodies, mainly the IAEA and the NEA. Finally, the policies associated with the decommissioning process of the main countries involved in nuclear energy will be detailed.

3.1.1. *Definitions of terms for end-of-life operations of a BNI*

Terminology concerning actions taken at the end of the life of a basic nuclear installation (BNI) is not yet fixed and accepted by all. In France, the ASN and IRSN instead use the term *démantèlement* (dismantling) for all clean-up and disassembly operations. Conversely, French project owners use different language; the CEA has the "A&D" (*Assainissement et Démantèlement* – Clean-up and Dismantling) program for both its civil and military installations, as does Orano, but EDF uses the term "deconstruction". The Americans have long developed "D&D" (Decontamination and Decommissioning) programs. The European Union, on the other hand, uses the term "decommissioning" to refer to all the administrative and technical operations that allow for a facility to be removed from the list of authorized installations.

Administrative operations include the preparation of decommissioning plans and the obtaining of permits and certificates of free release for the facilities and site.

Technical operations include decontamination, dismantling and waste management. Decommissioning is not intended to destroy buildings, but releases them from all obligations and controls corresponding to the class to which they belong. This is the final objective to be achieved [EU 05a].

3.1.2. *Stages in the life of a BNI*

The life of a basic nuclear installation can be subdivided into four periods: construction, operation, transitional phase and decommissioning phase.

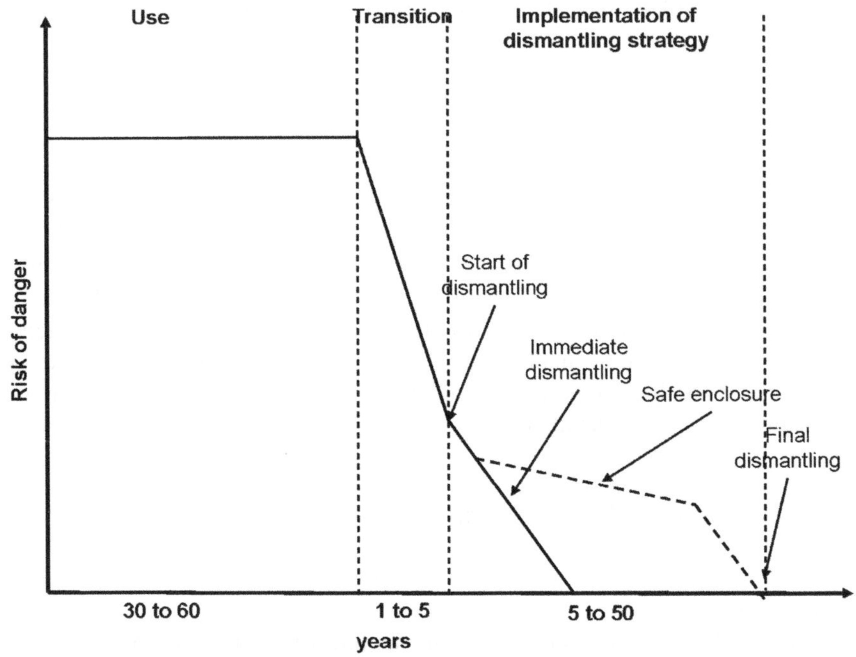

Figure 3.1. *The various phases in the life of a basic nuclear installation (source: [NEG 12])*

During the transitional phase between the abandonment of a nuclear facility and the commencement of final decommissioning, the planning, management and conduct of activities in preparation for final decommissioning take place [NEG 12]. After this transitional phase, decommissioning begins, which may be carried out according to several options (Figure 3.1).

The operational life of nuclear facilities is largely and primarily determined by economic and safety considerations (usually several decades) [EU 05a].

3.2. General principles of decommissioning and dismantling

The decommissioning and dismantling of nuclear facilities are obvious necessities from a health and environmental protection point of view.

International bodies such as the IAEA and the NEA have carried out much work on the subject, making it possible to issue recommendations to national authorities responsible for nuclear safety and security. States must make choices on both the decommissioning strategy and the desired end state of the nuclear site. This conditions the entire decommissioning process. It appears that the best solution for successful decommissioning is that it should be planned as early as possible in the life of the installation, and is best prepared for at the design stage of the BNI. Planning for decommissioning is an essential prerequisite, which should be reviewed and adjusted regularly. Decommissioning is a lengthy and costly process.

Abelquist [ABE 14] addresses all technical aspects required for the planning and execution of radiological studies in support of decommissioning.

3.2.1. *The necessity and goals of decommissioning*

At the end of a BNI's operation, a high level of radioactivity remains throughout the installation and this represents risks that must be controlled. The structures of an abandoned BNI deteriorate more rapidly than the risks related to ionizing radiation will diminish. Therefore, a BNI that has been shut down must be monitored. This is the main reason why it is necessary to initiate decommissioning operations.

The main risks concern workers exposed to strong irradiation, as well as risks of environmental contamination by the most mobile radionuclides. It is also essential to manage radioactive waste, which is produced in much larger quantities than during the operating phase. In all dismantling operations, there are conventional risks because some of these operations are similar to the deconstruction work carried out in the construction industry. There are also risks related to the loss of design and operating memory and maintenance of skills, and those related to inadequate monitoring that may lead to long-term pollution of the site or its environment.

The decommissioning and dismantling of a BNI have three objectives. The first objective is to stop the operation of the BNI and its role; this role would often have been to produce electricity or to contribute to an experiment, but could also be another. Second, the shutdown leads to the evacuation of dangerous substances and waste still present in the BNI. Finally, the end state will often be achieved by decontamination, then the evacuation of the equipment of the operating process, the clean-up of the premises, the possible deconstruction of the installation's buildings and the clean-up of the soil [IRS 17].

3.2.2. *IAEA recommendations*

The issues related to the decommissioning of nuclear power plants are taken into account by the IAEA, in the context of a growing number of activities, and led to the creation of the Decommissioning and Environmental Remediation Section (D&ERS) in 2018. The IAEA's International Decommissioning Network (IDN) was created to link professionals involved in decommissioning activities, and to enable them to exchange information and to collaborate. The ARTEMIS service (Integrated Review Service for Radioactive Waste and Spent Fuel Management, Decommissioning and Remediation), for example, provides expert advice on decommissioning, radioactive waste and spent fuel management, and environmental remediation programs [FRA 18e].

The life of a BNI passes through six stages: site selection, design, construction, commissioning, operation, and finally, decommissioning. Prior to construction and decommissioning, a study of each of these stages must be presented by the operator to the regulatory authority (e.g. the ASN in France which submits them for analysis to the IRSN). A formal agreement is required before any decommissioning and dismantling operations can begin. In the case of dismantling, the operator must also explain the procedure for managing the waste generated by all operations. This decommissioning stage is the last one and it removes all regulatory controls attached to this type of installation, with the exception of nuclear waste disposal facilities where decommissioning is replaced by closure.

Decommissioning should be considered at all previous stages. This important step consists of several sub-steps: planning the decommissioning, performing the decommissioning operations and lifting the decommissioning license.

Decommissioning planning begins at the design stage and continues throughout the life of the BNI. It includes the preparation of an initial decommissioning plan, the collection of information and data to implement the plan, the selection of a decommissioning strategy, the radiological characterization of the BNI, the cost estimate and the search for funding, the request for advice from the appropriate authorities and possibly public consultation.

The decommissioning operations themselves include project management, implementation of the final decommissioning plan approved by the authorities, management of radioactive and non-radioactive waste and demonstration that the BNI subscribes to the final stage criteria previously defined. All this work is carried out by or under the responsibility of the operator, and is overseen by the regulatory body.

States choose their own decommissioning strategy. Two options are possible: immediate dismantling or deferred dismantling. Both are possible for all facilities. Strategies combining both are also possible.

The IAEA [AIE 17] recommends 15 requirements to be met to ensure the best conditions for decommissioning a BNI.

The first three requirements relate to the safety and protection of workers, the public and the environment. First, the protection and safety of populations during decommissioning must be optimized. During this stage, the basic international standards for radiation protection of workers and the public are rigorously applied. Second, a step-by-step approach is adopted throughout the decommissioning stage on the possible radiological risks attributable to this stage. The third requirement is to plan for the safety of decommissioning.

There are three requirements regarding the responsibilities associated with decommissioning. It is the government of the state in which the BNI is located that is fully responsible for decommissioning. The same is true of the decommissioning regulatory body, such as the ASN for France, which regulates all aspects of decommissioning. Finally, the decommissioning licensee (usually the operator of the BNI) plans the decommissioning and carries out the associated operations according to the decommissioning authorization. The licensee is responsible for all aspects of safety, radiation protection and environmental protection during this stage.

The seventh requirement relates to decommissioning management and requires the licensee to have an integrated management system that covers all aspects of decommissioning.

The eighth requirement states that the licensee shall select the decommissioning strategy and ensure that it is consistent with the national policy on radioactive waste management.

The requirement for decommissioning funding is a reminder that this should be set out in national legislation.

The following two requirements relate to decommissioning planning during the lifetime of the facility. The licensee is required to develop a decommissioning plan and maintain it throughout the life of the BNI. In addition, the licensee develops a final plan and submits it to the regulator for approval prior to commencing decommissioning operations.

The following three are the requirements to be followed when carrying out decommissioning operations. The licensee shall implement the final decommissioning plan, including radioactive waste management, in accordance with national regulations. The licensee shall make provisions for emergency response and report all events that are pertinent to safety to the regulatory body. It also manages waste streams during decommissioning.

Finally, the fifteenth and final requirement is that the licensee demonstrates, upon completion of the decommissioning operations, that the end-stage criteria meet those described in the final decommissioning plan. The regulatory body verifies this compliance and decides on the lifting of the decommissioning authorization [AIE 17].

3.2.3. *NEA work and publications*

Within the Nuclear Energy Agency (NEA), the Working Party on Decommissioning and Dismantling (WPDD) was created. This group operates under the aegis of the Radioactive Waste Management Committee (RWMC).

At the end of 2016, two phases of work were completed. The first developed strategic guidance for decision-makers on the selection and adaptation of radiological characterization strategies, which provides an overview of good radiological characterization practices at different phases of the lifecycle of a nuclear installation. The second phase focused on strategies for the practical implementation of radiological characterization from the perspective of the final state of waste and materials [LAR 17b].

3.2.4. *Decommissioning and dismantling strategies*

There are three main strategies for decommissioning nuclear facilities: "immediate decommissioning", "deferred decommissioning" and "secure containment with sealing". These three principal strategies are referred to as DECON, SAFSTOR and ENTOMB by the European Union [EU 05a].

3.2.4.1. *Immediate dismantling or decontamination (DECON)*

Under DECON, all radioactive components and structures are cleaned or disassembled, packaged and shipped to a low-level waste disposal site, or they are temporarily stored on site. This task lasts five years or more but once completed, the regulator terminates the basic nuclear facility's license. As a result, this part of the site can be reused for other purposes.

Immediate decommissioning will result in higher radioactive waste activity levels, generated due to the lack of time during which decay can occur. Nevertheless, it will shorten the overall duration of the decommissioning project by reducing the amount of work required [WNN 19v].

3.2.4.2. *Deferred dismantling or secure storage (SAFSTOR)*

With the SAFSTOR option, the BNI is kept intact and placed in protective storage for decades. Depending on the installation, this period varies from 20 to 150 years. This method consists of isolating the facility containing radioactive materials from the environment and monitoring it with an on-site security team. Time is used as a decontamination agent by physical decay of radionuclides. This is only an advantage for short- or medium-lived radionuclides. Thus, in the case of a BNI authorized to remain inactive for 30 years, the radioactivity of cobalt-60 (a period of 5.27 years), for example, will be reduced to one-fiftieth of its level at the time of shutdown. After 50 years of cooling, the radioactivity will be only one-thousandth of its original level. After this monitored waiting period, and therefore once the radioactivity is considerably reduced, the installation can be dismantled according to the DECON strategy.

Delayed decommissioning will reduce the radioactive waste's activity level generated as a result of radioactive decay and may allow for more manual (as opposed to remote) handling. This will result in lower disposal costs, but potentially higher overall decommissioning costs due to the longer post-operational life [WNN 19v].

The Americans developed the Safestore concept in three main stages. The first comprises a provisional schedule, removal of spent fuel, engineering work to prepare for dismantling and operational management of potentially evacuable waste. The second stage consists of the decommissioning of the facility, construction of the sarcophagus and the monitoring, upkeep and maintenance of the site. The third stage is preparation for the dismantling of the reactor, the management of high-level waste, actual dismantling of the reactor and the clean-up and authorization of the site release. It has been applied in particular at Hinckley Point B in the UK [BAY 03].

Examples of this strategy of building a safe enclosure are numerous, such as the nuclear power plants of Dodewaard (the Netherlands), Vandellós-1 (Spain), Kozloduy (Bulgaria), Lingen (Germany) and Berkeley (UK) [THI 12].

3.2.4.3. *Containment or entombment (ENTOMB)*

This option consists of enclosing (encapsulating) all radioactive structures, systems and components in a time-resistant material such as concrete. The encapsulated BNI will be properly maintained and monitoring will continue until the radioactivity decreases to a level that allows for the termination of the facility's license. Most nuclear power plants will have radionuclide concentrations exceeding the limits of unrestricted use even after 100 years. Therefore, special provisions will be necessary for the extended monitoring period required by this option. To date, no facility owner has proposed the option of a sarcophagus or fallout for nuclear power plants undergoing decommissioning. This option is, in fact, similar to declaring the site as a shallow landfill [EU 05a].

Finally, entombment uses the existing structure to contain the decommissioning waste *in situ*. The resulting buried structure effectively becomes radioactive waste within its own disposal site, which requires ongoing institutional controls, proportional to the associated waste categories. Because of social, political and land scarcity problems, national policies for decommissioning civilian nuclear power plants do not normally support entombment [WNN 19v]. On the contrary, this solution has sometimes been adopted in the United States because its territories are vast. This is the case for the installations at the Hanford National Laboratory and Idaho sites, as well as the plutonium-producing reactor P at the Savannah River site [THI 12].

An intermediate solution has recently been developed with the storage of large components, or the segmentation of highly radioactive components, for the physical decay of short- and medium-lived radionuclides. The dismantling of the installation itself is thus distinct from the non-segmented components, their segmentation being planned for later, after decontamination or possibly clearance. This combines the advantages of the early and delayed decommissioning strategy [THI 12].

3.2.4.4. *The choice of strategy*

When choosing a decommissioning strategy many factors must be considered. These factors fall into three categories: political and socio-economic factors, technological and operational factors and long-term uncertainties. Most of these factors are not quantitative in nature and require subjective assessment.

Political and socio-economic factors are determined by the national and/or local situation. However, these national policies vary considerably from country to country and evolve over time. Technological and operational factors will influence the choice of strategy; even though decommissioning techniques exist, new ones emerge frequently. The complexity of the decision-making process explains the difference between national decommissioning strategies and the evolution of these strategies. Similarly, the methods of financing decommissioning activities differ from country to country [AEN 06].

The main factors influencing the choice of decommissioning strategy are national nuclear policy, the characteristics of the facility, management of the health and safety of workers on site and the surrounding population, environmental protection, the potential for radioactive waste management, the proposed future use of the site, improvements in decommissioning technologies that could be made in the future, the cost and availability of funds for the project and various social considerations.

The national nuclear strategy will be detailed in section 3.4. Among the characteristics of power plants, the main ones that influence the choice of decommissioning strategy are the type of reactor, the location of the facility and the total amount of radioactivity it contains. The amount and location of radioactivity are determined by the type of reactor and its operating history. For example, a Boiling Water Reactor (BWR) circulates steam containing radioactivity through the turbine circuit, whereas in a Pressurized Water Reactor (PWR), the radioactivity is contained in the primary cooling system and does not contaminate the turbine circuit [EU 05a].

3.2.4.5. *The end-state choice*

Decisions on end-state strategy will have a direct and major impact on decommissioning planning and associated costs. At the end of a nuclear facility's life, two end-state objectives may be chosen, either the "greenfield" or "brownfield" option.

The "brownfield" option involves maintaining some regulatory control, i.e. the site is reused with certain restrictions, the nuclear facility is removed, and the radioactivity that was present is removed as radioactive waste. Any remaining radioactivity is normally associated with underground structures and low levels of soil contamination.

The "greenfield" option removes all regulatory controls and the site is released from any regulatory oversight with no restrictions on future use. The site is cleared of all traces of radioactivity [WNN 19v].

Among the completed decommissioning projects, there are several successful strategies for reusing decommissioned sites for new purposes. For example, some former BNIs have been transformed into museums, new industries or waste disposal facilities [LAR 12b].

3.2.5. *Decommissioning planning*

The operator of a BNI provides a decommissioning safety case to the national safety authority, who will approve it (or not), allowing the process to begin. This decommissioning safety case for an installation is different from the safety case for its operation. It must be updated as changes and transformations occur in the situation of the installation, including its management. The decommissioning safety case will require a characterization of the installation, a safety assessment and, in most cases, an environmental assessment, as these data are specific to the installation in question [AEN 05b].

For proper planning and safe implementation of facility decommissioning, a good estimate of the amount and nature of the radioactivity present in the facility (i.e. the source term) is necessary. Most of this inventory for nuclear power plants, research reactors and gas accelerators is created by activation induced by neutrons or other particles (protons and ions) [IAE 19b].

Information on the structure of the project organization depends on the size and complexity of the project and the attributes that the project organization must have. It includes details such as staffing, organization of the management team, required workforce skills, change management, training and use of internal resources in relation to contractors [DEY 12].

3.2.5.1. *The decommissioning process*

Eight main steps can be distinguished when decommissioning a BNI. The first is the initial timeline for developing a strategic plan and procedures. The second stage consists of removing the spent fuel for nuclear reactors and reprocessing plants. The third is the inventory and characterization of radioactivity (according to Waste Acceptance Criteria (WAC)). The fourth step is decontamination after careful preparation. The fifth step is the actual dismantling, which must be rigorously planned. The sixth stage is the characterization, transport and management of nuclear waste. The seventh step is the conventional demolition of the remaining structures with their disposal in possible reuse channels. The last stage is the final state of the site which, in the best case, will be perfectly clean and suitable for other uses (Figure 3.2).

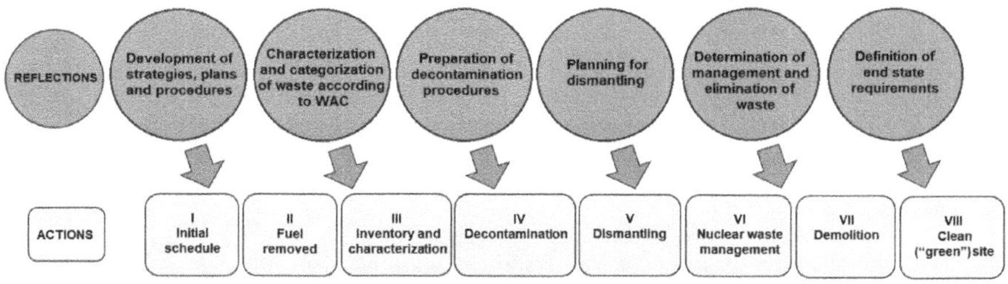

Figure 3.2. *The various stages of the decommissioning and dismantling of a nuclear installation (source: [BER 19]). For a color version of this figure, see www.iste.co.uk/amiard/disarmament.zip*

During the decommissioning of a nuclear installation, the dismantling operations are the most delicate and indispensable. They consist first of allowing the various components of the installation to cool down, i.e. to allow the physical decay of the different radionuclides to take place over time so that the irradiation decreases significantly. The second stage consists of dismantling and cutting up the various components of the installation. In a third step, the numerous components of the installation must be sorted and then stored in specialized landfill sites.

3.2.5.2. *The decommissioning schedule*

Careful planning is one of the key factors in the successful management of any large infrastructure project. There are several potential obstacles that can lead to cost escalation and unnecessary delays. These can be overcome with the experience that has now been accumulated and with real-world case-based solutions that have been developed.

The transition from an operating nuclear facility to the decommissioning phase is essential in the lifecycle of each facility. This is prepared for during the transition phase (Figure 3.3). A number of organizational and technical changes are required for the facility to meet the new objectives and requirements, and a number of activities must be initiated to accompany the transition and preparation for decommissioning of the facility. Careful preparation and planning are essential to the success of global decommissioning and dismantling projects, both to minimize undue delays and costs and to ensure a safe and efficient decommissioning process [NEA 18a].

The decommissioning sequence will be successful if two points are addressed. First, the definition of the desired end-state and future use of the site shall take place early in the lifecycle of the plant, i.e. during the design phase of the BNI. Second, the establishment of radiological, physical and chemical inventories shall be carried out as early as possible. These should be updated throughout the operational phase up to the closure of the facility.

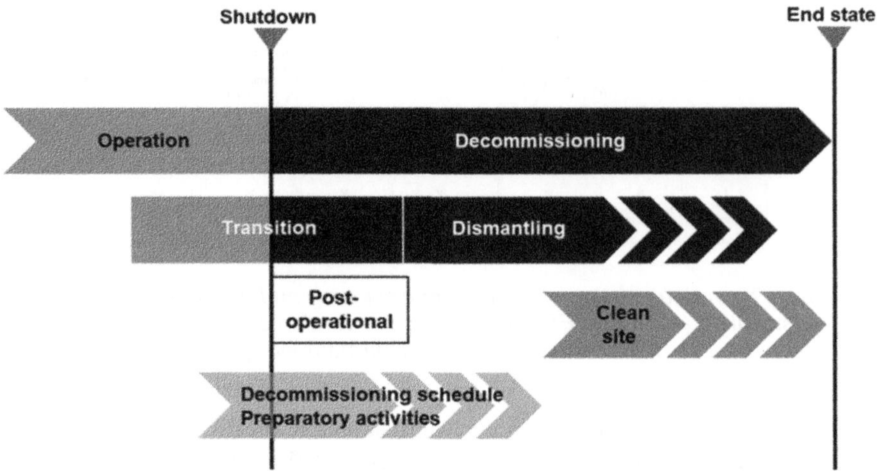

Figure 3.3. *The three phases of a nuclear facility: operation, transition and decommissioning (source: [NEA 18a]). For a color version of this figure, see www.iste.co.uk/amiard/disarmament.zip*

3.2.5.3. *The decommissioning and dismantling stage statuses defined by the IAEA*

A traditional definition of the stages of a decommissioning status has been proposed by the IAEA [IAE 95] and is widely used worldwide. It is based on the classification into three stages representative of the decommissioning state and dismantling site of a nuclear installation.

The first stage is called "storage with surveillance". It consists of minimal decontamination, draining of liquid systems, disconnection of operating systems, accompanied by physical and administrative controls to ensure limited access and continuous monitoring of the site (regular monitoring and inspection). This maintenance lasts for a predetermined period of time, usually a few years to a few decades. Before completing this stage 1 for nuclear reactors, the spent fuel must be removed from the installation. This stage does not legally modify the nuclear license of the site.

The second stage is referred to as "restricted site release". All equipment and buildings that can be easily dismantled are removed or decontaminated and made available for other uses. All remaining fluids are drained from the systems. In nuclear reactors, the biological shield is extended and sealed to completely enclose the reactor structure. In fuel cycle facilities, the primary radioactive building and equipment will sometimes be removed. Monitoring around the barrier may be reduced, but it is desirable to continue periodic spot checks and environmental monitoring.

Stage 3 is called "unrestricted site release". All buildings, equipment and materials that cannot be decontaminated below established release threshold levels are removed (at least for states accepting release thresholds) and the resulting waste is packaged and permanently stored or disposed of. The remaining parts of the plant and site are released for unrestricted use. In some cases, the facility is completely deconstructed and the site is returned to greenfield conditions. No additional monitoring is required [EU 05a].

3.2.6. *Duration of decommissioning*

The decommissioning of a nuclear facility can be completed in a relatively short period of time or it can be spread over 100 years or more. The choice of a long time frame, often defended by industrialists, poses a serious ethical problem. Indeed, generations that have not benefited from the advantages of the facility are left to manage the costly liabilities of the waste management.

Reactor	Type, power (MWe)	Operational dates	Duration
United States			
Big Rock Point	BWR (67)	1997–2006	9
Connecticut Yankee	PWR (560)	1996–2007	11
Fort St. Vrain	GCR (330)	1989–1996	8
Rancho Seco	PWR (873)	1989–2009	12
Trojan	PWR (1095)	1992–2005	13
Carolina Virginia Tube Reactor	PHWR (17)	1967–2009	4
Yankee Rowe	PWR (167)	1991–2007	16
Elk River	BWR (22)	1968–1974	6
Maine Yankee	PWR (860)	1997–2005	8
Saxton*	PWR (3)	1972–2005	18*
Shippingport*	PWR (60)	1982–1989	7*
Shoreham*	BWR (820)	1989–1995	6*
Pathfinder*	BWR (59)	1967–1992	3*
Germany			
Grosswelzheim	Proto (25)	1971–1998	15
Kahl	BWR (15)	1985–2010	22
Niederalbach	HWGCR (100)	1974–1994	20
Japan			
JPDR	BWR (12)	1976–1996	15

Table 3.1. *Duration of decommissioning of several nuclear power reactors (source: [LAR 18a]). *ISFSI (Independent Spent Fuel Storage Installation) only*

Table 3.1 summarizes the decommissioning time for nuclear reactors that have already been decommissioned. This duration is relatively short for American reactors but, for some, decommissioning is not complete and has remained at the temporary storage stage (ISFSI – Independent Spent Fuel Storage Installation). Table 3.2 summarizes the predicted decommissioning duration for several reactors in various states. The average decommissioning time is close to 25 years.

Reactor	Type, power (MWe)	Operational dates	Duration
United States			
Zion 1.2	PWR (1040)	1996 and 1997–2018	8
France			
Brennilis (EL-4)	HWGCR (70)	1985–2032	35
Bugey-1	GCR (153)	1994–2037	29
Chooz-A	PWR (305)	1991–2025	17
Creys-Malville	FBR (260)	1998–2027	21
Germany			
Stadium	PWR (60)	2003–2023	20
Italy			
Caorso	BWR (60)	1990–2028 or 2032	28–30
Garigliano	BWR (153)	1982–2024 or 2028	24–28
Latina	GCR (150)	1987–2023 or 2027	23–27
Trino Vercellese	PWR (260)	1993–2026 or 2030	26–30
Japan			
Tokai-1	GCR (137)	1998–2025	24
Spain			
José Cabrera	PWR (510)	2006–2018	9

Table 3.2. *Selected decommissioning projects (source: [LAR 18a])*

3.3. Lessons from the past

With more than 70 years of operational experience and societal contribution, in the form of debates on nuclear energy, about half of the world's 600 first-generation research and recycling nuclear facilities are already at decommissioning stage and the other half are approaching it. As of December 2018, 172 nuclear power facilities had been permanently closed, of which 72 are in the process of being decommissioned and more than 20 have had their licenses terminated, which means that these facilities are no longer considered to be nuclearized [WNN 19c]. For its part at the end of 2018, the CEA [CEA 19] listed 176 nuclear reactors that had been shut down and 447 reactors in operation. As of December 31, 2017, there were 50 PWR, 40 BWR, 38 GCR and 38 other nuclear power reactors in the world in shutdown, including 28 PWR, 18 BWR, 37 GCR and 22 other reactors for Western Europe [IRS 19c].

The IAEA has taken this dimension into account by launching various research programs such as DeSa (International Project on Evaluation and Demonstration of Safety for Decommissioning of Facilities Using Radioactive Material).

The main lessons that can be learned from past decommissionings are the need to better coordinate the efforts of companies working in this field, and the need for the license owner to be prepared to deal with the unexpected. Challenges include international harmonization of national legislation, financial assurance of the cost of decommissioning and the safe and permanent management of nuclear waste from dismantling.

3.3.1. *Experience in decommissioning*

Germany and the United States are the pioneers in the decommissioning (D&D, Decontamination and Decommissioning) of nuclear power plants and therefore the market leaders, with many projects in progress or completed. In Germany, 27 nuclear installations have so far been decommissioned and are in the D&D phase. Of these, 15 sites have final dismantling status and three facilities have terminated licenses. In the United States, 37 units have been closed, of which seven have the status of an independent spent fuel storage facility only and three have terminated licenses [WNN 19c].

In response to the International Plan of Action on the Decommissioning of Nuclear Facilities, the international DeSa project has provided a useful platform for the presentation and exchange of lessons learned in the development and review of decommissioning safety assessments. The project led to the development of a harmonized safety assessment methodology, a regulatory review procedure and recommendations on the application of the graded approach [IAE 13b].

In the case of several reactors of the same type, on the same site or on several sites, after an initial experiment, a simpler decommissioning plan can be drawn up for the installations. The characterization of the radioactive materials present and the establishment of the quantities and categories of waste are faster. This also makes it possible to optimize and better plan decontamination and dismantling activities [CRO 12].

3.3.2. *Structuring and organization of companies*

Worldwide experience has shown that companies involved in all aspects of a decommissioning project (planning, operation, maintenance, waste characterization, treatment, disposal, etc.) have had less difficulty planning and managing the project – including optimizing their human resources and skills – than those involving a number of separate organizations.

3.3.3. *Ongoing decommissioning challenges*

Several challenges must be overcome to ensure that the decommissioning of nuclear facilities is as rapid and safe as possible. One of the first challenges is the lack of regulatory parity between countries, which hampers multinational approaches to decommissioning and waste management. There are also the financial and technical challenges of decommissioning and associated waste management that need to be overcome. These are key elements in demonstrating to the public and policy-makers that nuclear energy is an essential and practical form of low-carbon generation [WNN 19v].

3.3.4. *Management of the unexpected in dismantling*

The IAEA [AIE 16] has identified the unexpected cases that arose during the dismantling of BNIs. The experiences come from several states such as the Czech Republic, France, Ukraine, the United States and the United Kingdom, in particular the nuclear reactor accident at the Windscale plant.

3.3.5. *The transmission of information*

Information and Communication Technologies for Education (ICTE) offer means for the collection and dissemination of science and technology content, particularly useful for supporting emerging industries and their associated ecosystems, especially for the decommissioning industry [GIR 16]. In this field, management and long-term preservation of information are particularly necessary [LEC 12].

3.4. The decommissioning and dismantling policies of the various states

Decommissioning and dismantling policies vary widely from state to state. Here, we detail the policies of for states with significant nuclear energy, and therefore states with many reactors to be dismantled in the more or less distant future.

It is important that each nuclearized state develops its own national nuclear strategy. In this regard, the most relevant questions are whether there is a national policy for the elimination of shutdown nuclear facilities and whether decommissioning is the responsibility of private industry or government. It is also essential to know the national policy for the management of the state's radioactive waste, as well as the national policies for the protection of workers and the public [EU 05b].

When choosing a decommissioning strategy, countries with a small D&D program, without a radioactive waste repository, generally prefer a safe enclosure, while no country with a large D&D program opts for deferred decommissioning as the preferred strategy. Several Western countries have even modified their D&D programs in favor of immediate decommissioning [THI 12].

In the European Union, at the end of 2016, 126 reactors were in operation and 90 had been shut down (26 in Germany, 4 in Bulgaria, 2 in Spain, 12 in France, 4 in Italy, 2 in Lithuania, 3 in Slovakia, 5 in Sweden, 30 in the UK, 1 in the Netherlands and 1 in Romania). Only three were completely decommissioned in Germany [BRU 19]. In its balance sheet at the end of 2018, the CEA [CEA 19] listed 94 reactors as shutdown in the EU.

Following the Chernobyl accident in 1986, the European Union decided to shut down high-power channel-type tube reactors (*reaktor bolshoy moshchnosti kanalnyy* – RBMK) and other first-generation, Soviet-designed nuclear reactors. Also, upon accession to the European Union, Bulgaria, Lithuania and Slovakia agreed to close the reactors at the respective sites of Kozloduy, Ignalina and Bohunice. The European Union then launched nuclear decommissioning assistance programs to comply with the accession agreements. The European Commission manages these programs by assigning implementation tasks to the European Bank for Reconstruction and Development in the three member states, the Central Project Management Agency in Lithuania and the Slovak Innovation and Energy Agency. The European Commission has proposed to continue supporting these programs in the period 2021–2027.

3.4.1. *US policy*

In the United States, the nuclear regulator is the United States Nuclear Regulation Commission (US NRC). This authority has created a website on decommissioning, which was updated in 2018 [NRC 18].

In the United States, D&D planning is volatile due to a mix of political, social and economic parameters, all of which have indirect impacts on the final cost per kWh. These parameters have even led to the postponement or cancellation of planned nuclear shutdowns, as seen with the Clinton, Byron and Fitzpatrick plants [WNN 19c].

The decommissioning process begins when a licensee decides to permanently cease operations. Licensees can choose from three decommissioning strategies: DECON, SAFSTOR or ENTOMB. Currently, only the first two, or a combination of them, have been selected by operators of nuclear facilities in the United States.

The decommissioning process is detailed for three types of nuclear facilities: power reactors, nuclear fuel cycle facilities and uranium recovery facilities.

There are three important steps in the process of decommissioning the facility. First, the so-called transition stage, which allows the licensee to submit to the NRC notification of the shutdown of the facility and a Post-Shutdown Decommissioning Activities Report (PSDAR) or a Decommissioning Plan (DP) for other types of nuclear facilities. This report provides a description of the planned decommissioning activities, a schedule for carrying them out and an estimate of the expected cost. In order to protect the environment, the licensee must prepare and publish its draft Environmental Impact Statements (EIS), a Record of Decision (ROD) and a final Safety Evaluation Report (SER).

The second stage includes the main decontamination and dismantling activities.

The third and final step is to complete the decommissioning process. The nuclear license owner must submit a License Termination Plan (LTP) detailing the implementation of the LTP and the completion of the decommissioning. In addition, the licensee must demonstrate that the applicable requirements of the License Termination Rule (LTR) will be met. Finally, the operator proposes a Long-Term Surveillance Plan (LTSP) for the site.

Most decommissioning plans consider making the site accessible to the public for unlimited use, meaning that any residual radiation would be below the NRC annual exposure limit of 2.5 mSv (25 millirems) and the site would no longer be under regulatory control.

Throughout the entire decommissioning process, the sorting and safe disposal of nuclear waste must be ensured by the operator. The US Department of Energy uses various technical analyses to assess whether certain radioactive waste by-products can be managed as Low-Level Waste (LLW), i.e. Waste Incidental to Reprocessing (WIR). When disposing of radioactive waste, the NRC offers a performance assessment. It corresponds to a quantitative evaluation of the potential releases of radioactivity from a storage facility into the environment and an evaluation of the resulting radiological doses.

The United States has created an Environmental Restoration Waste Management (ERWM) structure which is specifically responsible for site remediation and nuclear waste management.

The entire decommissioning process in the United States is subject to federal legislation (CFR – Code of Federal Regulations). Thus, for nuclear waste, performance criteria must follow Title 10, Part 61 of the Code of Federal Regulations (10 CFR, Part 61), "Licensing Requirements for Land Disposal of Radioactive Waste". These provisions are intended to protect the public, workers and the environment. NRC staff guidance for the review of decommissioning plan performance evaluation (DP – Decommissioning Plan) is described in Volume 2 of NUREG-1757 "Consolidated Guidelines for Decommissioning: Characterization, Review and Determination of Radiological Criteria". Similarly, licensees who decommission their facilities are required to demonstrate to the US Nuclear Regulatory Commission (NRC) that their proposed methods will ensure that the decommissioning can be carried out safely and that, at the end of the decommissioning activities, the facility will comply with the radiological criteria for license termination, as defined in Subpart E of Title 10, Part 20 of the Code of Federal Regulations (10 CFR, Part 20).

Currently, in the United States, 36 reactors have been permanently shut down [SCH 19].

3.4.2. *Russia's policy*

The policy of the USSR, and then Russia, has evolved over time with regard to decommissioning and dismantling. Russia's federal nuclear safety agency is Rostekhnadzor, which was established in 2004.

ROSATOM (Russian State Atomic Energy Corporation) was established by the Russian Federation as a state-owned company in December 2007. It manages more than 400 organizations and is responsible for the decommissioning of approximately

2,500 nuclear facilities (Table 3.3). Currently, in Russia, eight power reactors have been permanently shut down [SCH 19].

Type of reactor	Number of facilities to be decommissioned
Reactor	37
Plutonigenic reactor	10
Research facility	54
Storage center	609
Building	1,260
Nuclear icebreaker ship	14
Contaminated soils	466

Table 3.3. *Russian nuclear facilities in decommissioning stage (source: [NEA 18b])*

ROSATOM considers the creation of conditions for the timely and cost-effective decommissioning of nuclear and radiation facilities to be one of its business priorities. This must be done in accordance with modern international standards and, as far as possible, by excluding the transfer of the problem to future generations.

Currently, a complete system of new legislation for the decommissioning of facilities by ROSATOM is being developed. At the time of decommissioning, the safety rules are available for nuclear power plants (NP-012-99), for nuclear research facilities (NP-028-01), for nuclear ship reactors (RT-037-02) and for nuclear fuel cycle facilities (NP-057-04).

All Russian nuclear power plants are managed by ROSATOM's branch, JSC Concern Rosenergoatom. Currently, the power units that have been shut down and safely stored for several years are Novovoronezh Units 1 and 2 and Beloyarsk Units 1 and 2. Preparatory work is in progress.

The work on the Beloyarsk nuclear power plant is complicated by the need to remove spent fuel residues from the storage pools, some of which are damaged. For the practical implementation of decommissioning technologies of the Novovoronezh nuclear power plant with WWER reactors, there is an Experimental Demonstration Engineering Center on Decommissioning.

Other sites are in the process of being decommissioned, such as the 10 plutonigenic (uranium–graphite) reactors with a military function that are located on the Mayak, MCC and SCC industrial sites. Currently, tests of *in situ* technologies for decommissioning this type of reactor are being carried out in an experimental demonstration center. In addition, there are 14 research reactors that are subject to decommissioning, as well as a significant number of units and vessels of the nuclear fleet (naval icebreakers).

Priority work is associated with the closure, conservation and decommissioning of the open ponds that served as an outlet for the Light-Water Reactors (LWR) at Mayak, MCC and SCC. A particularly acute problem for the Mayak site is the volume of the storage ponds (especially the Techinsky cascade reservoirs) and the high radionuclide content of some of them (B-9 Karachai Pond and Pond-17). Work on environmental issues at this site is difficult and is currently being taken into account.

Preparatory work is being carried out to achieve an environmentally sound storage of residues from uranium extraction and enrichment plants [NEA 14c].

3.4.3. *Germany's policy*

Following the accident at Fukushima Daiichi in Japan, the German government asked its citizens if they wanted to continue using nuclear power. After a negative response, Germany adopted a policy of phasing out nuclear energy, as well as a voluntary approach to D&D, with immediate dismantling operations underway [WNN 19c].

By a decree of July 31, 2001 (German Atomic Energy Act – AtG), Germany decided to stop using nuclear energy and to dismantle the seven oldest reactors, as well as the Krümmel reactor, before 2022. To this end, Germany has published a detailed guide to assist decommissioning in the nuclear industry [GRS 16].

Currently, in Germany, 29 reactors have been permanently shut down [SCH 19].

The nuclear regulatory authorities in Germany are the Federal Ministry for the Environment, Nature Conservation and Nuclear Safety (BMU – *Bundesministerium für Umwelt, Naturschutz und nukleare Sicherheit*) for nuclear safety and radiation protection and the Federal Office for Radiation Protection (BfS – *Bundesamt für Strahlenschutz*) for radiation protection and waste disposal. In 2016, 9 reactors were shut down, 16 were being dismantled, 3 had been completely dismantled and 8 reactors were still in operation [BRE 16].

The shutdown of the Grafenrheinfeld nuclear power plant in June 2015 marked a new phase in the energy transition. This "exit" from nuclear energy should be seen rather as an "entry" into the decommissioning economy and, in particular, final disposal. However, with the closure of the Grafenrheinfeld nuclear power plant, major challenges are emerging, namely the safe decommissioning of nuclear power plants and the long-term storage of nuclear waste. There are still no reliable cost estimates. Nevertheless, immediate decommissioning seems to be the best choice.

Even today, six decades after the first commercial use of nuclear energy in Germany, there is still no storage, or even a planned storage site, for high-level radioactive waste. Moreover, decommissioning and final decommissioning involve an unusual political time scale of decades to centuries [VON 15].

3.4.4. *UK policy*

In 2005, the United Kingdom created a specialized decommissioning agency, the Nuclear Decommissioning Authority (NDA). Its objective is to decommission and clean up the UK's civil nuclear legacy in a safe and cost-effective manner and, where possible, to accelerate work programs that reduce risk. The NDA manages 17 nuclear sites covering 953 hectares and is responsible for decommissioning more than 800 nuclear buildings. This program involves 16,000 employees and will take place over more than 100 years.

The NDA defines an overall domain-wide strategy, decommissioning management priorities, allocates budgets and monitors performance at the 17 sites. Site License Companies (SLCs) carry out work at each site in accordance with the decommissioning plans agreed by the NDA [NAO 18].

Although the NDA itself employs only 300 people, its annual budget is £3.2 billion. The vast majority of the NDA's budget is spent through contracts with site license companies, which also subcontract to other companies that provide special services. The NDA aims to achieve this by introducing innovation and expertise from contractors through a series of competitions similar to the model that has been used in the United States. In this regard, in April 2017, the NDA lost a lawsuit in the Supreme Court regarding the awarding of a major contract for the decommissioning of 12 Magnox nuclear facilities. Indeed, an unsuccessful competitor challenged the NDA's choice and its appeal was accepted.

In February 2018, the Public Accounts Committee (PAC) of the British Parliament concluded that the NDA had "grossly underestimated" the costs and "completely failed" to procure and manage the Magnox Ltd contract.

In order to carry out its activities, the NDA is highly organized. Responsibility for site operations has been restructured into five site license companies (SLCs). The management of the SLCs is outsourced to different organizations specialized in decommissioning (PBOs – Parent Body Organizations), which often include several private companies. The five geographical zones are Sellafield, Magnox Ltd, Dounreay, LLW Repository Ltd and Springfield.

The Sellafield zone is managed by Sellafield Ltd, formerly a subsidiary of British Nuclear Fuels Limited (BNFL). This zone includes the Sellafield nuclear chemistry facility and the Calder Hall and Windscale sites. In contrast, the uranium enrichment plant at Capenhurst is now managed by Urenco. Since April 2016, NDA has directly controlled the Sellafield Ltd site. The main subcontractor is Nuclear Management Partners Ltd, a consortium of URS, Amec Foster Wheeler and Areva.

The area managed by Magnox Ltd includes Chapelcross, Hunterston A, Trawsfynydd, Wylfa and Oldbury (formerly Magnox North Ltd) and Berkeley, Bradwell, Dungeness A, Hinkley Point A and Sizewell A (formerly Magnox South Ltd). Subcontracting is provided by Cavendish Fluor Partnership Ltd, a consortium of Babcock International and Fluor.

The Dounreay site is managed by Dounreay Site Restoration Ltd, comprising the Dounreay site, with Cavendish Dounreay Partnership Ltd as subcontractor, a consortium of Babcock International, CH2M HILL and URS.

The operator of the low-level waste repository near Drigg (Cumbria) is LLW Repository Ltd, and the subcontractor UK Nuclear Waste Management Ltd, a consortium of URS, Studsvik, Areva and Serco.

The Springfields nuclear fuel production facility near Preston (Lancashire) is operated by Springfields Fuels Ltd with Westinghouse Electric Company, a subsidiary of Toshiba, as subcontractor.

When it was created, the NDA also took over the ownership of Direct Rail Services, the rail freight operating company created by BNFL in 1995 to transport nuclear materials, as well as International Nuclear Services, which manages and transports nuclear fuel.

The NDA also owns Radioactive Waste Management Ltd (RWM), which is responsible for implementing a geological disposal facility in the United Kingdom and providing radioactive waste management solutions.

In February 2017, the national archive for the British civil nuclear industry was brought together in Wick, in the county of Caithness, Scotland under the name Nucleus.

In order to coordinate fundamental and applied research in the field of decommissioning and dismantling, the United Kingdom created a National Nuclear Laboratory (NNL) in 2009, mainly comprising the British Technology Center at Sellafield and Nexia Solutions. The NDA is a stakeholder in this laboratory, owning the Sellafield site and funding most of the research required in the nuclear park. This national laboratory complements other initiatives to develop a sustainable workforce such as the National Skills Academy for Nuclear (NSAN), including the development of Energus in West Cumbria, and complementary research and development facilities such as the Dalton Nuclear Institute.

Currently, in the United Kingdom, 30 reactors have been permanently shut down [SCH 19].

3.4.5. *France's policy*

Unlike the United Kingdom, France has not created an agency specialized in decommissioning. Decommissioning and dismantling are the responsibility of the operators. In France, the operators of nuclear installations are EDF, ORANO (formerly AREVA) and the CEA. However, the ASN retains a central role in monitoring and advising. Each operator must draw up a decommissioning plan and submit it to the ASN.

Nuclear operators in France have all adopted the principle of immediate decommissioning. The strategy adopted in France is for the operator to prepare for the decommissioning of its facility as early as the design stage, to anticipate decommissioning prior to the shutdown of its facility and to send the decommissioning authorization application file before shutting down its facility, and for decommissioning operations to take place "as quickly as possible" after the facility is shut down. This is a period that can vary from a few years to a few decades depending on the complexity of operations. In addition, to cover the cost of decommissioning, a system has been created to secure the costs associated with the decommissioning of nuclear installations, the management of spent fuel and the management of radioactive waste [ASN 19a].

France's choice of immediate dismantling has a twofold objective: to avoid placing the burden of dismantling on future generations and to benefit from the knowledge of the teams present during the operation of the installation in order to dismantle it [SCH 15].

However, since 2015, EDF has reversed this strategy for first generation reactors (UNGG) and has postponed their final dismantling beyond 2100 (see Chapter 6 for more detail) [IRS 19c].

The decommissioning and dismantling of nuclear facilities in France follows the administrative approach shown in Figure 3.4.

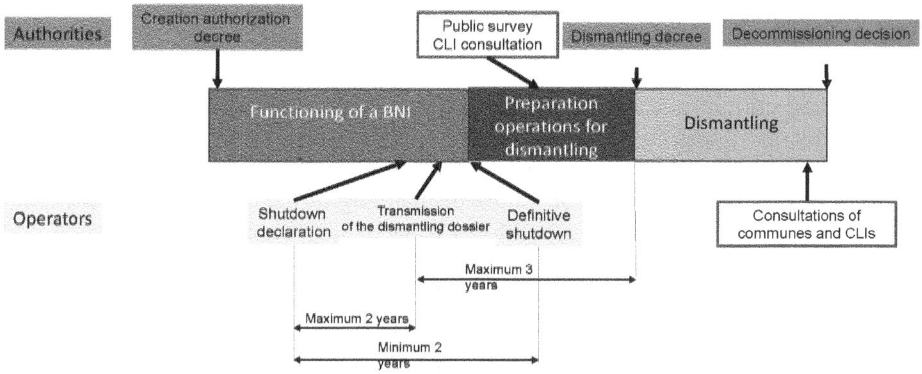

Figure 3.4. *The decommissioning and dismantling process of nuclear facilities in France (source: [ANC 17, ASN 19a]). For a color version of this figure, see www.iste.co.uk/amiard/disarmament.zip*

In 2019, 35 nuclear facilities of all types (power or research reactors, laboratories, fuel reprocessing plants, waste treatment facilities, etc.) were shut down or being dismantled in France, corresponding to approximately one-third of the basic nuclear facilities in operation other than power reactors. This figure will increase considerably in the near future [ASN 19a].

Currently, in France, 12 reactors have been permanently shut down [SCH 19].

The *Assemblée Nationale* has produced a report on the technical and financial feasibility of dismantling the BNIs [AUB 17]. The authors of this report estimate that decommissioning will take longer than expected, that the initial assumptions are not all respected, that the provisions are among the lowest in the OECD with no safety net in case of cost variance, and that technical feasibility is not fully assured. The authors recommend revising the rules for forecasting decommissioning costs, establishing a provisional schedule of reactors to be dismantled, relaxing the rules on very low-level waste (VLLW), facilitating competitive bidding for decommissioning and accelerating the decommissioning of gas–graphite reactors (UNGG).

3.4.5.1. *EDF*

All the nuclear reactors for the production of electricity are the property of EDF, and this company assumes full technical and financial responsibility for the dismantling of its nuclear power plants. EDF confirms its objective to dismantle its nuclear facilities in the shortest possible time. To this end, the Group relies on specific skills that it is committed to developing over the long term, both at its own facilities and at those of its main service provider partners. Thus, the Chooz A dismantling project is proceeding according to schedule and within budget. The experience feedback from the Chooz A dismantling site will benefit the nuclear power plant currently in operation, entirely made up of pressurized water reactors [EDF 17].

As a priority, EDF has scheduled the dismantling of the Saint-Laurent-des-Eaux silos, the EL4-D installation on the Monts d'Arrée site, Superphénix, Chooz A and Bugey 1, and, in a second phase, the dismantling of the five reactors of the UNGG (Uranium Naturel Graphite Gaz) sector of Chinon A and Saint-Laurent-des-Eaux A [ASN 19a].

In the field of decommissioning, EDF studies focus on three main areas: the preservation of the memory of installations through a program to conserve documents useful for their dismantling, the implementation of tools to facilitate the estimation of the volume of waste produced by these operations, and finally, the choice of a decommissioning strategy based on feedback from first generation reactors, as well as on similar projects conducted abroad [RON 15].

However, EDF is facing two major problems leading to considerable decommissioning delays. The first concerns the dismantling of the Brennilis reactor in the Monts d'Arrée. The second concern is the dismantling of first generation reactors where the quantities of contaminated graphite are significant and there is currently no safe storage solution at present.

3.4.5.2. *CEA*

The CEA owns a large number of nuclear facilities, many of which are experimental reactors. Some of these installations are civilian and depend on the *Direction de l'énergie nucléaire* (DEN); others are military and are managed by the *Direction des applications militaires* (DAM).

The DEN is piloting the dismantling of 22 CEA civil nuclear facilities, nearly two-thirds of the French facilities currently being dismantled. Because of the wide variety of facilities in operation (experimental reactors, chemistry laboratories, effluent and waste treatment plants, etc.) each project is a special case. Some dismantling has been completed, such as the dismantling of the SILOETTE,

MELUSINE and SILOE reactors in Grenoble, and the dismantling of the HARMONIE reactor in Cadarache. Dismantling is in progress at Saclay INB 35, LMA and ULYSSE; at Fontenay-aux-Roses INB 165 and 166; at Grenoble, LAMA and STED/STEL; and at Cadarache, RAPSODIE, PHEBUS, ATPu, ATUE and STED. Future dismantling programs are scheduled in Saclay, INB 72 and OSIRIS; in Marcoule, PHENIX and in Cadarache, PEGASE. To these must be added the military installations in Marcoule (APM, GI, UP1, Usine, SPF/AVM stripping, ASE, RCD bitumen and non-bitumen) [PIK 15].

The ASN has expressed doubts about the CEA's strategy. Indeed, it notes the CEA's difficulties in implementing the strategy of immediate and complete dismantling. In 2004, at the request of the ASN, the CEA submitted a 10-year dismantling plan for its various kinds of facilities. This plan was analyzed by the ASN authorities, who ultimately considered that the strategy adopted by the CEA and supported by the establishment of a dedicated fund, was generally satisfactory from a safety point of view.

However, in view of recurring delays and the postponement of several deadlines for dismantling operations at certain facilities, the CEA provided a progress report in 2011, at the request of the ASN, presenting an update of this strategy, justifying the deadlines selected and specifying the reasons, technical or otherwise, for the delays observed. This report was submitted in December 2016. The CEA has defined priorities, based primarily on an analysis of potential hazards, to reduce the risks presented by these facilities.

Despite the relevant organization recently put in place by the CEA to manage its dismantling program over the long term, the ASN and ASND are questioning the robustness of the CEA's action plan and the resources available, both human and financial, to deal as quickly as possible with all situations presenting the most significant safety or environmental issues. Among the weaknesses, the two authorities note in particular the cooperation between the various centers for the management of aqueous radioactive effluents or solid radioactive waste, leading to the availability of only one installation for certain operations [ASN 19a].

3.4.5.3. *Orano*

Orano (formerly AREVA NC) inherited a large number of nuclear facilities; therefore, the dismantling of old facilities is a major challenge. It must carry out several large-scale dismantling projects in the short, medium or long term (UP2-400 plant in La Hague, Eurodif Production plant, INBS facilities in Pierrelatte, etc.). The implementation of these projects is closely linked to the waste management strategy given the quantity and the non-standard and difficult to characterize nature of the waste generated during dismantling operations. Regarding the waste management strategy for the La Hague facility, ASN reported its position in 2006, which was generally good. The same applies to the ASND in 2012 concerning the waste management strategy for the AREVA NC INBS in Tricastin [ASN 19a].

3.4.5.4. *The French doctrine of decommissioning and dismantling*

The regulations relating to the dismantling of basic nuclear installations (BNIs) were clarified and supplemented from 2006 by the law no. 2006-686 of June 13, 2006 relating to transparency and safety in nuclear matters, then by the decree of November 2, 2007 now codified and the order of February 7, 2012. More recently, the law relating to the energy transition for green growth modified the legislative framework relating to the dismantling of the BNIs by imposing new constraints on the operator (declaration of the shutdown and dismantling file two years before final shutdown).

France distinguishes between two major phases in the life of a BNI: the operating phase of the facility, governed by the decree authorizing its creation (*Décret d'autorisation de creation* – DAC) and the dismantling phase, following the final shutdown of the facility, governed by the dismantling decree (*Décret de démantèlement* – DEM). The decommissioning application file must include all the stages, from final shutdown to final decommissioning, even though some stages will include "stopping points", i.e. periods of inactivity.

In the dismantling plan of a BNI provided by the operator from the creation of the installation, many pieces of information are mandatory such as the dismantling strategy and methods, the initial state at the final shutdown, the conservation of the history of the BNI and the characteristics of its specificity.

Preparatory operations for dismantling include the evacuation of some of the radioactive substances, fitting out of premises, preparation of worksites, training of teams, operations to characterize the installation (production of radiological maps, collection of elements from the history of the operation) with a view to dismantling.

Decommissioning allows the lifting of regulatory controls to which a BNI is subject. Depending on the final state reached, the ASN may make the decommissioning of a BNI conditional on the implementation of public utility easements. These may impose a certain number of restrictions on the use of the site and buildings or precautionary measures (radiological measures in the event of scouring, etc.) [ASN 19a].

The remediation of civil engineering structures (concrete walls, floors) of a nuclear waste area when the phenomena of activation and/or migration of contamination in these structures have taken place requires aggressive action on these structures. This can be, for example, abrasion to reduce their thickness. The aim is to permanently decommission the nuclear waste zone into a conventional waste zone. The development of the new waste zoning boundary is based on the use of three independent and successive lines of defense, the stacking of which guarantees a high level of confidence in the distinction between the nuclear and conventional waste zones [ASN 19a].

To assist operators, the ASN has published three guides, updated on August 30, 2016 [ASN 16a, ASN 16b, ASN 16c].

3.4.6. *China*

In China, the regulatory authority is the National Nuclear Safety Administration (NNSA) under the Ministry of Environmental Protection (MOE), while the China Atomic Energy Authority (CAEA) regulates the nuclear industry.

The NNSA was established in 1984 and merged in 1998 with the State Environmental Protection Administration (SEPA). It is responsible for safety regulations for civil nuclear facilities and for issuing technical nuclear safety standards. In addition, since December 2003, it has been in charge of the supervision of radioactive sources (production, import, export, sale, transport and storage).

The CAEA is the competent authority in the nuclear industry. It has overall responsibility for all nuclear facilities and activities associated with the nuclear fuel cycle and radioactive waste management. It is also responsible for emergency preparedness of China's nuclear facilities and activities. It conducts and promotes international cooperation and exchanges on behalf of China and is also responsible for the training and examination of nuclear facility operators and their qualifications [YAN 06].

Due to a large but relatively young nuclear park, China has no experience in decommissioning. Its first decommissioning took place in 2006 on an experimental reactor.

3.4.7. *Sweden*

Half of Sweden's original 13 nuclear reactors were shut down by 2020. Decommissioning these reactors was a challenge for all parties involved, including licensees, the waste management system, the financing system and the Swedish Radiation Safety Authority (SKI). Amft *et al.* [AMF 19] describe the administrative aspects of decommissioning, such as SKI's guidelines, the definition of fundamental concepts in the regulatory framework and a proposed revision of the licensing process under the Environmental Act. The Swedish approach to decommissioning is shown in Figure 3.5.

At the end of 2019 five reactors in Sweden had been permanently shut down [SCH 19].

OKG (Oskarshamns Kraftgrup) has three BWR reactors at the Oskarshamn site in Sweden, two of which are currently being shut down. Some difficulties are being encountered due to the fact that on the same site, two units are being decommissioned and one unit is in electrical operation. The company is currently involved in the Swedish regulatory framework with the definition and preparation of the Environmental Impact Assessment (EIA), the updates of the Safety Analysis Report (SAR) and the decommissioning plans.

Experts from the industrial owner OKG, Westinghouse and Studsvik (now Cyclife Sweden and Studsvik Consulting) were asked to develop a decision basis for a comprehensive strategy for the management of materials and waste resulting from the decommissioning of the two BWR nuclear power plants. This strategy ensured safe, timely and cost-effective management of waste during decommissioning [RAN 17].

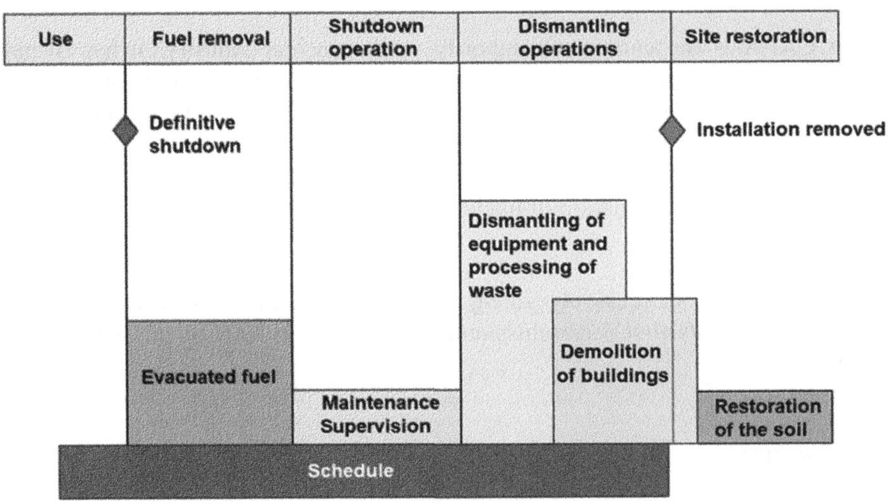

Figure 3.5. *Diagram of the decommissioning phases according to the Swedish licensing process (source: [RAN 16]). For a color version of this figure, see www.iste.co.uk/amiard/disarmament.zip*

3.4.8. *Japan*

Japan is facing an extreme situation following the multiple nuclear accidents at Fukushima in 2011. Thus, 24 reactors have currently been shut down in the long term and 27 have been permanently shut down [SCH 19]. This exceptional event demonstrated the importance of a nuclear authority that is truly independent of any industrial influence [AMI 19a]. The consequence was the replacement of NISA (the Nuclear and Industrial Safety Agency) by a new regulatory authority, the Nuclear Regulation Authority (NRA). It is an external agency under the Japanese Ministry of the Environment. The NRA is responsible for renewing the regulation of nuclear activities and monitoring its implementation in order to protect the population and the environment. Among the many nuclear power reactors to be decommissioned, the most serious concern for Japan is that three of them are seriously damaged (reactors 1, 2 and 3 at Fukushima), while nuclear reactor 4 suffered but its tank is intact and the spent fuel from the core, which had been put in a pool before the accident, has since been removed.

3.4.9. *Other states*

Other countries are faced with the permanent shutting down of their nuclear facilities due to aging, a change in policy or a nuclear accident damaging the facility.

3.4.9.1. *Brazil*

In Brazil, IPEN (*Instituto de Pesquisas Energéticas e Nucleares*) has played a decisive role in the development of nuclear science and technology since it was founded in 1956. The Institute is located in the city of São Paulo on the campus of the University of São Paulo in an area of nearly 500,000 m^2. Fuel cycle activities were carried out at the pilot plant scale and most of the facilities were built during the 1970s and 1980s. Nevertheless, radical changes in Brazilian nuclear policy in the early 1990s led to interruption of activities and the closure of the facilities.

Many nuclear installations have been dismantled, such as the pilot plant for uranium purification, the pilot plants for the dissolution of thorium sulfate and for the aqueous diffusion of UF_4, the laboratory for isotopic characterization, the pilot plant for UF_6, the pilot plant for the fabrication of UO_2 fuel pellets. Conditioning of radioactive waste was carried out. In 2011, the future decommissioning activities concerned the ADU precipitation pilot plant, UO_3 calcination, denitration, UF_4 production, thorium nitrate purification and the reprocessing laboratory [LAI 11].

The uranium hexafluoride plant produced 35 metric tons of this gas per year and was closed in 1992. Decommissioning studies were launched in 2002 and D&D began in 2005 [SAN 09].

The dismantling operations took place in two phases. From 2000 to 2001, the pilot plant for the dissolution of thorium sulfate and the production of UF_4 by aqueous diffusion in building 2 of the CQMA (Center of Chemistry and Environment) was dismantled. Between 2002 and 2003, the pilot plant for the dissolution of ADU (unpurified yellow cake) and the purification of uranyl nitrate was dismantled in building 1 of the CQMA [LAI 05]. Between 2005 and 2006, the pilot plant for conversion of uranium hexafluoride was dismantled. In 2007, decommissioning of the pilot plant for the manufacture of UO_2 pellets was in progress [LAI 11].

The decommissioning of nuclear facilities in Brazil depends on many challenges that are due to the fact that Brazil has a modest and recent nuclear program, no specialized decommissioning companies and limited expertise in this field. In addition, political authorities have been very hesitant about D&D's decision and priorities, and many staff have retired or dispersed. The storage capacities for radioactive waste in IPEN are limited, and there are no decontamination and treatment techniques adapted to the reduction of the volume of medium- and low-level waste [LAI 11].

3.4.9.2. *South Korea*

In South Korea, the regulatory authority is the NSSC (Nuclear Safety and Security Commission).

The ultimate goal of decommissioning is the unrestricted release or use of the site. Therefore, the Historical Site Assessment (HSA) is essential. This is the first step to completing the Final Status Survey (FSS) according to the South Korean procedure (RSSI – Radiation Survey and Site Investigation). The HSA process leads to a series of actions to collect, review and evaluate existing information describing the completeness of the site information from the beginning of the activity to the end of the license. The procedure is based on the MARSSIM guideline and allows prioritization of the factors to be considered, which will be useful for the description of the site. This is the first step in the decommissioning of Kori Unit 1 [YU 17].

Technically, decommissioning is a mature industry. Many steps and processes are similar to the maintenance, storage and transport procedures encountered during the operation. However, to minimize the hazardous and radioactive materials produced during decommissioning as much as possible, the issue should focus on the protection of workers, the public and the environment. There are eight factors that impact whether or not decommissioning will be successful and these must be evaluated and prioritized. The most important factors are the consideration of stakeholders and the management of spent fuel. Taking them into account should make it possible to continue the successful decommissioning of Kori-1 without problems [YUN 16].

In South Korea, one power reactor has been shut down in the long term and two have been permanently shut down [SCH 19]. According to the CEA [CEA 19], South Korea has only one reactor (PWR of 516 MW net) shut down in June 2016 (Kori-1) and 24 in operation.

3.4.9.3. *Canada*

The regulatory authority in Canada is the Canadian Nuclear Safety Commission (CNSC). It regulates the use of nuclear energy and materials to protect the health, safety and security of Canadians and the environment and to respect Canada's international commitments on the peaceful use of nuclear energy.

Currently, in Canada, one reactor has been shut down in the long term and six have been permanently shut down [SCH 19]. The CEA [CEA 19] lists only six shutdown reactors. In contrast, only one, the Gentilly-2 reactor, was in the process of being decommissioned in 2018 [RAD 18]. The cost of dismantling this CANDU reactor was estimated at 1.5 billion dollars in 2012 [COU 12].

The breakdown of the total cost of decommissioning was estimated in Canada. The permit termination phase, including the management of the decontamination and decommissioning project, and the management of the waste resulting from these operations, represents 65% of the total cost. The site remediation phase, including preparation for unrestricted use, accounts for 10% of the cost, while the management of used fuel, including long-term storage and management, accounts for 25% of the total cost [KPM 18].

3.4.9.4. *Switzerland*

Following the nuclear accidents in Fukushima in 2011, the Swiss Confederation decided in a referendum (approved by 58% of the Swiss population) to phase out nuclear power. Nevertheless, the world's oldest nuclear reactor (50 years old in 2020), Beznau, operated by Axpo, is expected to continue to operate following major investments (2.3 billion euros) to improve its safety. Similarly, the Gösgen power plant (1979, PWR, 985 MWe) is expected to operate until 2040 and the Leibstadt power plant (1984, BWR, 1220 MWe) will remain in operation until 2045.

Conversely, Mühleberg nuclear power plant (BWR, 335 MWe) is a nuclear power plant that has been permanently shut down since December 20, 2019. It is located in the north of the canton of Bern. It operated for 47 years under BKW FMB Energy Ltd. Decommissioning began at the beginning of 2020. The 16,000 tonnes of spent fuel will not be removed from the reactor until 2024, and currently no site has been selected for final storage.

3.4.9.5. *Italy*

In addition to Ispra-1, the public company Sogin (Società Gestione Impianti Nucleari) is responsible for the decommissioning of four nuclear power plants (Trino, Caorso, Latina and Garigliano) and five fuel cycle research plants (Eurex in Saluggia, Fn in Bosco Marengo, Opec and Ipu in Casaccia and Itrec in Rotondella) [WNN 19e].

Among the four reactors undergoing decommissioning in Italy, two are BWRs (Caorso, 860 MWe and Garigliano, 160 MWe), one PWR (Trino, 270 MWe) and one Gas–Graphite (Latina, 210 MWe) [GUE 06b].

3.4.9.6. *Taiwan*

In Taiwan, the Chinshan nuclear power plant has reactors at the end of their lives. The operating license for Unit 1 expired on December 5, 2018 and that of Unit 2 in July 2019. A broad public debate is taking place on the continuation of the nuclear power program. In a referendum in November 2018, voters categorically rejected the government's phasing out of nuclear power. The future decommissioning of the nuclear industry in Taiwan is pending future election results. However, the decommissioning of Nuclear Power Plants No. 1 and No. 2 is still in progress.

Taiwan's Environmental Protection Administration announced that it has approved a 25-year decommissioning plan for Taiwan's No. 1 nuclear power plant, also known as the Chinshan Nuclear Power Plant. According to the official plan, the decommissioning will be divided into four stages, including post-operational transition (8 years), decommissioning and dismantling (12 years), demolition of the reactor turbine buildings (3 years) and site reclamation (2 years) [CHU 19].

3.4.9.7. *Various other states*

In Lithuania, there were two RMBK 1500s that were operational from 1984 to 2009. Removal of the fuel is planned for 2022 and final decommissioning should be effective in 2027 [LAE 19].

In Belgium, the *Agence fédérale de contrôle nucléaire* (AFCN) is a public establishment with legal personality (a public interest organization under the supervision of the Minister of the Interior), established by the law of April 15, 1994 on the protection of the population and the environment against the dangers resulting from ionizing radiation. Currently, in Belgium, one power reactor has been permanently shut down [SCH 19].

At present, the nuclear power reactors that have been permanently shut down are in Spain, Bulgaria, Slovakia, Armenia, the Netherlands and Ukraine, numbering 3, 4, 3, 1, 1 and 4, respectively [SCH 19].

3.5. Conclusions

Many nuclear facilities around the world are reaching end-of-life stage and need to be decommissioned and dismantled. This has become an industry with many players, often private. Two countries, the United States and Germany, have acquired a great deal of experience in this field.

The main problems related to decommissioning come from the high levels of irradiation that require remote-controlled work, and the enormous amount of radioactive waste of various categories produced during dismantling. This is accentuated in cases where the installation has suffered a nuclear accident.

Two international bodies, the IAEA and the NEA, have worked intensively to draft recommendations to national authorities with nuclear facilities to ensure that decommissioning and dismantling stages take place under good conditions that guarantee the protection of the public, workers and the environment at reasonable cost and in the shortest possible time.

Decommissioning is the responsibility of each state with a regulatory body generally independent of the nuclear industry. The choice of decommissioning strategy, whether immediate or deferred, varies from country to country. Thus, the United States favors deferred decommissioning and Germany favors immediate decommissioning. Likewise, the end state is more of a "greenfield" approach for Germany and, conversely, a conditional release for the United States.

At the global level, progress could be made if harmonization of administrative procedures and the definition of radioactive waste were put in place.

4

Procedures and Technologies Involved in Decommissioning

4.1. Introduction

The decommissioning and dismantling of a nuclear facility is a lengthy and costly process. These operations must be rigorously planned long before they begin. It is also essential to plan their financing. Decommissioning is the source of a wide variety of radioactive waste, some of which is highly radioactive. Their management must therefore be precisely anticipated. The long process of decommissioning and dismantling is a source of radioactive risk for both for the environment and all living organisms, including humans. Dismantling is based on a large body of techniques which, with hindsight, are increasingly well controlled. Technological innovations for dismantling nuclear facilities are numerous in a large number of fields.

4.2. The cost of dismantling

The increase in the number of operations will be significant from 2001–2050, with approximately 473 nuclear facilities to be dismantled at a cost of about US$164 billion [GON 03]. Taking inflation into account, this sum corresponds to about €186 billion. This is probably a large underestimate as it was predicted in the 2000s, and does not include the 19,000 medical facilities for radiotherapy.

The cost of decommissioning nuclear power plants depends on many factors, the main ones being the sequence of chosen decommissioning steps, the schedule of each decommissioning step and the decommissioning activities achieved at each step. In addition, costs depend on country and site-specific factors, such as reactor type, waste management and disposal practices and labor rates [EU 05a].

As decommissioning costs have turned out to be much higher than expected, government agencies soon realized that long-term decommissioning liabilities represent a significant risk. As a result, states have called for funding mechanisms to be established and made clear. Gradually, various studies have been conducted and regulations followed that have slowly developed methods to forecast liabilities and regulate funding, without relying on the financial resources of the nuclear licensee [LAG 12].

4.2.1. *International recommendations*

Several international bodies such as the IAEA and the NEA have issued recommendations to standardize and ensure the estimation of the cost of decommissioning nuclear facilities. For example, the IAEA recommends the use of the ISDC (International Structure for Decommissioning Costing), which defines the principles and context of a costing methodology. This method is particularly intended for the preliminary stages of costing the decommissioning of research reactors and other small nuclear facilities. The general ISDC costing methodology has been modified in CERREX (Cost Estimate for Research Reactors in Excel), into a simpler version adapted to the preliminary costing steps, using Excel spreadsheets [IAE 13a].

The process for estimating the costs of nuclear decommissioning facilities has continued to evolve in recent years, with a general trend towards demonstrating greater levels of detail and a more explicit consideration of uncertainties, which may affect the costs of the decommissioning project.

For the decommissioning of nuclear power plants, an internationally developed cost structure has already been developed using the International Decommissioning Cost Structure (ISDC) for Nuclear Installations [NEA 12]. The ISDC identifies 15 cost categories that are quantified. Although the ISDC has been developed in a way that can be adapted to the decommissioning of other types of installations, only the IAEA has done so at the international level for research reactors, within the framework of the DACCORD program (Data Analysis and Collection for Costing of Research Reactor Decommissioning) [IAE 17]. To date, there are no internationally agreed cost structures for the decommissioning of other types of nuclear facilities, or for the provision of radioactive waste infrastructure [NEA 19].

The studies and calculations have mainly focused on estimating uncertainties (Figure 4.1).

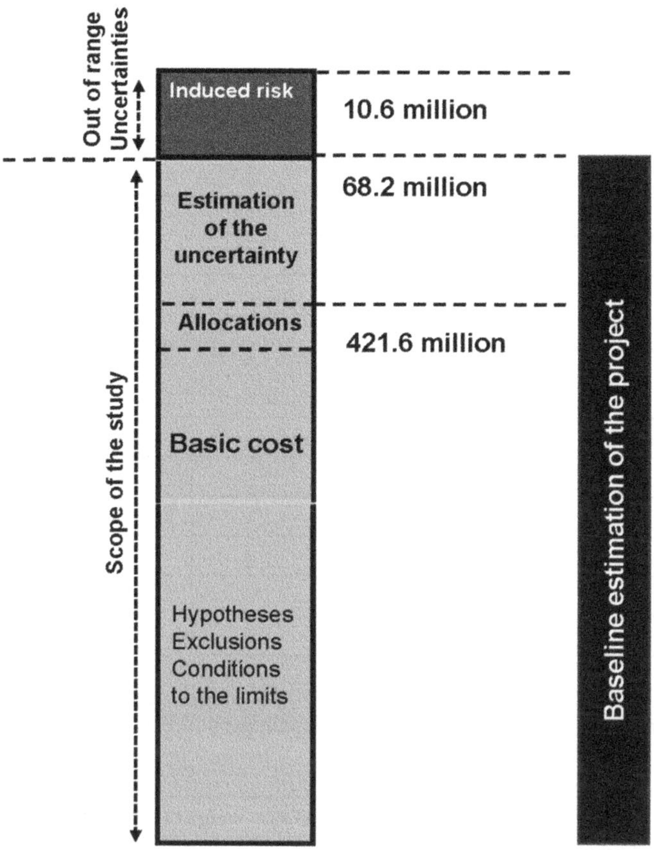

Figure 4.1. *Overview of the estimate by calculating the cost of decommissioning and the uncertainties on this estimate (source: [NEA 17b])*

Despite the variability in decommissioning costs, the study showed that their reported estimates remain below US$500 per installed kilowatt of electricity (kWe), for almost all water reactors. They are significantly higher for gas-cooled reactors (about US$2,500 per kWe (Table 4.1).

The cost of immediate decommissioning is estimated to be between US$93 and US$909 per kWe, and for deferred decommissioning, between US$178 and US$405 per kWe for PWRs. For WWER, the cost ranges from US$162 to US$506 per kWe, and for BWRs, from US$104 to US$2,300 per kWe (Table 4.2) [NEA 03]. The cost of nuclear waste management also varies greatly depending on the type of reactor, with PHWRs and GCRs generating much more waste than other reactors (Tables 4.1 and 4.3).

Reactor type	Cost of dismantling (%)	Cost of waste (%)	Average cost (standard deviation)
PWR	30	23	320 (195)
WWER	25	17	420 (100)
BWR	33	23	330 (1,150)
PHWR	34	43	360 (70)
GCR	25	43	>2,500 (–)

Table 4.1. *Contribution (%) of decommissioning and waste management phases to the total cost of decommissioning [NEA 03]*

Reactor type	Immediate dismantling	Deferred dismantling
PWR	93–909	178–405
WWER	162–367	218–506
BWR	104–1,644	76–2,300
PHWR	–	275–432
GCR	3,248	721–4,470

Table 4.2. *Decommissioning costs (in US dollars per installed kWe) of reactors by reactor type and decommissioning strategy (source: [NEA 03])*

Reactor type	Waste weight (t/MWe)
PWR	10
BWR	10
PHWR	13
WWER	17
GCR	100

Table 4.3. *Quantity of nuclear waste (in tons per installed MWe) by reactor type (source: [NEA 03])*

In addition, the European Union has taken a strong interest in the financial aspects of decommissioning, as well as in two issues related to it, which are radiation protection during this process and its environmental impact. Finally, the vast subject of the management of nuclear waste arises from its eventual decommissioning into common waste (release limits according to mass activities). Each European Member State has its own waste management strategies and policies, including material release criteria. The harmonization of waste management practices and clearance levels between Member States, and for all nuclearized states, would be beneficial, not only in terms of equivalent levels of safety in waste management and disposal, but also in waste minimization through disposal and recycling [NEA 01].

The NEA has made comparisons between the decommissioning cost estimates of various countries (France, Italy, Czech Republic, Spain, Sweden, Switzerland and the United Kingdom) [NEA 15].

The cost of the immediate dismantling of many US reactors ranges from \$307 million to \$800 million (2013). The cost for the dismantling of the Finnish Loviisa reactor, WWER-440, was estimated at €359 million in 2012. In the Netherlands, the dismantling of the reactor (BWR, 55 MWe) is expected to cost €180 million, plus 25% of transport and waste management costs. In Slovakia, the decommissioning of the Bohunice V1 reactor, WWER-440 model 230, is expected to cost €1.141 million. The estimated cost of decommissioning the Spanish José Cabrera PWR reactor is €258 million. In Switzerland, the decommissioning of the two PWR reactors and the two BWR reactors was estimated at a total of 2.974 million Swiss francs (or €691 million per reactor) in 2011. The total decommissioning of the British Magnox reactors is estimated at £15.4 billion (about €17.04 billion or €631 million per reactor) [NEA 16a].

4.2.2. *The American example*

The large number of INB decommissionings in the United States provides a wealth of experience on the cost of these operations. Thus, US spending on military nuclear activities in 2018 and 2019 and projected spending in 2020 are shown in Table 4.4 [DOE 19].

The estimated costs of dismantling Maine Yankee (900 MWe PWR) and Big Rock Point (67 MWe BWR) in 1998, were US\$577 million and US\$421 million, respectively [LAG 12]. Therefore, the costs per installed kWe differ by a factor of 10 (\$0.64 million and \$6.3 million, respectively).

The estimated costs of decommissioning US reactors are increasing, taking into account new expenditures. Thus, the cost of decommissioning US reactors (in millions of US dollars) was 508 (in 1997) and 577 (in 1998) for Maine Yankee, 407 (in 1999) and between 636 and 750 (in 2003) for Yankee Rowe, from 426 (in 1996) and from 850 to 931 (in 2003) for Connecticut Yankee, from 436 (2003) to 473 (2007) for Big Rock Point and from 504 (2010) to 517 (2012) for Rancho Seco [LAG 17].

	FY 2018 Enacted	FY 2019 Enacted	FY 2020 Request
Uranium Enrichment Decontamination and Decommissioning Fund	840,000	841,129	715,112
Atomic Energy Defense Activities. National Nuclear Security Administration			
Federal Salaries and Expertises	407,595	410,000	434,699
Weapons Activities	10,642,138	11,100,000	12,408,603
Defense Nuclear Nonproliferation	1,999,219	1,930,000	1,993,302
Naval Reactors	1,620,000	1,788,618	1,648,396
Total NNSA	14,668,952	15,228,618	16,485,000
Environment and Other Defense Activities			
Defense Environment Cleanup	5,988,048	6,024,000	5,506,901
Other Defense Activities	840,000	860,292	1,035,339
Defense Nuclear Waste Disposal	0	0	26,000
Total Environment and other Defense Activities	*6,828,048*	*6,884,292*	*6,567,840*
Total Atomic Energy Defense Activities	**21,497,000**	**22,112,910**	**23,057,840**

Table 4.4. *US Department of Energy 2018, 2019 and 2020 Nuclear Budgets in thousands of US dollars (based on [DOE 19])*

4.2.3. *The British example*

In 2005, the cost of decommissioning UK sites was estimated at £55.8 billion, with Sellafield requiring £31.5 billion [NDA 05]. However, in 2006, the NDA (Nuclear Decommissioning Authority) reported that the cost of cleaning up existing waste would be higher than expected, and gave a new decommissioning cost estimate of approximately £72 billion over a 100-year period. In 2008, the estimated decommissioning costs increased to £73.6 billion, but after taking into account discount rates, £44.1 billion. A 2006 estimate included £14 billion in compensatory revenue from fuel reprocessing at Sellafield. In 2009, the NDA sold land near three existing reactor sites, for expected new nuclear power plants, for more than £200 million (Table 4.5). These 2005 estimates were later revised upwards.

In 2013, a critical Public Accounts Committee report stated that the private consortium managing Sellafield had failed to reduce costs and delays. Between 2005 and 2013, Sellafield's annual operating costs increased from £900 million to about £1.6 billion. The estimated undiscounted lifetime cost of transactions with the Sellafield site increased to £67.5 billion [HOU 13]. Sellafield executives were forced to apologize when the projected cleanup costs exceeded £70 billion by the end of 2013. In 2014, Sellafield's decommissioning cost estimate was increased to £79.1 billion, and in 2015 to £117.4 billion (about €130 billion). The annual operating cost was £2 billion in 2016 (Table 4.6). By way of comparison, estimates for the dismantling of the La Hague plants were estimated at only 6 or 7 billion euros [COU 20]. This gives food for thought on the accuracy of the estimates.

Owners and sites	Total cost (millions of pounds sterling)	Owners and sites	Total cost (millions of pounds sterling)
Magnox Ltd		**UKEA**	
Berkeley	800	Culham	194
Bradwell	1.127	Dounreay	2.782
Chapelcross	1.661	Harwell	872
Dungeness A	1.161	Windscale	622
Hinkley Point A	1.128	Winfrith	454
Hunterston A	1.214	**British Nuclear Group**	
Oldbury	1.366	Digg	746
Sizewell A	1.236	Calder Hall	1.284
Trawsfynydd	1.116	Sellafield	23.698
Wylfa	1.708	Capenhurst	1.104
ES & T*	- 1.374	**Westinghouse**	
		Springfields	371
Total	43.270		

Table 4.5. *Estimated life cycle financial data for each UK site (based on [NDA 05]). *Revenue from the sale of generated electricity*

All of the cost estimates provided above clearly demonstrate the fragility of the estimates, which regularly increase sharply over the years. For example, the total NDA expenditures increased by a factor of 2.5 in 2015–2016 and by a factor of 1.5 in 2017–2018 (Table 4.6).

Year	Sellafield	Other NDA sites	Total
2009–2010	25.2	19.9	45.1
2010–2011	32.7	16.5	49.2
2011–2012	37.2	15.6	52.9
2012–2013	42.0	16.9	58.9
2013–2014	47.9	17.0	64.9
2015–2016	117.4	43.3	160.7
2017–2018	Not estimated	Not estimated	234

Table 4.6. *Evolution of NDA cost estimates for the decommissioning and cleanup of Sellafield and other UK nuclear sites (£ billion) (source: [NAO 15; NAO 18])*

4.2.4. *The French example*

According to the ASN, decommissioning expenses are divided into five categories [ASN 19a]. The first three categories are dismantling charges, charges for the management of spent fuel and charges for the recovery and conditioning of old waste. The fourth category includes the long-term management costs of radioactive waste packages produced by the first three categories. The last category includes charges for monitoring after the closure of facilities.

In France, the procedures for implementation by operators and control by the administrative authority are specified in Decree 2007-243 from February 23, 2007, on securing the financing of nuclear expenses, and supplemented by the Order of March 21, 2007. The law also created a National Commission for the Evaluation of the Financing of the Decommissioning of Basic Nuclear Facilities and the Management of Spent Fuel and Radioactive Waste, to assess the adequacy of provisions for future expenses and asset management. In this commission, the ASN provides technical support to the Minister in charge of energy, by contributing its knowledge of the scenarios and hypotheses adopted by the operators.

The ASN is very critical of the fourth triennial reports provided in 2016, criticizing EDF for its lack of detail on the costs, and the three operators for not taking the costs of soil and structural remediation into account. In addition, it recommends an external audit on the management of used fuel by EDF and on the resumption of the costs of old waste from La Hague by the CEA [ASN 19a]. For its part, the Court of Auditors emphasizes, that between 2013 and 2018, the estimates almost doubled for the corresponding dismantling for EDF (+4.5 billion $€_{2018}$) and increased by more than 25% for the CEA (+3.2 billion $€_{2018}$, excluding the effect of smoothing), as well as for Orano (+1 billion $€_{2018}$, excluding transversal risks) [COU 20]. For the three operators, the total amount of future dismantling costs amounted to €46.4 billion at the end of 2018. This organization believes that the provisioning of future expenses is not always based on the most realistic dismantling schedules.

4.2.5. *The Russian example*

ROSATOM's decommissioning cost estimation methodology is generally aligned with the International Decommissioning Cost Structure (IDCS) for nuclear facilities. The methodology and its application reflect current good practice in decommissioning cost estimation.

In order to gain confidence in the alignment of the methodology with international guidelines and best practices, ROSATOM asked the Nuclear Energy Agency (NEA) to conduct an international expert review, the so-called International Expert Feedback (IEF) [NEA 18b].

4.3. The production of radioactive waste

The dismantling of a nuclear facility generates waste and effluents that are very different from those produced during the operation of a BNI. In the same way, the quantities can be greater, especially for very low-level waste [IRS 18c]. Those responsible for the decommissioning of a BNI must strive to limit the volume of radioactive waste and release the soil for possible reuse. The first of these actions can use several solutions. One of these, which is often the most effective, is the release of materials from decommissioning.

Clearance is the removal of radioactive materials, or radioactive objects associated with licensed practices, from any subsequent regulatory control, by the regulatory body. The control referred to here is control for the purposes of radiation protection [AIE 07]. Conceptually, clearance is closely related to exemption, but differs, as exemption is a decision that certain sources and practices need to be uncontrolled. In the case of the exemption, the regulatory authority considers that a source or practice need not be subject to some or all elements of regulatory control, because the exposure (including potential exposure) from the source or practice is too low to justify the application of those elements [AIE 07]. A number of issues related to the concept of clearance and its relationship with other concepts, such as exemption and exclusion, have been addressed by the IAEA [IAE 04a]. The precise radiological characterization of the material to be dismantled has an important role in determining its clearance, storage and/or disposal potential [VAL 12].

The second solution to reduce the volume of nuclear waste is transformation. To do this, many techniques are available to the operator, such as super-compaction, incineration, evaporation and solidification (drying or cementing).

Radioactive waste management is successful when all of the waste has been released or has found a disposal route. Often, it is necessary to provide sufficient interim storage capacity for radioactive waste packages. An example is provided with the dismantling of the KKN reactor in Germany (Table 4.7) [VAL 12].

Waste	Volume (t)	Treatment	Future use
Inactive components	405	Clearance	Conventional waste
Dismantling tools, contaminated and activated steel	2,259	Melting, reuse in controlled areas	Restricted use (FZK)
Activated metals	522	Conditioning (FZK)	Temporary storage (FZK)
Contaminated or activated concrete	890	Conditioning (FZK)	Temporary storage (FZK)
Secondary waste	255	Packaging (FZK)	Temporary storage (FZK)
Liquid waste	1,540 m^3	Reprocessing (KK1)	Emissions via KK1
Building rubble	82,500	Clearance	Reuse

Table 4.7. *Quantities and fate of nuclear waste from the decommissioning of the KKN reactor (source: [VAL 12]). FZK is the Karlsruhe Research Centre in Germany*

In France, it is estimated that more than 700,000 m^3 of radioactive waste will be produced by the dismantling of nuclear installations by 2030 [DUT 15].

4.3.1. *The channels planned for the management of radioactive waste*

Waste is divided into 12 categories according to its radioactivity (four levels: very low, low, medium and high activity – VLA, LA, AA and HA) and physical period or half-life (three levels: very short, short and long life – VSL, SL and LL). Very short-lived waste (<100 days) is only managed by radioactive decay [IRS 17].

In practice, only five main categories of radioactive waste are retained, according to their level of radioactivity and their lifespan:

– VLA, very low activity (27% of the volume of radioactive waste produced in France);

– LAA-SL, low and medium activity with a short life (63% of this volume);

– LA-LL, low, long-life activity (7%);

– AA-LL, medium, long-life activity (3%);

– HA-LL, high-activity long-life waste, representing 96% of the total radioactivity of the waste but only 0.2% of its volume.

A dismantling site generates a large majority of VLA. For example, the BNIs at the CEA center in Grenoble produced 96.1% of VLA, 3.7% of LA, 0.1% of AA and 0.03% of HA [LOR 14].

In France, for example, the National Plan for the Management of Radioactive Materials and Waste (*Plan national de gestion des matières et des déchets radioactifs* – PNGMDR) has identified a channel for each category of radioactive waste, especially very low-level waste (VLLW). However, the existing disposal center will not be able to absorb all of the VLLW in the context of dismantling operations. It is therefore necessary to develop the management of very low-level waste. Alternative or complementary options to disposal are now being studied for this type of waste. They aim to optimize waste management and even limit its production. These approaches are in line with the PNGMDR, the environmental code and the positions of the French Nuclear Safety Authority and the IRSN [IRS 18c].

There is currently no solution in France for four categories of nuclear waste (LA-LL, AA-LL, HA-SL and HA-LL) [IRS 17]. In this case, developments in the management of long-lived low-level waste are ongoing. Investigations concerning the creation of a specific storage center at Soulaines-Dhuys, in the Aube, are continuing. However, an alternative would be to distribute LA-LL waste from the downstream end of the cycle into the storage centers already planned. As for the wastes with reinforced natural radioactivity (uranium waste from the Areva plant at Malvési, depleted uranium, and radiferous (containing radium) and thoriferous (containing thorium) wastes), they would be grouped in a common storage center [IRS 18c].

The British approach is slightly different. Waste that is considered "clean" is identified and removed from the site. "Exempt" materials are classified according to the SoLA (Substances of Low Specific Activity) exemption criteria under the Radioactive Substances Act (RSA93), i.e. activity less than 0.4 $Bq.g^{-1}$. The Very Low-Level Waste (VLLW) category requires specification and procedures for eventual disposal; however, this category is not designed for bulk decommissioning waste. Low-Level Waste (LLW) must be tested for acceptance criteria at existing disposal facilities and packaged in specific containers before being transported to disposal sites. There are currently no disposal facilities for HLW and ILW (High-Level Waste and Intermediate-Level Waste) in the UK, so packaging and interim storage are the only solutions used. The most radioactive HLW in the UK is encapsulated and stored for at least 50 years to allow cooling [BAY 03].

Due to the large volumes of waste generated by the dismantling of a BNI, it is essential to prioritize waste management. The management of materials and waste is a real challenge for the operator. Fundamental research and technological innovations are needed to better understand the long-term interactions between waste, packaging and storage environments [NEA 14b].

4.3.2. *Material release thresholds*

During the dismantling of a BNI, enormous quantities of materials are produced, creating large volumes of waste. However, a large proportion of this waste is non-radioactive and can therefore be released unconditionally, i.e. without clogging up the radioactive waste storage facilities (Table 4.8). Indeed, the vast majority of materials and scrap metal from decommissioning operations are neither contaminated nor irradiated, and can be recycled and reused or disposed of as conventional waste. On the contrary, radioactive materials are sorted, packaged and evacuated to final disposal centers (Centre de l'Aube in France, El Cabril in Spain, etc.) [NEA 10].

Material	Quantity (ton)	Unconditional clearance (%)
Metallic	2,010	70.8
Concrete	28,658	83.7
Heavy concrete	472	92.3
Other materials	187	44
Total	31,327	80.9

Table 4.8. *Quantities of material that can be released (based on [SEL 12])*

Clearance, or in other words, the lifting of regulatory control over radioactive materials, is now a mature concept. Guidelines for clearance thresholds have been developed internationally. This is the case for the IAEA [IAE 96] and the European Commission [EUR 14], thus promoting a good level of international harmonization of clearance. However, many countries have developed their own national regulations on clearance. Thus, the release thresholds, expressed in mass activity

($Bq.g^{-1}$), are either universal (whatever the material, its origin and destination) or dependent on the material, its origin and destination.

It should be emphasized that the majority of countries, particularly in Europe, practice a policy of releasing VLA waste on the basis of the mass activity thresholds below. It is considered that the management of the waste concerned does not require the implementation of radiation protection provisions. This waste is therefore managed in channels designed for non-radioactive waste. EURATOM proposes such thresholds (Table 4.9) [EUR 14].

Radionuclide	Release threshold ($Bq.g^{-1}$)	Radionuclide	Release threshold ($Bq.g^{-1}$)
^{3}H	100	^{14}C	1
^{60}Co	0.1	^{134}Cs	0.1
^{137}Cs	0.1	^{90}Sr	0.06
^{55}Fe	1,000	^{63}Ni	0.2
^{235}U	1	^{232}Th	1
^{238}Pu	0.1	^{239}Pu	0.1
^{241}Pu	10	^{241}Am	0.1
^{242}Cm	10	^{243}Cm	1

Table 4.9. *Proposed EURATOM release thresholds as a function of mass activity (based on [EUR 14])*

As part of the overhaul of its materials licensing program, the US Nuclear Regulatory Commission (NRC) and the Office of Nuclear Material Safety and Safeguards (NMSS) have consolidated and updated numerous decommissioning guidance documents into a three-volume NUREG [USN 06a, USN 06b, USN 12].

On the contrary, some states, such as France, do not have a release threshold in their regulations. The possibility of releasing very low-level waste (VLLW) is always a derogation. French regulations distinguish between radioactive waste and "conventional" waste resulting from nuclear activity, on the basis of a geographical zoning of the places where it is produced, based on an analysis of the operation of the installation and not on a radioactivity measurement [DOR 15]. Thus, a sorting mechanism exists, which separates waste that is "deemed" to be radioactive from that which is not (conventional waste). At the time of dismantling, the operator asked the ASN to establish a "waste zoning" that identifies two types of zones, "zones with possible production of nuclear waste" (ZppDN) and zones for "clean" waste. In the first zone, the waste produced must be managed as if it were radioactive and is subject to specific and reinforced management. This waste is directed to a dedicated storage facility, ANDRA's industrial center for grouping, storage and disposal (Cires). This is the only outlet in France for VLA waste. However, the capacity of the Cires will not be sufficient to store all of the VLA waste that will be produced in the coming years, due to the considerable development of dismantling programs. This calls for a rethink on the possible development of the VLA sector [IRS 17].

Today, there are mainly four factors limiting release. These are, first and foremost, the provisions within the regulatory framework of a particular country; second, the availability and cost of sites for the final disposal of very low-level waste (VLLW), which vary from one state to another; third, the opinion and attitude of the public, as well as certain industrial sectors, such as metallurgy, towards the origin of these materials; finally, the total amount of materials from the nuclear sector that, in the short and medium term, due to major decommissioning projects, will be accompanied by a high volume of materials that will require release [OSU 09].

4.4. The environmental and health risks of dismantling

During the various phases of decommissioning a BNI, the risks to the environment and to humans are significant and must be assessed in order to better control them.

In its concept of radiation protection, the IAEA considers that an intervention by the authorities is justified for annual individual exposure doses of 3 to 30 mSv, and that these interventions are always justifiable when the doses are higher than 30 mSv [GON 03, AMI 19b]. In France, the annual tolerable limit is only 20 mSv.

Radiological characterization is a key element for the planning, control and optimization of decommissioning and dismantling activities. This radiological characterization plays an important role in the decommissioning process of shutdown nuclear facilities, in order to ensure environmental and radiation protection. For this reason, at the end of 2013, the Working Party on Decommissioning and Dismantling (WPDD) of the OECD Nuclear Energy Agency (NEA) decided to extend the mandate of the Working Party on Radiological Characterisation and Decommissioning (WGRCD) and the Task Group on Radiological Characterization and Decommissioning for a second phase, focused on the characterization of nuclear facilities from the point of view of waste and end-state material. The first phase focused on global strategies for radiological characterization [LAR 16].

For sustainable sanitation, the NEA presented some recommendations for the integration of land and water sanitation, as well as performance assessment

[NEA 16c]. The integration of the remediation of contaminated land and groundwater into the decommissioning plan should be done at an early stage. A review should also be carried out by regulators of different approaches to achieving safety and environmental protection objectives, including those applied in non-nuclear industries. A case study approach and sharing of experiences could be facilitated by using the International Atomic Energy Agency (IAEA) web portal CONNECT.2.

The development of performance indicators for the assessment of the sustainable option, is a long-term strategic approach to the design and operation of the life cycle of nuclear facilities. Examples of end states are provided by the NEA for various states (Germany, Canada, Spain, the United States, France, the United Kingdom and Russia), as are examples of site remediation and selected end states for the United States, Russia, the United Kingdom and France [NEA 16c].

4.4.1. *Assessment of environmental and health risks during decommissioning*

The IAEA proposes a method for assessing risk during the decommissioning of a BNI [IAE 19c]. Risk assessment should be performed at two levels, at the planning stage (RMSL – Risk Management at the Strategic Level) and during the execution of the decommissioning operations (RMOL – Risk Management at the Operational Level). Planning risk assessment primarily focuses on managing the uncertainties surrounding key assumptions and strategic decisions during the decommissioning planning phase, from the Initial Decommissioning Plan (IDP) to the Final Decommissioning Plan (FDP). The IAEA is coordinating an international project on this topic (DRiMa International Project on Decommissioning Risk Management) [IAE 19c].

A risk analysis begins by arbitrarily assigning risk scores based on the likelihood of the occurrence of those risks (Table 4.10). It then assigns an impact score based on the cost of the budget and the duration of the program (Table 4.11).

Risk score	Probability (%)	Scale
1	0–20	VL
2	21–40	L
3	41–60	M
4	61–80	H
5	81–100	VH

Table 4.10. *Example of a probability scale for the use of risk analysis (modified from [IAE 19c]). VL: very low; L: low; M: medium; H: high; VH: very high*

Impact score	Scale	Impact criteria	
		Cost (% of budget)	Duration of the program
1	VL	<1%	<1%
2	L	1–5%	1–5%
3	M	6–10%	6–10%
4	H	11–20%	11–20%
5	VH	>20%	>20%

Table 4.11. *Example of an impact scale for the use of risk analysis (source: [IAE 19c]). VL: very low; L: low; M: medium; H: high; VH: very high*

By multiplying the risk (probability) and impact scores, a scale of occurrence ranging from 1 to 25 is obtained (Figure 4.2). The strategy to adopt will depend on the score, which is made more visible by using color. When the occurrence score is between 20 and 25 (red), it is essential to change the downgrading plan (Avoid). When the score is between 6 and 16 (yellow), an action is required to reduce the probability of the risk, or transfer the risk to another party (Mitigate). When the score is between 1 and 5 (green), the risk is acceptable (Accept) [IAE 19c].

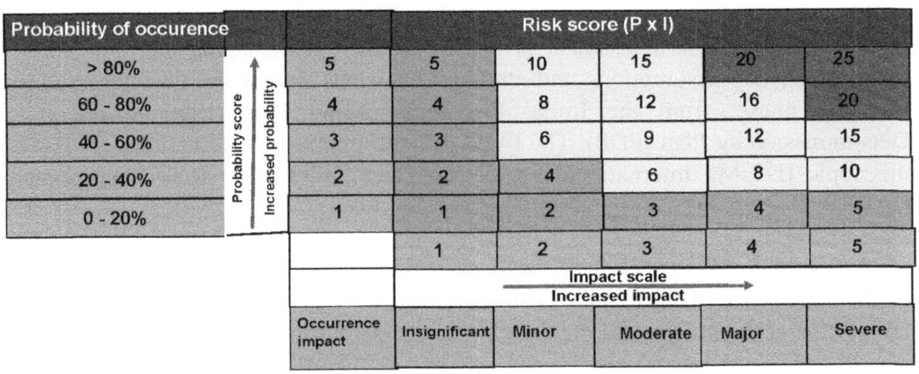

Figure 4.2. *Example of a probability versus impact diagram (source: [IAE 19c]). For a color version of this figure, see www.iste.co.uk/amiard/disarmament.zip*

Other examples of environmental and health risk assessments are detailed in the book *Risques chimiques environnementaux* [AMI 17].

4.4.2. *Environmental impact studies*

The operator must provide an Environmental Impact Assessment (EIA) with each application to decommission a BNI. Over the next few years, a large number of nuclear power plants will be decommissioned as they approach the end of their lives, and proposals for decommissioning will be submitted for various EIAs.

During the normal operation of a BNI, various radioactive, thermal and chemical releases occur. These releases are either into the atmosphere, into a liquid medium (river or sea), or onto the surrounding soil. They depend on the type of installation, as well as on its aging. In addition, BNIs, especially reactors, are greedy in taking water samples from the environment.

When shutting down a BNI, the operator is required, by regulation, to provide an environmental impact assessment of the operations related to decommissioning. The main difficulties take place during dismantling and mainly concern noise, physical and chemical pollution. These difficulties may have direct or indirect impacts, in the short or medium term, on the environment and human health.

In France, according to the Ministry in charge of ecology, an EIA is a "systematic study, formalized by regulations and laws, of the consequences of a project on the environment in the context of normal operation of the installations concerned. It is mandatory for each application for authorization of a classified facility or development meeting certain criteria" [MIN 19].

Impact studies must contain certain information. After a description of the project and an analysis of the initial state of the site and its environment, the EIA must provide an analysis of the effects of the project on health and the environment. It must also analyze the cumulative effects with other known projects in its vicinity. The reasons for the choice of project must be explained, as well as the compatibility of the project with land use and its articulation with the plans, schemes and programs of other ongoing projects. The operator must detail the measures envisaged to avoid, reduce and compensate for the effects of their project. They must also make a presentation of the methods used and any difficulties encountered in carrying out the impact study, and demonstrate the link between the impact study and the risk management study. The ASN has published a guide to assist the operator on this subject [ASN 16a].

EIAs are generally very large documents, especially since the BNI is large. For example, Exhibit 7 of the impact study for the dismantling of the Phoenix reactor is more than a thousand pages long [CEA 10].

EIAs are reviewed by several bodies, such as the French Environmental Authority (*Autorité Environnementale* – AE). For example, in the case of the dismantling of the Georges Besse plant of EURODIF, the AE [AE 16] was very critical and therefore recommended that the impact study be updated, with a new public inquiry held prior to each decision on the treatment of metal waste, the deconstruction of buildings and the remediation of contaminated soil. Similar criticisms are found for the decommissioning of BNIs in other states, such as the Canadian experimental reactor WR-1. It is believed that the EIA does not provide an explanation for the inexperienced reader [BAU 17]. More importantly, the *in situ* disposal proposed by Canadian Nuclear Laboratories (CNL) for WR-1 does not meet its strategic requirement to "contain radioactive contamination in such a way that risks to the public and the environment are kept ALARA (As Low As Reasonably Achievable)".

They are also subject to public consultation, with a public inquiry that is reported by the French Commission of Inquiry (*Commission d'Enquête*).

The French approach mentioned above is reflected in the decommissioning policy of many BNIs around the world. In Sweden, for example, environmental impact studies are examined twice by the authorities (Figure 4.3).

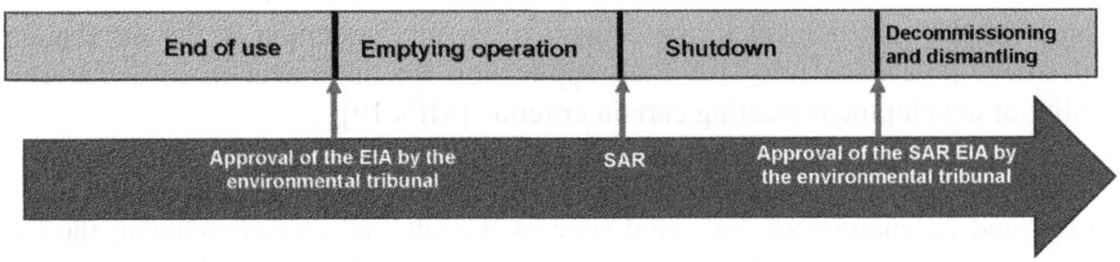

Figure 4.3. *Decommissioning operations and environmental and health impact assessments in Sweden (based on [RAN 16]). SAR: Safety Analysis Report*

In order to improve the proposals submitted, three EIAs, in which public participation was associated with this activity in three European Union countries, Germany (Greifswald), Spain (Vandello's Vandellòs) and the United Kingdom (Trawsfynydd), were examined in detail [BON 04]. These reviews attempted to identify elements of good practice that could be applied, in order to provide future decommissioning projects with good records. Five "success factors" summarize the main aspects of good practice to be applied. These include: acceptance by the authority of the value of public participation; integration of participation in all EIA activities; participation activities beginning before the final strategies are adopted; and transparency of the decision-making process and provision of sufficient information to the public to facilitate participation [BON 04].

In the case of European environmental impact assessments, although the issues of concern are similar in all countries, the different legislative processes from one country to another do not lend themselves to a coordinated approach. On the contrary, the ethics of dealing with these issues with respect to all parties involved, should be universal [LOV 12].

4.4.3. *Total remediation of the site*

Residual contamination on a nuclear licensed site must be removed, so that the site can be delicensed [BAY 03].

The choice of sanitation techniques is large. For example, *ex situ*, the operators have excavation, surface stabilization, soil washing, chemical treatment and magnetic separation at their disposal. Among *in situ* treatment techniques, the most frequently used are natural decay, electroremediation, phytoremediation, bioremediation, vitrification, capping or cryogenic barriers. The treatment of contaminated groundwater must also be addressed [FEL 12].

During the last phase of a site decommissioning project and the end of a BNI's life cycle, regulatory approaches that aim to terminate the nuclear license and release the site are of great importance, but unfortunately vary from one state to another [ATY 12].

The decommissioning of nuclear facilities and corresponding corrective measures are underway worldwide to allow for the reuse of sites, or parts of sites, for other purposes. In the nuclear field, the term "remediation" refers to measures taken to reduce the impact of contamination on land and associated groundwater.

The NEA's Decommissioning and Dismantling Working Group (WPDD) has gathered information on experiences, approaches and cleanup techniques that minimize risks to workers and the environment, as well as the costs and disruptions of decommissioning programs at various nuclear sites.

Environmental remediation is a multi-phase activity, which consists of identifying environmental problems, gathering information to develop a range of solutions to solve the problems, evaluating options and selecting the preferred solution, carrying out the remediation project that will solve the problem, and then verifying and documenting that the solution has been successful. A nuclear site remediation project is generally divided into seven phases: problem definition, investigation, resolution, planning, remediation, project closure and possibly institutional control [NEA 14a].

Experience in the renovation of nuclear sites is often weak or moderate, except in the United States and Germany, which have carried out several projects. NEA working group members have selected 12 case studies. They provide an in-depth understanding of the remediation practices used and the main problems encountered. The 12 sites studied are in France, STED of Grenoble (CEA) and Monts d'Arrée (Brennilis); Spain, PIMIC and "Lenteja" (CIEMAT); the United Kingdom, Windscale; Korea, a uranium conversion plant (Deajeon); Germany, an MOX fuel assembly plant (Hanau); the United States, groundwater contamination and waste burial sites at Hanford and a permeable wall for treatment *in situ* and Canada, a laboratory at Chalk River Laboratories [NEA 14a].

4.4.3.1. *International projects*

In this area, the International Atomic Energy Agency (IAEA) has taken the initiative to launch its Network on Environmental Remediation and Management (ENVIRONET) program. Indeed, many legacy (so-called orphan) sites around the world require decontamination and remediation, including mining, fuel cycle, military and research sites, as well as sites contaminated by naturally occurring radioactive materials (NORM).

Since 1995, the Agency's Technical Cooperation Department has supported 31 national decommissioning and environmental remediation projects, including 14 to assist member states with decommissioning and 17 with environmental remediation. It has also supported 16 regional projects, including seven for decommissioning and nine for environmental remediation, and two interregional projects.

During the period 2020–2021, the IAEA's technical cooperation program plans to support 11 new decommissioning and environmental remediation projects, with a total estimated budget of €5.9 million ($6.4 million). These include nine national projects and two interregional projects. It also plans to support 16 new radioactive waste management projects, with a total estimated budget of €5.8 million. Among the projects, two are relatively representative, Chernobyl and Tanzania.

In Chernobyl, a project to restore a cooling pond heavily contaminated by radionuclides (after the 1986 accident at the Chernobyl nuclear power plant), received financial means from the IAEA. During the operation of the plant, the pond, which was located between the plant and the Pripyat River, covered an area of almost 23 km^2 and contained 150 million cubic meters of water. After the closure of the last Chernobyl operating unit in 2000, such a body of water was no longer needed, and the water supply to the pond from the river was interrupted in 2015.

Tanzania is home to the uranium mine project located on the Mkuju River. The deposit was discovered in 2007. Unfortunately, the Selous Game Reserve, in which the Mkuju River is located, was classified as a World Heritage Site in 1982. This prompted the government of Tanzania to request an IAEA mission to evaluate uranium production sites [WNN 19e].

The United Kingdom has developed the SAFEGROUNDS project, which defines the principles for managing contaminated land. This program is divided into 32 steps. The essential steps are to clearly define objectives, identify all options, assess constraints, identify and agree on criteria, use a good assessment methodology and evaluate all options. The remediation of the historic sites of Harwell and Dounreay is detailed [BAY 03].

4.4.3.2. *Soil release thresholds*

Site clearance is one of the last steps in the decommissioning of nuclear facilities, but, to date, few facilities have reached this stage. The competent authority of the country where site clearance is to take place needs a correct assessment of the radioactive contamination of the soil and subsoil in order to make its decision. Appropriate measurement techniques are available, which can be used in conjunction with statistical methods. Based on these contamination results, a final radiological assessment of the site is possible. This allows a plan for the release of the site to be drawn up.

This type of approach has been carried out in the United States. The MARSSIM (Multi-Agency Radiation Survey and Site Investigation Manual) is a multi-agency consensus document that was developed collaboratively by four federal agencies with authority and control over radioactive materials, the Department of Defense (DoD), the Department of Energy (DoE), the Environmental Protection Agency (EPA) and the Nuclear Regulatory Commission (NRC). The objective of MARSSIM is to describe a consistent approach to planning, conducting and evaluating end-state investigations of land surfaces and basements to meet dose or risk-based release criteria, while promoting the efficient use of resources [MAR 00]. Some release threshold values for US soils are reported in Table 4.12.

A manual provides information on the design and application of the code (RESRAD – RESidual RADioactive) to estimate residual radioactivity. Version 6 contains a total of 145 radionuclides (92 main radionuclides and 53 associated radionuclides), and the half-life limit for the associated radionuclides has been reduced to one month [YU 01]. To complete it, the RESRAD-BUILD computer code is a pathway analysis model designed to assess the potential radiological dose to an individual working or living in a building contaminated by radioactive materials [YU 03].

Radionuclide	Threshold of release ($Bq.g^{-1}$)	Radionuclide	Release threshold ($Bq.g^{-1}$)L
^{3}H	4.1	^{14}C	0.44
^{60}Co	0.14	^{134}Cs	0.21
^{137}Cs	0.41	^{90}Sr	0.06
^{55}Fe	373	^{63}Ni	0.2
^{228}Th	0.1739	^{232}Th	0.04
^{235}U	0.3	^{238}U	0.52
^{238}Pu	0.09	^{239}Pu	0.08
^{241}Pu	0.27	^{241}Am	0.08
^{242}Cm	5.92	^{243}Cm	0.12

Table 4.12. *Soil release thresholds for US sites (NRC Derived Concentration Guideline Level (DCGL) values as a function of soil contamination (in Bq.g-1)) (based on [ABE 14])*

This type of modeling has been applied in the United States for the Cintichem site, in order to determine the release thresholds applicable to surface soil contamination for the radionuclides specific to this site, more precisely [AEN 06].

This approach is also used in Europe. Germany proposes release thresholds for its BNI sites based on surface soil contamination (Table 4.13). An application to the Spanish Vandellòs reactor site has also been carried out [AEN 06].

Radionuclide	Release threshold ($Bq.g^{-1}$)	Radionuclide	Release threshold ($Bq.g^{-1}$)
^{3}H	3	^{137}Cs	0.06
^{14}C	0.04	^{131}I	0.2
^{55}Fe	6	^{239}Pu	0.04
^{60}Co	0.03	^{241}Am	0.05

Table 4.13. *Prescribed site release thresholds adopted by Germany [STR 01]. The values apply to soil (according to [AEN 06b])*

The IAEA carried out a bibliographical synthesis of the various recommendations [IAE 96]. For example, for ^{60}Co, the proposed release thresholds for soils ranged from 0.02 to 1 $Bq.g^{-1}$ and for ^{137}Cs from 0.008 to 1 $Bq.g^{-1}$.

Soil reuse was addressed by the NEA at a relatively early stage. As early as 1992, the NEA's Co-operative Programme on Decommissioning (CPD) (NEA 11) established a Recycling and Reuse Working Group to examine the practice of recycling and reuse in the nuclear industry and identify barriers to the recovery of waste from decommissioning activities. The CPD mandated a new Task Group on Recycling and Reuse of Materials (TGRRM) in 2014, to update the 1996 report based on the decommissioning experience since 1996.

In 1996, the International Atomic Energy Agency (IAEA) recommended unconditional release levels for cobalt-60 and cesium-137 (two very important isotopes in nuclear power plants) of 0.3 $Bq.g^{-1}$ [IAE 96]. In scrap metal, the proposed release thresholds are 1 $Bq.g^{-1}$ for ^{60}Co and ^{137}Cs [NEA 96]. The European Commission estimates that the unconditional clearance levels are 0.1 $Bq.g^{-1}$ for ^{60}Co and 1 $Bq.g^{-1}$ for ^{137}Cs [EC 00]. The Basic Safety Standard (BSS) Directive (2013/59/Euratom [EUR 14]) proposes an unconditional clearance level of 0.1 $Bq.g^{-1}$ for ^{137}Cs [NEA 17a].

The main challenge remaining, is that an international agreement on clearance levels for materials resulting from decommissioning should be created [SEL 12].

4.4.4. *Health impacts of decommissioning*

The aspects of radiation protection related to the decommissioning of nuclear facilities are very important and are taken into account by the safety standards (SAR – Safety Analysis Report). The form and content of these safety standards are described in an IAEA guide for all life stages of a BNI [IAE 04b].

The safety reference system corresponds to the set of rules that govern the operation of an installation. These rules are modified as the facility enters the remediation/dismantling phase. The new standard, which is subject to authorization by the French Nuclear Safety Authority (in France, the ASN or ASND), specifies and frames the main stages of the project, the deadlines, the final state aimed at, the means implemented, the risks involved (discharges, dosimetry, etc.), the waste produced and the associated treatment processes [LOR 14].

The installation's safety reference system must therefore be adapted and developed to take into account whether or not the risks persist and whether the associated control measures need to be maintained or developed. Hence, the operator must produce a new safety reference system each time the situation changes. This is particularly obvious and necessary for dismantling operations that are spread out over a long period of time and include stopping points. This reference system must be sufficiently precise to allow the competent authority to authorize the passage to another phase of dismantling, until the final state is reached [IRS 19c].

Each stage in the life of a BNI presents a series of risks that develop over time (Figure 4.4).

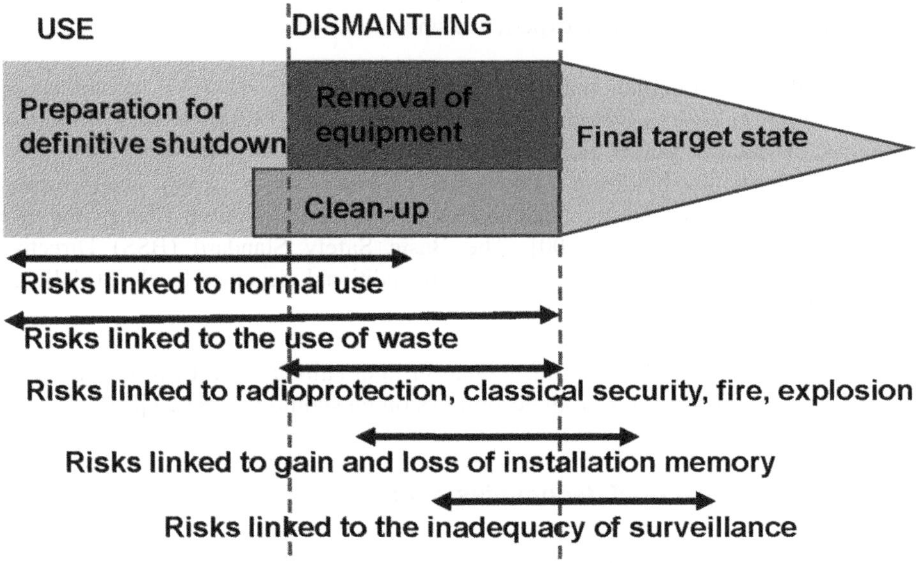

Figure 4.4. *Risks associated with the various phases of decommissioning (source: [AEN No. 08])*

4.4.5. *Social impacts of decommissioning*

Decommissioning is an industrial operation that is different from the commissioning of a nuclear installation. Its impact on workers is obvious; however, this impact can also extend to the public.

All personnel working on the dismantling site are subjected to additional irradiation which must be controlled and limited to the maximum. In addition, to successfully complete a decommissioning operation, operators must rely on well-trained personnel. Furthermore, the employment of personnel with temporary contracts from subcontracting companies must be strongly limited, and the latter must comply with the rules laid down by the nuclear regulatory authority, for example, the ASN in France. Employees must benefit from training, supervision and good working conditions. It is not enough to enact rules; their application must be monitored. In particular, radiation protection supervision must be rigorous, with monitoring of personnel, detection equipment adapted to the installations, analysis of workstations by radiation protectionists and recognition of occupational diseases [CCS 07].

In terms of the public, two aspects must be taken into account, a rigorous estimate of the additional doses of ionizing radiation delivered to local residents as a result of the dismantling, and informing the concerned public about this operation. For many states, during the decommissioning process, public consultations on the decommissioning project are planned. This is done so that, mainly the neighboring population, can express their doubts about the radioactive risk of the site and their desires for the future use of the site. This is the case in the United States, with public meetings concerning decommissioning plans (DPs), such as those for West Valley. Indeed, the NRC has long been accustomed to carrying out its regulatory responsibilities in an open manner and keeping the public informed of regulatory, licensing and monitoring activities at decommissioned or decommissioning sites [NRC 18]. This is also the case in France, although several organizations [CCS 07, NCA 17] consider that efforts in this direction are not sufficient. The ANCCLI, in its White Paper [ANC 17], proposed to amend the draft law on Energy Transition for Green Growth, in order to strengthen this aspect.

These social aspects remain remarkably under-studied. Although the communication, involvement and commitment of stakeholders are generally considered key catalysts for the success of a project, the needs and preferences of local communities are often neglected and a participatory form of dialogue is avoided.

Nuclear Decommissioning Projects and Programs (NDPs) are under discussion, but are progressing at Kozloduy-1 and 2, Ignalina-1 and 2, Superphénix, Greiswald-1 to 8 and Scanzanio Jonico. On the contrary, such programs have been accepted and are progressing at Vandellòs-1, Barsebäck-1 and 2, José Cabrera, Trawsfynydd, Onkalo and Olkiluoto [INV 17a].

Nuclear decommissioning projects and programs are threatened by several risks, the two main ones being the length of the schedule and significant costs with uncertain estimates. An innovative approach has been proposed to compare decommissioning projects, both in the nuclear and non-nuclear industries, in the UK and worldwide [INV 17b]. It is possible to compile a list of the main characteristics of NDPs and statistically test their correlation with project performance. Ultimately, the development of guidelines to improve the selection, planning and delivery of future NDPs is possible [INV 17b].

The European Union has also reflected on socio-economic, political and public perception issues, which are the main non-technical issues influencing decommissioning projects. The relatively negative public perception of nuclear activities is one of the main problems that needs to be solved by the authorities and industrialists involved in this field [BIS 01].

4.4.6. *Regulatory provisions*

The regulatory provisions for decommissioning and dismantling (D&D) procedures are specific to each state. In addition, they change regularly. The NEA has summarized recent developments in D&D authorization for five countries (Germany, Spain, the United States, France and Italy) [AEN 08].

In Spain, the organization in charge of radioactive waste management (ENRESA – *Empresa Nacional de Residuos Radioactivos*) covers dismantling operations.

Germany has developed and taken into account the aspect of stakeholder relations well (Figure 4.5).

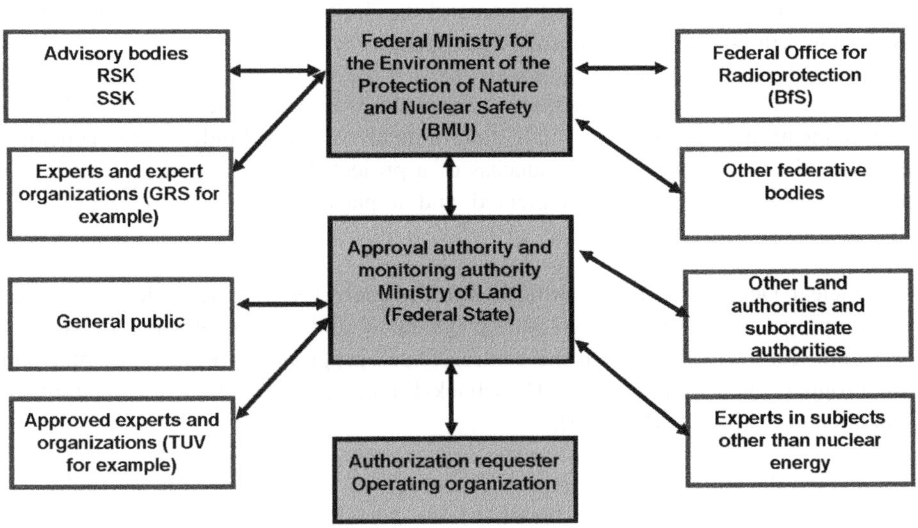

Figure 4.5. *Relationships between the parties involved in the decommissioning and dismantling process in Germany (according to [AEN 08])*

4.5. Nuclear decommissioning techniques

The main techniques of dismantling are cutting, decontamination and demolition techniques, as well as sanitation processes and the various possible robotizations.

4.5.1. *Cutting techniques*

Cutting operations are numerous and varied during nuclear decommissioning. These operations allow for the separation of the components of various circuits, as well as the gauging of large parts, so that they can be packed in the various containers used. The chosen process must meet these objectives by integrating, as well as possible, the constraints related to the intervention, which are radiological, environmental and dimensional, in particular. Cutting processes can be divided into three main families: mechanical, thermal and electrothermal processes and a fourth family of various processes.

In mechanical processes, cutting is obtained by shifting the mechanical properties of the material (load or breakage) without the effect of temperature, which remains close to that of the environment. With thermal and electrothermal processes, cutting is obtained by the melting or local vaporization of the material at high temperatures. Heat sources include plasma torches, flame cutting, thermal lances or lasers.

Other processes are used much less. These are pyrotechnical, chemical, electrochemical, electromagnetic and high-pressure water jet processes. Pyrotechnic processes are based on explosives (detonating cord, explosion and implosion) and are mainly used in building demolition. The chemical and electrochemical processes act by chemically attacking the material, or by electrolysis. Electromagnetic processes are obtained by using an intense pulsed electromagnetic field, which generates a high pressure in the conductive materials. This pressure is used for forming or punching. Ultra-high pressure water jet cutting is carried out by a very fine water jet (a few tenths of a millimeter in diameter) driven at a high speed resulting from the high injection pressure (3,000 to 5,000 bars). This action is reinforced by the addition of abrasive products in the water [CCS 07].

The performance of the various cutting techniques is summarized in Table 4.14. But sometimes, the best method is the enemy of the good. For example, in Brazil, where a power supply may be lacking on site, lightweight hand tools, mainly those powered by rechargeable batteries, were used during the dismantling of several IPEN-CNEN/SP buildings [LAI 16].

Cutting technique	Material	Thickness (mm)	Cutting speed	Manual use	Secondary waste control effort
Flame	Carbon steel	5–600	Fast	Yes	Average
Flame with powder	Metal	>100	Fast	Yes	High
	Concrete	<1,000	Fast	No	Very high
Thermal lance	Concrete	<2,000	Medium	Yes	Very high
	Metal	>500	Medium	Yes	High
Plasma arc	Metal	<130	Fast	Yes	Average
EDM*	Metal	>10	Slow	No	Low
CAMC*	Metal	<200	Fast	No	Average
Laser	Metal	<10	Fast	Yes	Average
Sawing	Metal	10–3,000	Medium	Yes	Low
Shaving	Metal	<10	Fast	Yes	Low
Grinding	Metal	<5	Slow	Yes	Average
	Concrete	<500	Medium	No	High
Diamond wire	Concrete	>300	Medium	No	Average
	Metal	>50	Slow	No	High
Water jet with abrasives	Metal	<300	Slow	No	Average
	Concrete	<500	Medium	No	High

Table 4.14. *Performance of various cutting techniques (source: [STE 12]). *EDM: electric discharge machining; CAMC: contact arc metal cutting*

4.5.2. *Decontamination techniques*

The implementation of decontamination is complex and can be carried out using a variety of techniques. The choice will depend on the techniques available, the desired dosimetric improvement and economic data.

The choice of the selected technique will depend on several assessments, such as the implementation constraints, the expected improvement of working conditions, the analysis of acceptability and the quantities of solid waste and liquid discharges generated by the decontamination for a safe operation. The dosimetric constraints are quantified by the evaluation of the total dose of the site and the expected dosimetric gain, considering that the decontamination operation is a source of external or internal exposure to the workers itself. The economic constraints are estimated by carrying out the economic balance sheet of the operation, taking into account the cost of the treatment of liquid effluents, as well as the conditioning, transport and storage of the solid waste generated by the operation.

There are approximately six categories of decontamination techniques, which are: chemical, electrolytic, thermal, mechanical, wireless and methods combining many of the other categories. Chemical processes use a chemical decontamination reagent, such as concentrated acids or bases, surfactants or strong oxidants (e.g. cerium IV). Electrolytic methods consist of imposing an anodic polarization on a metal part, which causes electrolytic oxidation (controlled erosion). Thermal methods use melting, incineration, distillation or drying. Mechanical methods allow for the mechanical removal of labile contamination, or the removal of the surface layer of the substrate. They are obtained by brushing, high-pressure spraying or bush hammering. Hertzian methods use ultrasound, microwaves or lasers. These methods can be combined to improve the effectiveness of decontamination [CCS 07]. Several of these methods generate aerosols that must be kept undispersed. It would therefore be useful to know about the performance of different techniques used when cutting steel or concrete in the production of aerosols, and a comparison between cutting underwater and in the air.

4.5.3. *Automation or remote operation*

Automation, robotization or remote operation involves introducing and moving a tool, performing a remote operation and removing the dismantled element for another treatment, when it is inaccessible, or when the operation is not humanly feasible for dosimetry reasons.

Two types of remote operation are distinguished, according to whether the vision of the operation is direct or indirect. In direct vision operations, the operators remain close by and are protected by a screen (water and leaded glass). Additional means of vision can also be added. During remote operations with indirect vision, the operators are placed in a control station. In this situation, the means of vision (video cameras) are of great importance [CCS 07].

4.5.4. *Remediation processes for civil engineering structures*

The cleanup of civil engineering structures can be carried out using four different processes: shaving, bush hammering, hydraulic rock breaking and diamond wire sawing.

The flush sanitation technique consists of removing the surface layer of concrete (often 6 mm maximum, per period). The basic element consists of diamond discs forming a roller. The discs are spaced a few millimeters apart and pulverize the concrete with their rotation. The feed rate and depth can either be set manually, or controlled by programming.

In bush-hammering processes, the tool is equipped with heads, with each head being fitted with small chisels arranged radially and free to rotate. The whole is pneumatically propelled at a rate of 1,200 strokes per minute with a head stroke of about 10 cm.

The hydraulic breaker (HBR) consists of a fixed part (the body) containing two moving parts (the piston and the tip). The piston, actuated by a hydraulic circuit, strikes the tip, generating a back and forth movement. By percussion, the concrete blocks are split into small blocks. This assembly can be mounted on several types of carriers.

Diamond wire sawing is considered when sanitizing thick concrete (several tens of centimeters). This is the case, for example, for concrete activated by the neutron flux of a reactor, such as the reactor vessel shaft of a PWR reactor, or the vessel of a UNGG reactor [CCS 07]. This method generates waste water and sand (which serve to increase the abrasive force of the cable), and thus produces radioactive sludge that will be collected and conditioned.

4.5.5. *The main demolition techniques*

Five main tools are frequently used in demolition operations: concrete tongs, hydraulic excavators, HBRs (hydraulic rock breakers), explosives and jacks. Concrete clamps act in two modes of action, either crushing the concrete or cutting the rebar (a reinforced steel rod in the concrete). Hydraulic excavators with large arms are used for the demolition of high structures (buildings, silos and water towers). An HBR is used to demolish reinforced concrete structures or foundations by percussion. The use of explosives in demolition is suitable for all types of structures with a few exceptions, such as sites that are still partly active. Jacking (the lifting of heavy elements by jacks) is a system of demolition by expansion, used for the bursting and fragmentation of walls and concrete blocks of great thickness, containing few iron rods or made of rock [CCS 07].

The NEA has conducted a very detailed study to compare the various methods and techniques for the dismantling and decontamination of concrete structures [NEA 11].

4.6. Technical innovations in nuclear decommissioning

The decommissioning and dismantling of nuclear facilities is a rapidly growing field with strong competition from companies around the world. Several attempts have been made to address various research and development (R&D) needs related to the challenges of decommissioning, such as characterization and studies to be carried out prior to dismantling, decommissioning technologies, on-site decontamination and remediation, nuclear material and waste management, site characterization and environmental monitoring.

A 300-page NEA report with more than 700 references identifies the seven main decommissioning challenges [NEA 14b]. These are the use of remote sensing and satellites, the use of robotics, mobile radionuclide modeling, statistical modeling of sampling, prioritization of waste management and site remediation. But above all, the first challenge is preliminary characterization (Predemo), which must identify the physical properties of the system, its structure and components, as well as the radiological risks of the system, which is a combination of historical information and studies. Finally, the sampling and analysis of the materials of the installation must be carried out [LAU 17].

The European Union has nine Technology Readiness Levels (TRLs), depending on the technological concept available. These are in increasing order of certitude: the basic principles observed, technology concept available, the proof of concept experimentally established, a technology is validated at the laboratory scale, or validated in a relevant industrial environment, a technology demonstration carried out in an operational environment, a prototype system demonstrated in an operational environment, a complete and duly qualified system is available and a real system has been tested in an operational environment ([EC 15], in [LAR 18]).

4.6.1. *Research and development policies in the field of decommissioning*

The effort of states and organizations in basic and applied research in the field of decommissioning and dismantling of nuclear facilities is significant. Only the example of European policy will be developed below.

The European Union does not intervene in national policies for the decommissioning of nuclear installations. On the contrary, it does finance research in this field. Since 1979, DG Research of the European Commission has carried out four successive five-year research and development programs on the decommissioning of nuclear installations, carried out under shared-cost contracts with European Union organizations. The main objective of these programs was, and still is, to establish a scientific and technological basis for the safe, socially acceptable and economically affordable decommissioning of obsolete nuclear installations.

In the early 1990s, four decommissioning pilot projects were selected to compare the different approaches. These were a fuel processing plant (AT1 in La Hague), a gas-cooled reactor (WAGR at Windscale), a boiling water reactor (KRB-A Gundremmingen in Germany) and a pressurized water reactor (BR-3 in Belgium) [EU 05b]. In 1996, a WWER reactor (Greifswald, Germany) was added to this European list of pilot decommissioning sites.

The dismantling of the WAGR, for example, served as a model for the future dismantling of gas-cooled graphite reactors. Innovative dismantling techniques were deployed, such as a computer-controlled remote dismantling machine using stereoscopic television cameras to assist in the dismantling process, acoustic cleaning of electrostatic pre-filters; an ultraviolet laser to decontaminate vital parts of the machine prior to maintenance; and a video–gamma camera to identify hot spots and sort radioactive materials.

The dismantling of the BR3 reactor in Belgium successfully focused on the development of dry and underwater cutting techniques for the highly activated internal parts of the nuclear core.

The decommissioning project for the WWER reactors in Greifswald, one of the world's largest nuclear power plants, started stage 3 of the decommissioning for five commercial WWER-440 reactors in Greifswald and one WWER-70 reactor in Rheinsberg. The remote-controlled dismantling of the first pressure vessel and the internal reactor components, using a new, specially developed robotic system, began in 2001.

For the 250 MWe boiling water reactor KRB-A (Germany), the dismantling of the internal core components, the heat exchanger, the activated concrete biological shield and the reactor pressure vessel were completed.

The AT1 reprocessing plant in France has successfully completed its decommissioning period and in 2000, the site was being cleaned up for future use [BIS 01].

Under the EC program, two decommissioning databases have been created – EC DB TOOL, for the collection of technical performance data and EC DB COST, for the collection of data on waste produced and doses. The two databases are now merged into one – EC DB NET. In the field of information exchange, the Commission has decided to support the creation of a "thematic network on decommissioning" (www.ec-tnd.net) as an effective instrument to facilitate these objectives [BIS 01].

The Fifth Framework Plan (FP-5) of the European Commission had a project in the field of decommissioning. The IRDIT (Innovative Remote Dismantling Techniques) project dealt with a particular aspect of the dismantling process, which was the remote dismantling of the reactor pressure vessel (RPV) and its internal components. For this purpose, two projects were selected: the remote cutting of the Russian type WWER reactor pressure vessels of units 1 and 2 of the Greifswald nuclear power plant (Germany), and the cutting of the RPV and the internal components of the BR3 pressurized water reactor in Mol (Belgium) [BIS 01].

This policy continued during the Sixth (2002–2006) and Seventh Framework Program for Research and Technological Development (FP – Framework Program) (2007–2013). In the European Framework Program, Horizon Europe 2014–2020, the Euratom 2019–2020 called for proposals, including 17 topics covering six areas. In the field of "decommissioning and environmental remediation", the NFRP-09 topic was to "promote innovation in the decommissioning of nuclear facilities". The total budget for this call was approximately €63.60 million for 2019 and €70.30 million for 2020.

4.6.2. *Industrial organization in the field of nuclear decommissioning*

Lessons from the past clearly demonstrate that in the field of decommissioning and dismantling, the organization of industrial companies is the best solution to achieve successful decommissioning.

The nuclear decommissioning ecosystem is a sector under construction. In France, the *Pôle de Valorisation des Sites Industriels* (PVSI) has been created. This hub provides access to data and information, which is essential, because in this context of cooperation and competition between companies in the sector, data becomes a strategic issue [SAM 16].

For several years, a number of companies have been created to take care of the dismantling. These new companies are often agreements and cooperations between two or more companies that pool their human, material and intellectual resources to solve a particular problem. For example, a joint venture for reactor decommissioning using graphite technology was created by EDF and Veolia. This new company, Graphitech, hopes to obtain contracts to help dismantle graphite reactors in France, Italy, Japan, Lithuania, Spain and the United Kingdom [WNN 19q].

In some cases, it is a large group operating in the nuclear industry that creates a subsidiary that is specialized in decommissioning. Examples are numerous, such as PreussenFlektra, a subsidiary of Eon, AVR GmbH, a subsidiary of EWN GmbH, among others. Some subsidiaries are also created specifically for a single state or a single project, such as ORANO Decommissioning and Services GmbH for ORANO in Germany, Cyclife Sweden for EDF in Sweden, to cite a few.

4.6.3. *Management of radioactive contamination*

Radiological characterization is an essential activity for planning and implementing the decommissioning of nuclear facilities. Effective characterization allows the extent, location and nature of contamination to be determined and provides crucial information for the decommissioning of facilities, management of materials and wastes, protection of workers, the public and the environment, and estimation of associated costs [NEA 17c].

4.6.3.1. *Quantification of radioactive contamination*

The quantification of the radioactive contamination of a site must be the subject of a serious study, with the definition of a strategy of the selected characterization process. This strategy will define the following steps, with a planned number of campaigns. The beginning of a campaign will be preceded by a scoping document. The planning must include several plans: characterization, sampling and analysis. The implementation is coupled with recording, an evaluation report and a characterization report. At the end, a summary must be drawn up [NEA 17c].

With the increasing number of nuclear facilities, particularly nuclear reactors, that have been shut down in recent years, it is becoming increasingly necessary to conduct assessments of the inventory and distribution of radioactivity in various materials, in order to make decisions for the management of the waste produced. The quantitative determination of various radionuclides present in these materials is therefore a priority. These radionuclides are extremely varied. They are mainly the neutron activation products of components and impurities in materials used in nuclear facilities, such as ^{3}H, ^{14}C, ^{36}Cl, ^{41}Ca, ^{60}Co, ^{55}Fe, ^{63}Ni, ^{79}Se, ^{93}Mo, ^{93}Zr, ^{94}Nb, ^{126}Sn, ^{133}Ba, ^{152}Eu, ^{154}Eu and some transuranics (isotopes of Pu, Np, Am and Cm), present in construction materials. In the event of a leak of spent nuclear fuel, long-lived fission products such as ^{90}Sr, ^{99}Tc, ^{129}I, ^{135}Cs and ^{137}Cs are the main concern. While gamma emitters are easily detectable, alpha and beta emitters require the separation of the sample matrix, which is time-consuming and expensive. Radiochemical analysis is the only way to implement the determination of these difficult to measure radionuclides. Unfortunately, appropriate methods are not always available for the characterization of various decommissioning materials. This is due to several reasons, particularly the fact that the sample matrix varies a lot. This includes concrete, graphite and ion exchange resins, which require different radiochemical methods for different sample matrices and target radionuclides. New radiochemical measurement methods are available, such as mass spectrometry, in particular, ICP-MS, and accelerator mass spectrometry (AMS) for ^{41}Ca, ^{36}Cl, ^{99}Tc, ^{129}I, ^{135}Cs and transuranics. The measurement time can be significantly reduced from days to minutes, allowing the rapid measurement of a large number of samples. This is interesting because some of these radionuclides are difficult to measure, as they are beta or X-radiation and are therefore easily absorbed into the material. Also, little radiation emerges from thick parts and thc ambient "background noise" makes on-site measurement very difficult. However, these radionuclides play an important role in radiation protection because of their very long half-life (15.7 million years for iodine-129 or 2.3 million years for cesium-135).

Several examples of analytical methods have been presented, such as the rapid determination of ^{3}H and ^{14}C in solid materials; the determination of ^{36}Cl and ^{129}I in graphite, concrete and metals; the simultaneous determination of multi-nuclides such as ^{41}Ca, ^{55}Fe, ^{63}Ni, ^{90}Sr and transuranics; as well as the difficult determination of low concentrations of ^{93}Mo, ^{93}Zr, ^{135}Cs, ^{79}Se and ^{126}Sn in decommissioning waste [HOU 16].

Dismantling operations are often made difficult due to the ignorance *a priori* of the position, nature and radiological characteristics of radiation sources. Operators can use dose rate modeling tools. Three tools are available. They are the CARTOGAM gamma camera and the MERCURAD dose rate modeling tool, developed in cooperation with the CEA and COGEMA, as well as the ISOCS gamma spectrometry system developed by Canberra. These three tools have been put into operational use in Europe, America and Asia [CHI 03].

4.6.3.2. *Radioactive decontamination*

A review of decontamination processes commonly used in nuclear decommissioning programs, as well as some emerging techniques has been presented [NOY 12]. Among the latter, the author details the decontamination of metal components and equipment, segmented metal parts and building surfaces, mainly with regards to their reuse and recycling.

The emerging techniques are mostly mixed techniques combining two or more of the chemical, electrochemical, biological, mechanical, laser ablation or ultrasonic methods [NOY 12]. The example of decontamination methods will be developed in Chapter 7.

4.6.4. *Numerical simulations in the field of nuclear decommissioning*

Numerical simulation is used in a very large number of fields. This is explained by the importance of having the most accurate and correct possible vision of the future, in order to eventually remedy and redirect this future. The success of numerical simulation is due to its increasingly obvious accuracy and precision. This has largely been made possible by previous experiences, but above all by the computational capacities made possible by the new tools at our disposal. In the field of decommissioning, numerical simulation is of interest in modeling the mobility of radionuclides in the environment and in artificial environments, organizing sampling plans, simulating dismantling operations or managing the various stages of decommissioning [AMI 17].

4.6.4.1. *Modeling the mobility of radionuclides*

Whether in the natural environment or in the anthropogenic environment, radionuclides are found in all three possible physical forms (gaseous, liquid or solid) and in all compartments. Their mobility between the various compartments varies according to several factors. One of these, and a primordial one, is the radionuclide itself; some radionuclides, such as tritium, are extremely mobile, others are much less so, or may even be immobile. The second factor is the nature of the medium. Generally, solid media provide greater immobility than media that are themselves mobile, such as water or air. Many studies propose numerical models that simulate, more or less efficiently, the mobility of radionuclides [AMI 13].

Decontamination and remediation cover a wide range of topics. These include a better understanding of the chemical interactions and mobility of contaminants with substrates, as well as the chemical and physical processes that can be used to decontaminate decommissioning materials, such as concrete, metals, graphite, tanks, groundwater and soil. Three-dimensional modeling allows a non-intrusive sample characterization [NEA 14b].

4.6.4.2. *Numerical simulation of dismantling operations*

Jeong *et al.* filed a patent for an invention that concerns a simulation of a nuclear installation decommissioning operation [JEO 18].

Similarly, current methods for simulating cutting processes are unable to do so flexibly. Therefore, the automatic generation of cutting products in the modeling procedure, throughout the cutting process was proposed [KIM 16].

It is also possible to simulate accidents due to human error during decommissioning operations [JEO 16a]. Similarly, a simulation can be used to establish a risk reduction methodology for the decommissioning of nuclear facilities, as well as a mathematical estimation method [JEO 16b].

An ergonomic assessment of the best working position (comfort) during the decommissioning of nuclear facilities was determined by computer. This methodology should make it possible to establish the ergonomic decommissioning plan for nuclear facilities. The safety of decommissioning will be improved and the costs can also be reduced [JEO 16c].

The working hours during decommissioning have a great effect on safety and costs. It is therefore interesting to be able to estimate the working time of each activity station, in order to optimize the trajectories of workers during the decommissioning of nuclear installations. This may reduce decommissioning costs and improve safety during decommissioning [JEO 16d].

Similarly, a system for assessing the risk of human error during decommissioning can be developed. This system should be used as a training tool to improve workplace familiarity and prevent worker accidents [JEO 16d].

The choice of the management option for the most globally relevant solution for nuclear decommissioning, and the identification of the criteria used to assess this relevance, can be obtained using numerical simulations. This communication tool will be used by all stakeholders, operators, regulators and so on [JEO 19].

In addition to radiochemical measurements (see the section above), another approach can be used to make an inventory of the radionuclides present in the various components of a nuclear installation to be decommissioned. This approach is based on neutron transport and radionuclide inventory calculations. However, a good estimation of the sources of uncertainty is required [LEL 19].

There are many factors that influence the decision-making process when decommissioning a nuclear facility. Using the study of 162 cases of permanently shut down nuclear facilities, these factors were identified and quantified [SUH 18]. It emerged from this vast study that the main factors are the periods of operation, the Health Development Index (HDI), the financing of decommissioning, public acceptance, public tolerance, the level of experience in the decommissioning process, the type of reactor, the operating history, the availability of radioactive waste facilities, the presence of several units on the site and the technological capabilities of the operator. However, two key factors remain: first, the way in which the decommissioning work is carried out, i.e. decontamination and immediate (DECON) or deferred (SAFESTOR) dismantling, and second, the choice of the end state of the site, either a clean site without any restriction of future use (greenfield) or a less clean site (brownfield), where there are restrictions on future use (parking, industry, etc.) [SUH 18].

4.6.4.3. *Computerized decommissioning management*

In South Korea, KAERI (Korea Atomic Energy Research Institute) is the only group of experts with decommissioning experience. This group is trying to develop a computer code to converge all of the data that has been accumulated during the decommissioning of the KRR-1 and 2 reactors and of the uranium conversion plant [HON 17].

On a research reactor (Fugen, ATR – Advanced Thermal Reactor), a computer support system was developed for decommissioning by adopting new technologies, such as three-dimensional computer-aided design (3D-CAD) and a virtual reality (VR) called DEXUS [IGU 17]. The 3D-CAD data is very useful for planning the decommissioning of highly radioactive materials, such as the reactor core. It also helps in coupling with robotics when data creation is costly. The VRdose simulation system predicts the dose received by workers and assists in the planning of manual work.

Augmented reality (AR) is one of the useful technologies for field application. The Knowledge Management System (KMS) for optimizing the decommissioning, transfer and education of lessons learned to the next generation looks promising [IGU 17].

	Early life phase	**Mid-life phase**	**End-of-life phase**
Nuclear installation	Design and construction of the BNI	Operations Maintenance	Decommissioning
Nuclear waste	Waste design according to the selected evacuations	Waste production by type and geographic area	Evacuation
Decommissioning scenario	Scenario design and validation by authorities	Scenario monitoring	Feedback

Table 4.15. *The different "products" according to the life cycle of a nuclear installation from its design to its end of life (according to [CHE 18])*

In a context where decommissioning activities are becoming more and more strategic for the nuclear sector, data information management is of primary interest. A PLM (Product Lifecycle Management) approach is proposed and tested as the information backbone for the decommissioning activity and the support of the whole process design. The life cycle of a nuclear installation has three essential phases, the beginning, middle and end of life. Each phase corresponds to different activities and actions to be undertaken for nuclear waste management and preparation for final decommissioning (Table 4.15). PLM tools include features to improve collaboration, a workflow engine to automate processes and approaches to manage product variants and versions [CHE 18].

4.6.5. *Cutting of large parts*

Cutting, particularly of nuclear reactor vessels, was very quickly a concern during dismantling. Solutions were proposed very early on. This was the case with a wire saw using diamond balls [KUB 94]. The main cutting methods are mechanical (saws, shears, nibblers, drilling, milling and hydraulic jackhammer). There are also thermal cutting (flame, plasma, thermal lance, electro-discharge and laser) and some other methods (expansive grout, high-pressure water and explosive cutting) [BAY 03]. However, currently, for the cutting of large parts, two tools are preferred: diamond saws and lasers, provided that the laser is not too thick.

4.6.5.1. *Cutting with diamond saws*

An example of cutting with a diamond saw is provided by the decommissioning of the Argonne Fast Source experimental reactor. It was built in the 1950s at the Materials and Fuels Complex of the INL, to create neutron fields that exposed materials to a radioactive environment. The reactor ceased operations in the early 1980s. Fluor Idaho and Battelle Energy Alliance worked together to remove the old reactor, while limiting the impact on research schedules and other ongoing laboratory operations. This included cutting and removing the 1.4-meter shield from the reactor. The demolition team decided that a diamond wire saw would be the most effective way to remove the shield from the structure, which was not contaminated by radiology. The teams cut the reactor into 56 blocks, each weighing between 1,360 and 2,270 kg, and removed them with a gantry crane [WNN 19p].

4.6.5.2. *Cutting with a laser*

However, laser cutting is currently the preferred method on most construction sites. The Chemical Oxygen–Iodine Laser (COIL) is a high-power fiber tool that can be used for a number of different industrial applications. In previous research, the University of Illinois at Urbana-Champaign (UIUC) identified the decommissioning and decontamination (D&D) of nuclear facilities as a primary objective of COIL technology. The use of a robot-driven fiber optic cutting and ablation tool in contaminated areas promises to reduce the risk to workers for the D&D mission. In addition, the high cutting speed of COIL will significantly reduce the time required to cut contaminated equipment, thereby reducing costs. The high power of COIL will allow for the dismantling of thick piping and equipment, as well as reactor vessels. The COIL laser holds great promise for the removal of material from contaminated surfaces, perhaps at depths greater than one inch (25.4 mm). Laser cutting and ablation minimizes dust and fumes, reducing the number of High Efficiency Particulate Accumulator (HEPA) filters required, thereby reducing the costly disposal of nuclear waste [SOL 06]. Great strides in power have been made and the thicknesses listed in Table 4.14 are now exceeded. For example, a laser with a power of 30 kW can cut a 1,000-mm thick biological shield wall at a cutting speed of 10 mm per minute [NOM 00]. 3D assistance in alignment is offered for laser cutting [KIM 17].

A patent has been proposed in which, after cutting the head of the vessel, some of the nuclear components of the reactor are introduced into the vessel and the vessel is hermetically packaged (encapsulated) before being stored in the long term [LES 18].

During laser cutting, particle and aerosol emissions must be minimized. Overall, it has been found that the mass of aerosol per unit of cutting area (by laser cutting) decreases with increasing laser power and cutting speed. When cuts are made underwater, the production of aerosol particle mass per unit area of cut is reduced by a factor of 10 and limited to less than 70 $g.m^{-2}$ [PEI 17].

4.6.6. *Automation in the field of nuclear decommissioning*

Among the innovative applications that can be used in the decommissioning of nuclear facilities, the NEA identifies robotics and remote sensing, telemetry and satellite technologies for characterization and monitoring [NEA 14b].

4.6.6.1. *Visual inspection*

The monitoring and inspection of highly radioactive areas are difficult, so the development of a camera resistant to strong radiation (several MGy) would be of great use. The CAMRAD camera has semiconductor components (CMOS – Complementary Metal Oxide Semiconductor), no shielding and works without significant degradation [GOI 18].

4.6.6.2. *Robotization*

Many decommissioning activities can benefit from remote handling equipment and robotics. These include the radiological characterization of a site, the recovery of contaminated objects, the dismantling, demolition and decontamination of buildings and nuclear equipment [MIC 12].

However, in order to assist the nuclear waste management mission and the decommissioning of BNIs, progress still has to be made in the detection and perception of ionizing radiation by nuclear robotics [MAS 17].

As part of the program to decommission the power demonstration reactor in Japan, remote handling technologies were developed. The systems developed are double-arm systems associated with a control mechanism. Operations are carried out in four successive stages: preparation, carrying, cutting and transport. This technique could be successfully applied to the dismantling of the main components of BNIs [TAC 00].

In 2024, CERN's Large Hadron Collider (LHC) will be upgraded to increase its "luminosity" (i.e. the number of collisions) by a factor of 10 (HL-LHC – High Luminosity-Large Hadron Collider). The ATLAS (ITk, Inner Tracker) indoor detector will be upgraded at the same time. It has suffered the most radiation damage, as it is the closest section to the line of light, and particle collisions are more frequent. Due to the risk of excessive radiation doses, human intervention to take the indoor detector out of service will be limited. Therefore, a study and evaluation of the radiation tolerance of a robotic finger were carried out at the Birmingham cyclotron facility. The finger was part of Shadow Robot Company's Shadow Grasper, which uses a set of Maxon DC engines [FRE 18].

4.6.6.3. *Satellite remote sensing*

At the end of the 1990s, spatial imagery developed considerably, making it possible to observe small geographic spaces from space. It is a tool used by Google to update its Google Earth and Google Map software, but the main client is the United States Department of Defense. The maximum resolution of GeoEyes 1 launched in 2008 was 41 cm and that of GeoEye 2 was 36 cm. However, for security reasons, it was reduced to 50 cm and only the military have access to this type of data. In addition, images of sites considered strategic, such as nuclear power plants or military bases, are controlled [ANO 20]. The applications are numerous, such as the monitoring of nuclear non-proliferation or military activities [JAS 99], the detection of nuclear waste in space [ULI 17], the estimation of radioactive contamination of the sea, using correlations with physical parameters of the ocean, measured by satellite [MAV 16], the state of conservation of buildings [HAR 19] or the measurement of the irradiation dose suffered by a satellite in space, using apparatus installed in a satellite [CUI 18, KOC 19]. The use of remote sensing for the decommissioning of nuclear installations is thus conceivable.

4.6.7. *Estimating radiation doses*

During the decommissioning of nuclear facilities, a large number of cutting and demolition activities are carried out, resulting in frequent structural changes and numerous irregular objects. Thanks to virtual reality and a flexible method, it is possible to evaluate the dose received from objects with various geometries and the evolution of this dose during the cutting of a part or demolition operations [CHA 18].

4.7. Conclusions

Feedback on the decommissioning and dismantling of nuclear facilities is abundant. It provides a better estimate of the cost of these operations.

In order to take this subject further, the CEA has published two monographs: one on remediation and dismantling [CEA 17] and the other on instrumentation and measurement in nuclear environments [CEA 18].

It is indeed essential to accurately assess the radiological, physical and chemical state of the installation to be remediated or dismantled. Then, the containment of the remediation–dismantling sites must make it possible to preserve both the environment and the health of workers [CEA 17].

Operations and interventions in hostile environments use robotics and 3D simulations. Similarly, laser cutting is an emerging technology for remote-controlled dismantling sites. During dismantling, the treatment and decontamination of structures, soils and effluents use various methods and techniques [CEA 17].

The remediation–dismantling wastes are varied and include organic wastes, graphite wastes, magnesian wastes, special wastes such as mercurial wastes, muddy or powdery high activity wastes and tritiated wastes. It is necessary to find the appropriate treatment [CEA 17]. Nitrated wastes should also be mentioned. Thus, in France, ORANO buys uranium in the form of U_3O_8 (a yellow powder called "yellow cake") that is dissolved in nitric acid in order to rid the uranium of various impurities consisting of stable elements (carbon, fluorine, sodium, calcium, potassium, etc.) or of radioactive isotopes (produced by the decay of uranium since its conversion to U_3O_8). The result is nitrated solutions representing several thousands of cubic meters, which are stored and are soon to be treated.

Feedback in remediation–dismantling is beginning to be consistent. Thus, for France, let us cite the feedback from the UP1 spent fuel treatment plant in Marcoule, the experience of dismantling sodium-cooled fast reactors and the example of the dismantling of the SILOÉ reactor in Grenoble [CEA 17].

5

The Dismantling of Military Nuclear Facilities

5.1. Introduction

The decommissioning of military nuclear facilities frequently presents special problems. This is due to a combination of several phenomena. First of all, the installations are often old and from a time when notions of security and safety were not as developed as they are today. Thus, the main Manhattan Project sites, such as Hanford, Oak Ridge or Savannah River, Soviet sites, such as Mayak and British sites, such as Windscale were created either during World War II or just after it. During this period, qualified as the Cold War, the goal of these states' military was to develop atomic weapons as quickly as possible and in large numbers. Thus, all the indispensable operations, such as the enrichment of fissile materials and the reprocessing of spent fuel, were carried out on a large scale with few safety and security concerns or rules. This resulted in numerous incidents and accidents on these sites, with environmental and health consequences, on the site itself for the former and off-site for the latter. The soil and subsoil at these sites, as well as the aquatic environments, were frequently heavily radioactively contaminated.

Three types of nuclear installations are indispensable to the development of atomic weapons: high-grade uranium enrichment plants, plutonium-producing reactors and spent nuclear fuel reprocessing plants, the last separating the plutonium-239 created in large quantities in the reactors, thanks to their design and directed operation. Alongside these three types of installation, a multitude of other buildings house various types of production of fissile materials, as well as design offices, test laboratories, etc.

The number of military nuclear facilities that are no longer operational is significant. For example, in the United States, the number of military buildings to be decommissioned in 2016 was 2,349 military nuclear plants, and the cost was estimated at US$32 billion [DOE 16]. The Manhattan Project developed mainly at five sites: Hanford, Oak Ridge, Los Alamos, Livermore and Savannah River.

Four of the five NPT nuclear-weapon states (United States, Russia, United Kingdom and France) have declared that they have permanently ceased the production of fissile material for nuclear weapons. France, for its part, ceased the production of weapons-grade plutonium in 1992 and that of highly enriched uranium in 1996. The plants at Marcoule and Pierrelatte have been closed and are in the process of being dismantled. China, on the other hand, has not formally declared a moratorium. It has suspended production, but there is no evidence that it is clearly committed to not increasing its stock of fissile materials, with a view to developing its nuclear arsenal [CHE 10].

The production of highly enriched uranium ceased in 1964 in the United States and in 1989 in the USSR, and production of weapons-grade plutonium ceased in 1988 for the United States and in 1994 for Russia. The United Kingdom declared a moratorium on the production of fissile material for nuclear weapons in 1995. In these three countries, some installations for military use have been shut down, while others have been converted for civilian use [CHE 10].

India continues to produce weapons-grade plutonium, but does not enrich uranium for military purposes, except for its naval propulsion program. Pakistan produces highly enriched uranium and weapons-grade plutonium. Israel has the capacity to produce weapons-grade plutonium with its Dimona reactor, but it is not known whether it is pursuing activities in this direction. North Korea has resumed its reprocessing and plutonium separation activities at its Yongbyon plant and has reached the "*final stage*" of enrichment [CHE 10].

5.2. The decommissioning of military plants at the beginning of the nuclear fuel cycle

5.2.1. *The decommissioning of military uranium enrichment plants*

One of the ways of making an atomic bomb requires highly enriched uranium. This explains why one of the first priorities of the American Manhattan Project was to build an enrichment plant. This was done as early as 1945 in Oak Ridge (the Y-12 plant). Since then all countries have followed this path. Originally, the enrichment system was based on gaseous diffusion. Since then enrichment by centrifugation has been more efficient and less costly. The first countries in possession of the atomic bomb, with the exception of Russia, have closed their military uranium enrichment plants (Table 5.1). On the other hand, India, Pakistan and North Korea still have military enrichment plants (Table 5.2), as does Iran.

According to Kurtz *et al.* [KUR 19], the Chinese factory in Heiping is still operational with an annual production of 230 ton separative work units (tSWUs).

Country	Factory	Operational dates
United States	*Oak Ridge, Y-12, TN*	1945–1947
	Oak Ridge, TN	1945–1985
	Portsmouth, OH	1956–2001
	Paducah, KY	1952–2013
United Kingdom	*Capenhurst*	1954–1982
France	*Pierrelatte*	1967–1996
	Georges Besse	1979–2012
China	*Lanzhou*	1964–1998
	Heping	1975–1987

Table 5.1. *Closures of military uranium enrichment plants (source: [FEI 14])*

Country	Factory	Type	Status	Security	Capacity (tSWU.yr^{-1})
India	*Rattehalli*	Military	Operational	No	(15–30)
North Korea	*Yongbyon*	?	Operational	No	8
Pakistan	*Kahuta*	Military	Operational	No	(15–45)
	Gadwal	Military	Operational	No	?

Table 5.2. *Military enrichment plants in operation in 2013 (source: [FEI 14]). Capacity supplied in ton separative work units (tSWU)*

In the USSR, then in Russia, uranium enrichment activities took place on four principal sites. The Siberian Chemical Combine of Seversk produced highly enriched uranium (HEU); currently, it produces low-enriched uranium (LEU) and converts HEU metal to oxide (from nuclear weapons), as well as fluorination. The Mayak Production Association in Ozersk performs the conversion of HEU metal to oxide. The Zelenogorsk Electrochemical Plant produced HEU and now LEU and fluorination of HEU and LEU. The Urals Electrochemical Plant in Novouralsk produces HEU and LEU but is expected to gradually reduce this production [POD 09].

5.2.1.1. *US military enrichment plants*

These plants are on the Oak Ridge site.

The Y-12 plant in Oak Ridge

The Manhattan Project needed enriched uranium for the production of atomic weapons. The construction of building 9201-4 (Alpha-4) at Oak Ridge in 1943 was originally used to house a process of separation of uranium isotopes based on electromagnetic separation technology. However, with the start-up of the Oak Ridge K-25 gaseous diffusion plant, the Alpha-4 site was put on standby in 1947. In 1953, the process of uranium enrichment was discontinued in this building. Subsequently, the building housed the Colex process, which uses a mercury solvent and lithium hydroxide as the feed material.

The Alpha-4 building had a floor area of more than 46,000 m^2 and contained 15,000 tons of electrical equipment and 23,000 tons of insulation, mortar, bricks, flooring, handrails, ducts, utilities, fuels and sludge. Since much of this equipment and building material was contaminated with elemental mercury, a clean-up was required [OER 94].

The decontamination and decommissioning (D&D) program at the Oak Ridge Y-12 mill is part of the Environmental Restoration (ER) and Waste Management (WM) programs. The objective of the ER program is to provide Y-12 with the ability to meet applicable environmental regulations through facility development activities and site remediation. The WM Program supports the ER program [OAK 92].

K25 closure and demolition in Oak Ridge

When it was built in 1944, the four-story K-25 gas diffusion plant was the largest building in the world with a surface area of more than 152,000 m^2 and a volume of 2,760,000 m^3. The K-25 became a prototype for other gaseous diffusion plants established in the early post-war period. Other gaseous diffusion enrichment facilities on the same site were the 34,700 m^2 K-27, completed in September 1945, the 6.1 ha K-29, completed in 1951, the 8.1 ha K-31, also completed in 1951 and the 13 ha K-33, completed in 1954. Other gaseous diffusion facilities were constructed in Paducah, Kentucky, in 1952 and Portsmouth, Ohio, in 1954. Nowadays the separation of uranium isotopes is generally carried out by the process of ultra-centrifugation, which requires less energy and is less costly. Centrifuge cascades began operating at Oak Ridge in 1961. Because of a lower demand for enriched uranium, K-25 and K-27 ceased production in 1964. However, in 1969,

K-25 began to produce enriched uranium again, but now only enriched by 3–5%, for civilian use in nuclear reactors.

The US Department of Energy (DoE) contracted with British Nuclear Fuels Ltd (BNFL) in 1997 to decontaminate and decommission the facilities K-29, K-31 and K-33. Various subsidiaries have been working on this project over the years. The demolition of K-29 began in January 2006 and was completed in August of the same year. The demolition of K-33 began in January 2011 and was completed in September of that year. It was followed by the demolition of K-31, which began on October 8, 2014, and was completed on June 26, 2015. Bechtel Jacobs was contracted to dismantle and demolish the K-25 facility in September 2008 and the operation was completed on December 19, 2013, at a projected cost of $1.48 billion. The last debris was removed in 2014. Demolition of K-27, the last of five gas diffusion facilities in Oak Ridge, began in February 2016 and was completed on February 28, 2017.

Two buildings, K-131 and K-631, which once supported the uranium enrichment process by gaseous diffusion in Oak Ridge, Tennessee, are the most contaminated buildings remaining on the 890-hectare site. K-131 was originally constructed to supply uranium hexafluoride to the uranium enrichment cascade, while K-631 was used to remove depleted uranium from the cascade in gaseous form, convert it to liquid and transfer it to transport cylinders. Both buildings were constructed in 1945, were part of the Poplar Creek facilities on the site and were closed in 1987. Demolition began with building K-131 (five stories), with K-631 (two stories) to be demolished later. The demolition of both buildings is being carried out by OREM (Oak Ridge Office of Environmental Management) and the clean-up contractor, UCOR, and is expected to be completed in 2020. The site will then be converted into a mixed-use industrial park [WNN 19a].

5.2.1.2. *French military enrichment plants at Pierrelatte*

The Pierrelatte gaseous diffusion plant (UDG, *usine de diffusion gazeuse*) consisted of four units of decreasing size (large, medium, small and very small), each corresponding to a level of natural uranium enrichment (low, medium, high and very high). The low plant, the first to be put into operation and the largest, received a flow of natural uranium and delivered the medium plant the material enriched to 2% to the medium plant; put into operation in 1965, the medium plant enriched the flow received to 6% and sent it to the high plant; from 1966 onwards, the two last units, the high and very high units, brought the enrichment to the level required for military applications.

The process was that of gaseous diffusion. It was a question of passing uranium, transformed into gas, several thousand times through the barriers, which were membranes. The plant extended over an area of 12 hectares and some buildings were more than 900 m long. The plant operated until 1996 and produced about 30 tons of highly enriched uranium-235.

The preparation for dismantling took place from 1996 to 2002. The objective of this preparation phase was to evaluate the different possible decommissioning scenarios, according to technical performance, the fate of the radioactive waste, costs and deadlines. By the end of 2008, all the equipment for the enrichment process had been removed. Operations continued irreversibly with the decontamination and crushing of the enrichment stages and associated piping. The waste produced was conditioned in accordance with the acceptance criteria for final disposal.

As planned in 2008, the dismantling program of the Pierrelatte UDG was completed at level 2 at the end of 2010. It led to the dismantling and crushing of 4,000 diffusers, 1,300 tons of diffusion barriers, 1,200 km of piping and 100,000 valves. Nine million diffusion barriers were dismantled, crushed, packaged and disposed of as waste [CEA 15].

The dismantling of this plant produced 20,000 tons of nuclear waste, including 17,000 m^3 of TFA, which were stored, ready to be evacuated to ANDRA (compressors, valves, motors, etc.); 2,000 m^3 are evacuated per year to ANDRA, to which we must add the 4,000 m^3 of "regulatory" TFA [DUP 13].

The cost of decommissioning is estimated at 1 billion euros [MER 09]. Funding for decommissioning was provided by the ARDEMU program (1997–2010).

5.2.1.3. *British military enrichment plants*

In the United Kingdom, the military uranium enrichment plant was built in the 1950s and located in Capenhurst, Chester. Enrichment was achieved by gaseous diffusion to produce highly enriched uranium that could be used for atomic bombs. In the 1960s, the plant was modified to produce low-enriched uranium that could be used to supply nuclear power plants.

It was a huge factory (1,200 x 150 x 30 m) with 4,800 diffusion stages and 1,800 km of gas transport piping with a diameter of up to 55 cm. It was operated between 1951 and 1964 with an annual processing capacity of 300 tons of fuel, managed by Urenco. It was shut down in 1973 and intensively decontaminated. Of the 160,000 tons of metal and concrete, 99% was recycled and reused. In 2005, the Capenhurst plant was completely dismantled and the buildings were demolished [AEN 05].

5.2.2. *Decommissioning of plutonium-producing reactors*

The second way to make an atomic bomb is with a supply of plutonium-239. To do this, various armies have operated special reactors that are described as "plutonigenic", because they were oriented to the production of this radionuclide in large quantities.

5.2.2.1. *American plutonium-producing reactors*

Total DoE plutonium acquisitions from 1944 to September 30, 1994, were 111.4 tons. Of the 111.4 tons of plutonium acquired, 104 tons were produced in government reactors (103.4 tons in production reactors and 0.6 tons in non-production reactors). In addition, 1.7 t were acquired from American civil industry and 5.7 t from foreign countries [FAS 20].

The US plutonium-producing reactors were located at two sites, Hanford and Savannah River. Most of the Hanford reactors were shut down between 1964 and 1971, with an average individual lifetime of 22 years (Table 5.3). The last reactor (N) continued to operate until 1987 as a dual-purpose reactor, as both a power reactor used to supply the civilian electrical grid via the Washington Public Power Supply System (WPPSS) and a reactor for the production of plutonium for nuclear weapons. Since then, most of the Hanford reactors have been placed in a sarcophagus ("cocooned") to allow the radioactive material to decay, and the surrounding structures have been removed and buried. Reactor B, on the other hand, has remained accessible to the public during occasional guided tours. It was included in the National Register of Historic Sites in 1992, and some historians have recommended that it be converted into a museum. Reactor B was also designated a national historic monument by the National Park Service on 19 August 2008 [SIN 11].

Reactor name	Start of operation	Reactor shutdown	Initial power (MWth)	Final power (MWth)	Status
Reactor B	September 1944	February 1968	250	2,210	(1)
Reactor D	December 1944	June 1967	250	2,165	2001
Reactor F	February 1945	June 1965	250	2,040	2004
Reactor H	October 1949	April 1965	400	2,140	2005
Reactor DR	October 1950	December 1964	250	2,015	2002
Reactor C	November 1952	April 1969	650	2,500	1998
Reactor KW	January 1955	February 1970	1,800	4,400	(2)
Reactor KE	April 1955	January 1971	1,800	4,400	(2)
Reactor N	December 1963	January 1987	4,000	4,000	2012

Table 5.3. *American plutonium-producing reactors at the Hanford site (source: [FAS 20]). (1) Non-contaminated, risk mitigation authorized; (2) 2008 preliminary plans*

Five plutonium-producing reactors were built at the Savannah River site (SRS) (Table 5.4). Reactor P and reactor R at the SRS were successfully decommissioned in 2011 using the *in situ* decommissioning (ISD) method. ISD has proven to be a viable alternative to standard demolition and disposal practices. In particular, the ISD cost of the two reactor complexes saved at least $400 million, as the actual cost of the ISD was $70 million, compared to an estimated cost of more than $250 million for the demolition of just one reactor complex and the transportation of the debris to a disposal facility (excluding burial costs).

The development and deployment of the ISD approach have been considered for other reactors and nuclear facilities at the SRS, as well as elsewhere in the United States and around the world. Nuclear facilities containing nuclear fuel are often robust structures with closed contaminated equipment and process systems, including miles of piping and tons of contaminated material. This innovative solution uses the existing structure as a disposal site. It permanently buries and stabilizes residual contamination and debris in the underground areas of the facilities of reactor P and reactor R at the SRS, while leaving the above ground structures intact. It also avoids the potential hazards associated with the production of approximately 137,000 tonnes of nuclear waste and its disposal at a cost in excess of $500 million [SRN 13].

Reactor name	Start of operation	Reactor shutdown	Initial power (MWth)	Final power (MWth)	Status
Reactor R	December 1953	June 1964	500	2,500	ISD
Reactor P	February 1954	August 1988	500	2,500	ISD
Reactor K	October 1954	July 1992	500	2,500	Put to sleep
Reactor L	July 1954	June 1988	500	2,500	
Reactor C	March 1955	June 1985	500	2,500	

Table 5.4. *US plutonium-producing reactors at the Savannah River site (source: [FAS 20])*

In a reactor as large and complex as the N reactor, decontamination is mandatory, both to reduce staff exposure to the lowest possible level and to minimize costs. During decontamination, quantities of nuclear waste are produced. Thus, 271 Ci of ^{60}Co, 69 Ci of ^{59}Fe and 51 Ci of ^{51}Cr were produced [KRA 80].

5.2.2.2. *Soviet plutonium reactors*

In the USSR, plutonium production was carried out at three sites.

The Mayak Production Association in Ozersk was specialized in the production and separation of weapons-grade plutonium. At the end of the Cold War, five plutonium production reactors, five tritium production reactors, several reprocessing plants and a plutonium metallurgy plant were in operation at the Mayak site. The radiochemical plant involved in the separation of weapons-grade plutonium was shut down in 1987, and all production from eight of the reactors was stopped in 1991. In 2009, plant 20 of the Mayak site was the only facility that remained involved in the production of fissile components (Pu and HEU) for Russia's nuclear weapons program (this activity was to end in 2014). Two of its operating reactors, Russlan and Lyudmila, were primarily involved in the production of various isotopes, but they also maintained a tritium production capacity. These two reactors were also fueled by HEU. Mayak was also involved in the storage of HEU and plutonium and the dismantling of fissile components [POD 09].

The Siberian Chemical Combine in Seversk produced and separated weapons-grade plutonium. In principle, this separation ended in 2010. The site also produced fuel for military-quality plutonium production reactors and stored plutonium and uranium. The production of plutonium and HEU components for nuclear weapons resumed at this site in 2014. There is a store of weapon-grade plutonium oxide on the site which is under American supervision.

The Zheleznogorsk Mining and Chemical Combine produced and separated weapons-grade plutonium. In principle, production stopped at the end of 2009 and separation in 2012. Since then, the site has been used for the storage of military-quality plutonium and uranium and military-quality plutonium oxide, which is under American supervision [POD 09].

The shutdown of the last three plutonium-producing reactors located in Seversk (Tomsk region) in the Siberian Chemical Combine (SCC) was the subject of agreements within the framework of a nuclear non-proliferation program between the Americans and the Russians as early as March 2003. Two of these reactors were shut down in 2008, on April 20 (ADE-4) and June 5 (ADE-5). These reactors operated from 1960 to 1993 and produced 800 kg of weapons-grade plutonium per year. After 1993, the reactors provided heat and electricity to Seversk. The ADE-2 reactor, in Zheleznogorsk, was shut down before 2010 [WNN 08a].

5.2.2.3. *British Windscale plutonium reactors*

The Windscale site, located on the northwest coast of England in Cumbria (now known as Sellafield), has two plutonigenic reactors (Windscale Piles). They are called piles because of the stacking of graphite bricks. Each reactor had a tall chimney. These two reactors were built by the British Ministry of Supply and were operational from 1950 to 1957. They were air-cooled and moderated with graphite, and their objective was to produce weapons-grade plutonium for use in atomic weapons. Pile no. 1 became operational in October 1950, followed by pile no. 2, in June 1951. They were supposed to last five years and actually served for seven.

These reactors were shut down following a fire in pile no. 1 on October 10, 1957. This fire was due to the Wigner effect, which had displaced carbon atoms from the crystalline mass of the graphite, and a sudden release of heat as a result. This energy ignited the graphite, which caused the fusion of the fuel into uranium metal.

With a view to decommissioning the two piles, graphite samples were taken in the late 1990s. The program initially planned was "safe storage" (cocooning) of pile no. 2 and dismantling pile no. 1. However, detailed data were needed to support this program and to verify that the 15 tons of graphite contained in pile no. 1 could not lead to a criticality accident. The reactors (in particular, pile no. 2) still contain significant Wigner energy and elaborate precautions were applied to ensure that intrusive operations could not lead to energy releases. In view of the results, there do not seem to be any obstacles to the continuous storage of pile no. 2. For pile no. 1, the data were used to support the safety case being developed for disassembly and defined acceptable procedures for handling, storage and disposal of the graphite blocks [MAR 98].

Phase 1 decommissioning was already carried out on pile no. 1 during the 1980s. This decommissioning included sealing the air outlet ducts connecting the central space to the chimney; installing a ventilation system to ensure containment via the depression; removing the fuel cartridges[1], air outlet and outlet ducts; isolating the water pipe from the cooling tank; cleaning, draining and sealing the water pipe and, finally, basic surveys to collect data to begin Phase 2. The methodology to be used by the UKAEA ensured that a fire-damaged reactor core would be safely decommissioned with doses below regulatory limits. Conventional risks to the workforce were kept as low as reasonably achievable, and the best available methods were used to limit releases to the environment. Upon completion of decommissioning, the total risk associated with the Windscale site was significantly reduced, with the waste stored in a safe, stable and secure form awaiting final disposal [POM 00].

1 An assembly of fuel rods in which natural or enriched uranium metal was found. Their sizes are around 50 cm, or eight times shorter than the UO_2 assemblies.

After the end of Phase 1, 7% of the spent fuel remained in pile no. 1. Pile no. 2 has been shut down and maintained since the late 1950s, although there have been various modifications and additions. All the fuel has been removed from pile no. 2 [HSE 02].

The 50th anniversary of the accident led to increased pressure to clean up the site permanently. Contaminated filters were removed after the fire, and the chimney in pile no. 2 was partially demolished in 2001. Since the contaminated chimneys could not simply be knocked over, they had to be systematically demolished from top to bottom with core drills. A robot was used to remove the chimney linings. Some 5,000 tons of concrete, steel and bricks had to be removed. The Pile Fuel Storage Pond (PFSP), where irradiated fuel cartridges were previously left to cool, was decommissioned in 2013.

Plans call for the removal of spent fuel from the batteries by 2030. Although work began in the 1980s, nuclear decommissioning operations are expected to last beyond 2040 [HSE 02].

5.2.2.4. *French plutonium-producing reactors (G1, G2, G3)*

France has built three plutonium-producing reactors: G1, G2 and G3, at Marcoule.

The G1 reactor was built in 1955 and commissioned in 1956. It is an enormous concrete cylinder, 20 meters in diameter, which stands under a 55-meter-high metal nave. The front face of the reactor is pierced by more than a thousand small horizontal channels that permitted the introduction of uranium into the core [MER 09]. The core consisted of 100 tons of uranium, distributed in more than 2,000 rods. The maximum thermal power of G1 was 30 megawatts, later increased to 46 megawatts. From 1958, G1 made it possible to produce 10 to 15 kg of plutonium per year, extracted in the UP1 plutonium extraction plant (Plutonium Plant No. 1) at Marcoule. The extraction was carried out by the Purex chemical process, the irradiated uranium was then reprocessed and the plutonium was extracted. The reactor was definitively shut down on 15 October 1968.

Reactors G2 and G3 are more powerful than reactor G1. In addition, they are cooled by carbon dioxide (CO_2) under pressure in a closed circuit, whereas G1 was air-cooled. This had the disadvantage of activating various impurities present in the air as well as argon (production of argon 41 from stable argon 40), a rare gas that is normally present in the air. These reactors contain 12,000 fuel rods inserted in a block of 1,200 tons of graphite pierced by about 50 vertical shafts provided for the passage of control and safety rods. Their thermal power was 250 megawatts each. The G2 reactor was put into service on July 21, 1958, and was definitively shut down in 1980. Reactor G3 was put into service on June 8, 1959, and was definitively shut down in 1984.

With G1 and G3, G2 produced about 100 kg of plutonium per year. In the end, the production of plutonium was about 3,500 kg in G1, G2 and G3 between 1959 and 1977. The production of enriched uranium by the three plutonigenic reactors (G1, G2, G3, until 1984) was 3 to 4 t, the tritinogenous reactors Celestin-1 and -2 (until 2009) 0.75 t and the Phénix reactor (1973–2009) 0.35 t, for a total of 4.6 t with an uncertainty of 0.5 t. The UNGG electronuclear reactors (Chinon A1, A2, A3, Saint Laurent A1, A2, Bugey and Vandellòs) produce more than 1.7 t. The analysis suggests that France's military stocks of plutonium and highly enriched uranium greatly exceed the needs of its present nuclear arsenal [GLA 11]. According to Merchet [MER 09], French plutonium reserves are about 5 tons.

Clean-up and dismantling operations for the three reactors G1, G2 and G3 began in 1969, for G1, and 1986 for G2 and G3, a few years after their final shutdown.

The first phase of dismantling began in 1969 and was completed in the 1980s. This first phase of dismantling consisted of removing all external circuits, particularly the cooling system, and ensuring the containment of the reactor blocks. The fuels were unloaded and evacuated to the UP1 reprocessing plant.

After 1996, all auxiliaries, filters, coolants and air circuits were then dismantled and cut out, as well as all circuits downstream of the reactors. These pipes were cut as close as possible to the reactor blocks. The containment of the internal structures of the reactors is today ensured by metal plugs that close all the openings of the concrete structure.

Between 1991 and 1996, the 4,000 tons of metal waste produced by the dismantling of the CO_2 pipelines were treated by melting. The ingots were evacuated to the surface storage center for very low-level radioactive waste in Morvilliers, in the Aube [CEA 15].

Between 1996 and 2010, dismantling work continued with the demolition of the reactor control buildings. Today, the G1, G2 and G3 reactors are monitored to ensure the safety of these facilities until dismantling operations resume.

The last step will be to remove graphite from the reactor blocks after their dismantling. This final phase of the complete dismantling of the three reactors will not resume until the future national storage facility for irradiated graphite waste is opened. This period of time will allow the cobalt-60 that has contaminated the internal metallic structures of the reactor block by activation to decrease in concentration by about a factor of 1,000, since its half-life is short (5.27 years). The program should be completed in the 2040s [CEA 15].

5.2.3. *Decommissioning of tritium-producing reactors*

Military tritium production is carried out by seven countries (China, France, India, Israel, Russia, the United Kingdom and the United States). China has had two facilities since 1968 and 1979, but its production is unknown. The United States had the N reactor (4,800 MWth) at Hanford, which operated for a short period (1967) and produced 6,250 g of tritium, and the K reactor (2,400 MWth) at Savannah River, which operated from 1953 to 1988, producing 6,300 to 11,500 g of tritium. France used two specific reactors, the Célestins, commissioned in 1967 and 1968, which produced 1,200 g of tritium. India used two reactors, Cirus (1960, 40 MWth) and Dhruva (1988, 100 MWth), producing 120 g and 300 g of tritium, respectively. Israel operates a reactor in Dimona (IRR-2, 150 MWth), commissioned in 1963, producing 500 g of tritium. Russia has reactors at the Ozersk site where there were five graphite reactors (total 6,565 MWth), which were shut down at the end of the 1990s, and two light-water reactors (LWR, 1,000 MWth each) with a total production of 4,055 g of tritium. In the United Kingdom, four Magnox-type reactors (4 x 220 MWth), operating since 1958–1960, have produced a total of 800 g of tritium. In 1995, out of a total of 19 reactors dedicated to production, only five were shut down [KAL 95].

In France, tritium releases, due mainly to the military installations at Marcoule, were in liquid form of 20 $TBq.yr^{-1}$, from 2003 to 2007, and in gaseous form of 500 $TBq.yr^{-1}$ [IRS 10]. For other countries, the quantity of tritium released into the atmosphere from reactors dedicated to tritium production and other research facilities was approximately 400 TBq for Canada for the year 2006. Releases were of the order of 10,000 $TBq.yr^{-1}$ in the 1990s for the United States [IRS 10].

In France, two reactors intended specifically for the production of tritium, Célestin I and II, are military nuclear reactors located at Marcoule. Moderated and cooled with heavy water, they are used for the production of tritium and plutonium for military use. The Célestin I reactor was commissioned on May 15, 1967, followed by Célestin II in 1968. In 1967, the Workshop for the Extraction of Tritium from Targets, also located at Marcoule, began to operate. Previously, from 1962 to 1976, a French laboratory at the CEA center in Saclay was equipped to produce large quantities of tritium [BYR 01].

At the end of 1993, the Célestin reactors stopped their production of plutonium. In 2008, the production of tritium for thermonuclear bombs continued from the operation of the Célestin I and II reactors. Tritium releases into the environment from these installations are significant [GAZ 10].

According to an American source, the Célestins were shut down on December 23, 2009. The future production of tritium for military use will be carried out in the land-based test reactor for naval propulsion at Cadarache [GLA 11]. The final shutdown of Célestin I in 2009 was confirmed by a parliamentary source. Dismantling began immediately. The schedule for dismantling is planned until 2050 [GRA 11].

5.2.4. *Decommissioning of fissile fuel fabrication plants*

There have been numerous factories throughout the world for the manufacturing of highly enriched uranium and plutonium for use in the assembly of nuclear weapons. One of the emblematic plants was the Plutonium Finishing Plant (PFP) at the Hanford site in the United States.

The plutonium finishing plant is also nicknamed Plant Z, because it was the last stage of the plutonium production process at Hanford, in operation from 1949 to 1989. This complex contained more than 90 buildings. The plant produced plutonium pellets (or buttons), the size of a hockey puck (about 7.6 cm), which were shipped to nuclear weapons manufacturing facilities. Since the end of the Cold War, highly contaminated equipment has been left on the site [HAN 20].

At the end of its activity, after 40 years of service, the plant contained mainly plutonium and plutonium-uranium compounds, metals and alloys that have been stored for several years. The 1994 inventory listed about four tons of plutonium and about the same quantity of uranium. In 1994, this inventory was declared surplus for national defense programs and should be eliminated. This fissile material was to be stabilized and conditioned for disposal. Metallic plutonium and high-dosage plutonium oxide items (plutonium oxide materials containing more than 85%, by weight, of plutonium) are packaged in long-term storage containers for disposal. Low-plutonium content materials (those containing less than 30% plutonium by weight), such as incinerator ash and sand, slag and crucible (SS&C) material, are packed in WIPP (Waste Isolation Pilot Plant) certified containers and transferred to the Hanford Central Waste Complex, awaiting shipment to the WIPP [LIN 02].

Veolia Nuclear Solutions – Federal Services (VNSFS) treated approximately 200 items, including glove boxes, 2.4 km of ventilation piping, contaminated equipment, asbestos and other hazardous waste. In 2020, after nearly two decades of clean-up and preparation, workers celebrated the completion of the demolition of the main PFP processing facility. Over the next few months, work continued with the removal of rubble from the Plutonium Recovery Facility, coring the ground and covering the building slabs with a ground cover. The DoE will use its Enhanced Controls Place and Cover Disposal methodology to accomplish this [VEO 20].

5.3. The decommissioning of military spent fuel reprocessing plants

Military plants for reprocessing spent fuel to recover plutonium and uranium 235 were very numerous in the first five states possessing the atomic bomb. A large number have been shut down (Table 5.5). On the other hand, new states have retained this type of plant, albiet with reprocessing capacities that were limited (Table 5.6). According to Kurtz *et al.* [KUR 19], a second plant, both civil and military, exists on the site of Tarapur II, with an annual production of 100 tons of irradiated fuel.

Country	Plant	Operational dates
United States	*Hanford, T and B Plants*	1944–1956
	Hanford, REDOX Plant	1952–1967
	Hanford, PUREX Plant	1956–1990
	Savannah River, F Canyon	1954–2002
Russia	*Mayak, Ozerk, BB Plant*	1959–1987
	Seversk (Tomsk-7)	1956–2012
	Zhedeznogorsk (Kranoyard-26)	1964–2012
United Kingdom	*Sellafield, B204*	1952–1973
France	*Marcoule, UP1*	1958–1997
China	*Jiuquan*	1970–1984
	Guangyuan	1976–1990

Table 5.5. *Closures of military reprocessing plants (source: [FEI 14])*

Country	Plant	Operational	Security	Capacity
India	*Trombay*	Yes	No	50
	Tarapur	Yes	No	100
	Kalpakkam	Yes	No	100
Israel	*Dimona*	Yes	No	40–100
North Korea	*Yongbyon*	Assumed	No	100–150
Pakistan	*Nilote*	Yes	No	20–40
	Chashma	In construction	No	50–100

Table 5.6. *Worldwide military reprocessing plants in operation in 2013 (source: [FEI 14])*

5.3.1. *The dismantling of US reprocessing plants*

The first reprocessing plants were built in the United States at Hanford, located in Area 200 West (three plants: T, U and Redox) and 200 East (two plants: B and PUREX). Today the U and REDOX plants are shut down, and in 1957, the T plant was converted into a decontamination and repair plant.

The Plant T complex was the first chemical treatment unit built at the Hanford site and the first of its kind in the world. Construction began in 1943 and was completed in October 1944. Plant T was originally built to extract plutonium from spent reactor fuel. The material for the world's first atomic explosion, the Trinity bomb test in July 1945, and the Nagasaki weapon test were extracted in this plant.

The U plant was also built during World War II to process plutonium. The partial demolition of this plant began with a plan called the Canyon Disposition Initiative, which involved demolishing the buildings surrounding the plant and remediating the surrounding underground dumps. The plant then collapsed on itself and was buried under a mound.

The REDOX plant is located at the southern end of the 200 West Zone. It is surrounded by numerous tank farms built to receive waste from the various plants.

Plant B is a World War II installation that treated reactor fuel to recover plutonium, from 1945 to 1952. The plant was modified and restarted in 1968 to separate the radioactive elements cesium and strontium from the waste from the tanks generated by the reprocessing of nuclear fuel. In 1998, plant B, which completed a major deactivation project, was shut down, except for periodic surveillance controls.

The plant for the extraction of plutonium and uranium, commonly called PUREX (Plutonium and Uranium Refining by EXtraction), chemically treated the irradiated fuel to recover the plutonium and uranium. It is an advanced version of the T, U, B and REDOX plants. It was built between 1953 and 1955 and is considered a third-generation transformation plant. PUREX operated between 1956 and 1972 and between 1983 and 1988. PUREX was closed in 1988 and completely deactivated in 1997. This plant processed about 75% of all the plutonium that Hanford produced [ANO 07a].

5.3.2. *The UP1 plant in Marcoule, France*

The plutonium plant no. 1, or UP1, was used for the dissolution of nuclear fuels, the separation, in a concentrated nitric solution, of fissile materials (uranium and plutonium), fission products and actinides produced in the fuels. The fissile elements were transformed into uranyl nitrate, plutonium oxide or metallic plutonium. The plant was put into service in January 1958, and the production of the first gram of plutonium took place on July 21, 1958, while the first ingot of plutonium was obtained on February 20, 1959. In 1965, the first treatment of fuel from an EDF reactor was carried out in the plant. The year 1994 marked the definitive end of military reprocessing. The last fuel dissolution was carried out on September 30, 1997, in the UP1 plant, which, in total, treated nearly 20,000 tons of spent fuel.

The UP1 plant in Marcoule (Drôme, France) was associated with numerous production workshops that were grouped into six groups. The first group includes the three decladding facilities, one specific to the G1 reactor, used for the decladding and storage of magnesium for the first fuels of the G1 reactor (then serving as a warehouse), another is the decladding workshop whose functions include the reception, storage and preparation for dissolution of irradiated fuels, treatment and reshipment of external transport packages, storage of fuel structural waste and water treatment waste. The third decladding unit is the MAR 400 installation whose functions, identical to those of the decladding workshop, are used for fuel to be reprocessed from the UNGG sector.

The fission products liquid storage facility includes the tanks containing the concentrated nitric acid solutions of the fission products and their various equipment, and the glass vitrification and storage workshop (AVM, *Atelier de vitrification de Marcoule*) where the vitrification process is carried out; the glass obtained is stored in concrete pits inside ventilated shafts.

The production workshops with their ancillary equipment are made up of about 60 buildings and structures of various sizes. The so-called support workshops are made up of five installations. These are the Liquid Effluent Treatment Plant (STEL, *Station de Traitement des Effluents Liquides*), which manages all the liquid effluents produced on the site; the Solid Waste Conditioning and Storage Workshop (CDS, *Atelier de Conditionnement et d'entreposage des Déchets Solides*), which collects, controls, conditions and stores and/or sends to ANDRA all the solid waste produced at Marcoule; the Equipment Decontamination Workshop (ADM, *Atelier de Décontamination du Matériel*), which is used to collect and treat the contaminated equipment of the establishment; the Laundry Decontamination Workshop (ADL, *Atelier de Décontamination du Linge*), which treats the clothes of all the personnel; and the laboratory for control, analytical chemistry of processes and assistance in industrial chemistry, which is used for the activities of the two reprocessing units of the site (the UP1 plant of COGEMA and the Marcoule Pilot Workshop of the CEA/VALRHO, Rhone Valley) [BAT 97].

The permanent shutdown program (MAD, *mise à l'arrêt définitif*) began on January 1, 1998. Remediation work carried out since that date has resulted in the removal of nearly 100% of the radiological activity. This allowed the dismantling program (DEM, *démarrage du programme de démantèlement*) to begin.

The fuel decladding workshops had a high level of irradiation, requiring remote operation work. At the end of 2014, all the equipment in this zone was dismantled.

Plutonium dissolution and extraction workshops are characterized by a great diversity and complexity of components, materials and equipment present in the installations (chemical dissolvers, extractor batteries, evaporators, tanks, etc.), associated with a high irradiation risk, due to the presence of fission products. The majority of clean-up operations before dismantling are carried out by remote operation. The dismantling of the spent fuel dissolvers was completed in 2012. The dismantling of fission product extraction lines was completed at the end of 2013. The equipment for separating uranium and plutonium was dismantled in 2010, while the dismantling of the high-radiological activity part of the plant was finished by 2020.

In the high activity effluent treatment facilities, the radioactive effluents from the rinsing of the plant's tanks and pipes were entirely vitrified in the Marcoule Vitrification Workshop until the end of 2012, with a view to their final storage.

Support units (laboratories, waste conditioning workshops, effluent treatment plants) support the sanitation and dismantling operations. They were the last to be dismantled.

The dismantling of the main components of the UP1 plant was completed in 2020. The recovery of old waste will be completed around 2040, with a view to sending it to deep geological disposal. Already, 8,000 of the 60,000 drums of bituminous waste have been reconditioned [CEA 15].

The dismantling of the UP1 plant produced 27,000 tons of nuclear waste, and the total cost of dismantling for Marcoule is estimated at 5 billion euros [MER 09]. The military expenditures of the CEA amount to 11 billion euros. Of this total, the Marcoule pilot workshop represents an estimated expenditure of 726 million euros, including contingencies of 199 million euros at the end of 2018 [COU 20].

5.3.3. *The dismantling of other military reprocessing plants around the world*

In the United Kingdom, the B204 primary separation plant was built between 1947 and 1951, and used to recover plutonium from Windscale Pile fuel and to reprocess fuel from Calder Hall and other Magnox reactors. It operated from 1952 and 1964 to reprocess metallic fuel. The south line was then converted to reprocess oxide fuel from 1969 to 1973. It was the precursor prototype of the thermal oxide reprocessing plant (THORP). In 1973, a major accident forced its definitive closure. It was an explosion of radioactive gas in the building and the chimney, contaminating 35 workers, largely by ruthenium-106, causing contamination of the skin and lungs [COR 17].

Decommissioning followed an integrated strategy that was used to sequentially dismantle each of the six piles and export the waste via a central waste conditioning and control facility [CHA 92].

In 2005, a seismic evaluation of the primary separation plant and its associated chimney was conducted and it was found not to meet modern seismic standards. Ten years later, in 2015, the chimney was described by the owner of the Nuclear Decommissioning Authority site as "posing a significant seismic risk to the Sellafield site". The chimney was therefore demolished using a self-climbing platform, held in place by friction and capable of moving at 5 mm per second. This removed the estimated 600 tons of concrete and reinforcement and some 25 tons of stainless steel from the old pile. This operation was carried out from top to bottom, at a rate of about 1 meter per week [COR 17].

Information on other military reprocessing plants in other countries is more incomplete. There is no information on the dismantling of Chinese reprocessing plants (Jiuquan and Guangyuan), nor on the closed Russian plants of Mayak (BB Plant), Seversk and Zhedeznogorsk.

5.4. Decommissioning and decontamination of military sites

The United States and the USSR built large military research and manufacturing centers on their territory, as early as 1943 for the former and at the end of World War II for the latter. These sites were covered by defense secrecy and covered vast areas. As already mentioned, security and safety conditions were minimal. This resulted in radioactive contamination of the sites and even, frequently, of the environment around these sites.

5.4.1. *US military facilities*

The Manhattan Project was carried out at three main US sites: Oak Ridge, Hanford and Savannah River. Since operations took place during a war and knowledge about the fate of radionuclides and their radiological harmfulness was still in its infancy, precautions were limited. This resulted in the release of large amounts of radioactivity into the environment, both in subsoils and in nearby watercourses (Table 5.7).

Site	Estimated quantities of radioactivity released (Ci)	Type of discharge
Oak Ridge	1,000,000	Injection of waste into the subsoil
Savannah River	900,000	Liquid spills in rivers and infiltration basins
Hanford	700,000	Liquid discharges to soils and surface ponds

Table 5.7. *Environmental contamination in the United States from nuclear facilities for the production of fissile materials (source: [BRA 96])*

5.4.1.1. *The Hanford site*

The Hanford site occupies 1,518 km^2 of Benton County, located near Richland. On the site, there are three nuclear reactors (105-B, 105-D and 105-F) and three plutonium fabrication units, each 250 meters long, as well as four groups of 64 single-shell underground tanks (241-B, 241-C, 241-T and 241-U). Many Nobel Prize winners, including Enrico Fermi, worked at this site, which was known as "the city of plutonium" [AMI 18, AMI 19a].

The Hanford site is organized into specialized geographical Areas. Nuclear reactors are in Area 100, fuel reprocessing plants and nuclear waste storage facilities in Areas 200 East and 200 West, and fuel fabrication plants and research laboratories in Area 300. The Hanford site is the largest waste storage area in the United States. It contains 177 underground storage tanks, 67 of which are at the end of their lives. After the end of plutonium production on the site in 1987, clean-up was the greatest challenge of an environmental project [HAN 00]. Thus, the total contamination of the Hanford use only Area was 37,250 Ci or 1,378 TBq [WAH 80]. The main radionuclides present in the soils of this Area are U, ^{99}Tc, ^{129}I and nitrates.

During the Cold War, the Hanford site remained very active, with the construction of new nuclear facilities, including new plutonium-producing reactors and a plutonium finishing plant. An important synthesis on the history of the site has been published [MAR 02a], as well as a census of numerous documents [BRI 01].

Historically, at Hanford, plutonium production decreased as reactors B, D and F aged and were poisoned by fission products. In addition, the graphite (moderator) swelled because of the Wigner effect, which damaged the load tubes where the uranium was irradiated to produce plutonium, making them unusable. In order to maintain the supply of plutonium, production was reduced in the reactors and the oldest unit, Reactor B, was shut down, so that at least one reactor would be available in the future. The process of extracting plutonium with bismuth phosphate left the unused uranium in a state from which it could not be easily recovered. Thus, a new methodology was sought by DuPont and Metallurgical Laboratory. They ended up developing a process of extraction by a redox solvent.

Prior to 2001, the Department of Energy (DoE) sponsored limited public tours of the Hanford site, but discontinued the program after the September 11th terrorist attacks in the United States. Public tours of Hanford resumed in 2004. Specific tours for various categories of the population are organized, as well as separate tours for Reactor B, the world's first large-scale nuclear reactor [INS 11].

Currently, the Hanford site is divided into four main parts, with contrasting future functions (Hanford Reach National Monument, River Corridor, Central Plateau and Inner Area). The estimated cost of remediating the site between 2014 and 2090 is $113.6 billion [TRY 14]. The remediation policy for the Hanford site is detailed in a USDoE document [USD 13].

Cleaning up the Hanford site is a complex and difficult undertaking. Three main Areas need to be cleaned: the River Corridor, the Central Plateau and the liquid waste tanks. Each of these three Areas is in itself a complex and difficult undertaking involving multiple projects and contractors and requiring several years and billions of dollars to complete.

The goals are the result of more than 20 years of dialogue between tripartite agencies, tribal nations, the State of Oregon, stakeholders and the public. The Columbia River is a critical resource for the people and ecology of the Pacific Northwest. The approximately 80 km stretch of the river known as the Hanford Reach, is the last free-flowing section of the river in the country.

2015 vision

Significant thought has been given to reducing the footprint of radioactive pollution through active clean-ups. It resulted in the "2015 Vision". This vision calls for eliminating hazards near the Columbia River by cleaning up most of the River Corridor, treating contaminated groundwater near the Columbia River and demolishing the site's main plutonium facility, the Plutonium Finishing Plant. The work will reduce the 200 km^2 footprint at the center of the site through active clean-ups, reduce overhead costs and allow the ministry to focus its efforts and resources on the large-scale clean-up of the Central Plateau of the site. In some areas, groundwater remediation systems will need to continue to operate for longer periods of time to achieve the desired quality objectives. The Hanford Reach National Monument lands (approximately 750 km^2) surrounding the Hanford site will be managed primarily to preserve natural and cultural resources.

Cleaning of the River Corridor

Cleaning up the River Corridor has been one of Hanford's top priorities since the early 1990s. This emergency is due to the proximity of hundreds of contaminated waste sites and facilities to the Columbia River. The disposal of sludge from the K West Basin, which is close to the river, also remains a high priority.

The River Corridor portion of the Hanford site includes Areas 100 and 300, along the south shore of the Columbia River. Area 100 contains nine decommissioned plutonium production reactors. These Areas also contain numerous support facilities and disposal sites for solid and liquid wastes that have contaminated the groundwater and soil. Area 300, located north of the City of Richland, contains fuel manufacturing facilities, nuclear research and development facilities and disposal sites for associated solid and liquid wastes that have contaminated the groundwater and soil.

For sites within the River Corridor, the objective of the remediation is to restore groundwater to drinking water standards where possible and to achieve ambient groundwater quality standards prior to discharge to the Columbia River.

The River Corridor has been divided into six geographical areas for the purpose of making decisions regarding water sources and groundwater rehabilitation. Upon completion of the clean-up, the federal government will retain ownership of most of the lands in the River Corridor and implement long-term stewardship activities to ensure the protection of human health and the environment.

Cleaning of the Central Plateau

The Central Plateau consists of two Areas, the Inner Area (about 26 km^2) and the Outer Area (about 168 km^2). The Inner Area contains the main facilities for treatment, waste management and disposal of nuclear fuel. The Inner Area will be devoted to the long-term management of waste and the containment of residual contamination. The Outer Area will be remediated to protect human health and the environment. The clean-up of the Outer Area was completed in 2020.

Cleaning up the Central Plateau was a very complex activity due to the large number of landfills, surplus facilities, active treatment and disposal facilities and areas of deep soil contamination. Previous releases of more than 1.5 billion m^3 of liquid waste and cooling water into the ground have resulted in the contamination of approximately 155 km^2 of soil and groundwater. Today, some plumes extend well beyond the plateau. The overriding priority for authorities remains to contain and remediate these plumes. At the end of clean-up efforts, hazardous and radioactive residual contamination remains, both in surface disposal facilities and in underground environments in parts of the Central Plateau. Figure 5.1 summarizes the major sources of contamination in the Central Plateau.

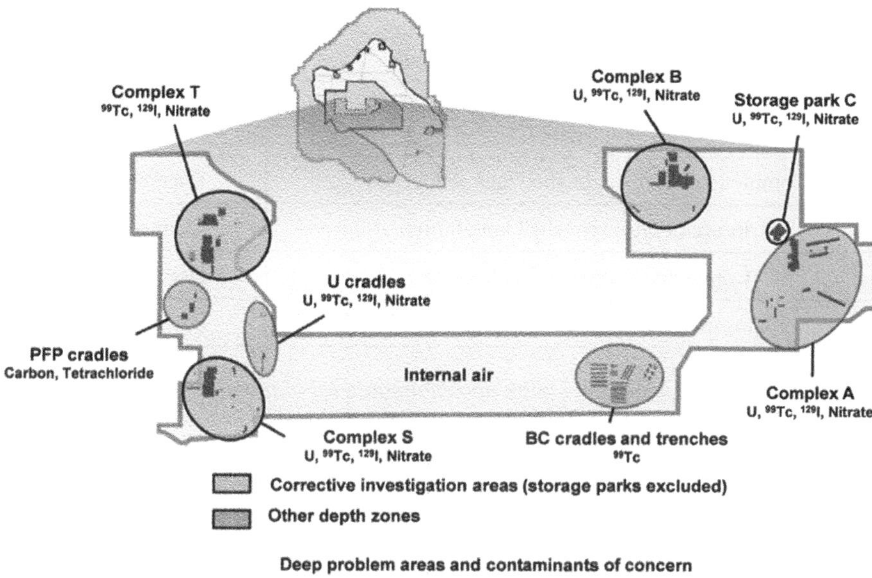

Figure 5.1. *Internal zone problems with the various study areas and contaminants of concern in each zone (source: [USD 13]). For a color version of this figure, see www.iste.co.uk/amiard/disarmament.zip*

Cleaning of waste tanks

This cleaning component is located in the inner area of the Central Plateau and is one of Hanford's most difficult legacies to solve. The stockpiles contain approximately 0.2 million m^3 of radioactive waste, stored in 177 underground tanks. Of these 177 single-shell tanks, 67 have, or are suspected to have leaked up to about 3,800 m^3 of waste. In some areas, discharges from some of the single-tank farms have reached the groundwater.

A key step in reducing the risk to human health and the environment from tank waste is to recover as much waste as possible from single-shell tanks and place it in double-shell tanks. The waste must then be sent to the waste treatment plant to be treated and converted by a vitrification process into solid waste (glass). The stockyard remediation program is presented in Table 5.8.

Metrics	Completed in
Complete recovery of tank farm C (10 tanks)	September 2014
Closure of waste management area C	June 2019
Start-up of the waste treatment plant	December 2019
Waste treatment plant, initial operations	December 2022
Completion of the recovery of nine single-shell tanks from storage park C	September 2022
Completion of all single-shell tank recoveries	December 2040
Closure of all single-shell tank farms	January 2043
Complete treatment of tank waste	December 2047
Closure of all double-shell tank farms	September 2052

Table 5.8. *Dates and measures for cleaning waste in the tanks (source: [USD 13])*

Hanford Reservoir waste treatment is managed by DoE-ORP (Office of River Protection). This program is primarily based on the WTP (Waste Treatment and Immobilization Plant), which uses a proven technology, vitrification, to immobilize chemical and radioactive waste from the tanks in a solid form, that of glass, and thus isolate it from the environment. The problem of storage of this waste remains to be solved.

Future activities after cleaning

A large part of Hanford's 943 km^2 is for conservation and preservation and is consistent with the uses of national monuments. However, more than 97 km^2 near the town of Richland are reserved for industrial purposes [USD 13]. The 2015 vision for the future Hanford site is shown in Figure 5.2.

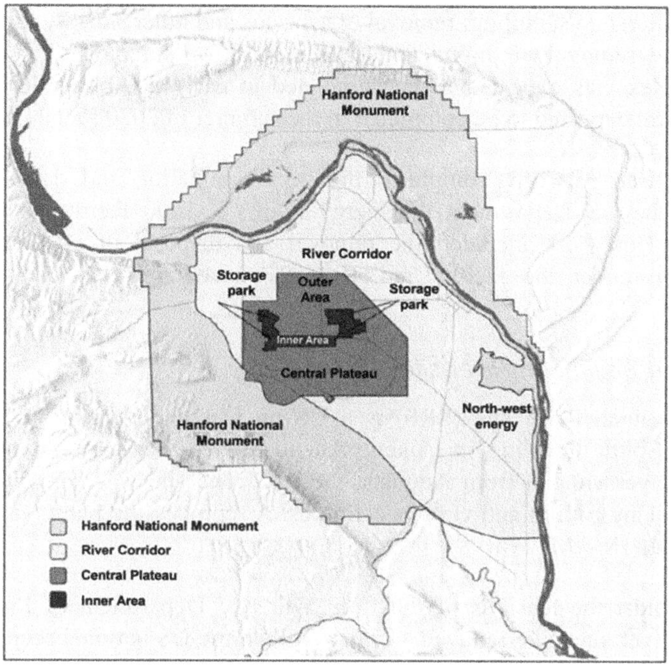

Figure 5.2. *Main elements of the Hanford clean-up completion framework, the River Corridor, the Central Plateau and the waste storage tanks (the River Corridor includes the south bank, which is part of the Hanford Reach National Monument). For a color version of this figure, see www.iste.co.uk/amiard/disarmament.zip*

5.4.1.2. *The Oak Ridge site*

Oak Ridge National Laboratory (ORNL), located in Oak Ridge, Tennessee, is a multi-disciplinary laboratory for science and technology. This site hosted the gaseous diffusion plant (designated under code K-25) where the two isotopes of uranium ^{238}U and ^{235}U were separated [AMI 18, AMI 19a]. The complex extends over more than 21,800 m^2, and some buildings reach 55 m in height. It was built in stages to develop and test the technology of uranium enrichment by centrifugation.

Recently, several buildings of the Manhattan Project have been renovated, such as K-1004-J and K-1600. Workers are demolishing part of the complex that includes laboratory K-1004-J, a research and development center built in 1944 as part of the Manhattan Project, around which the rest of the complex was built in the 1970s. Teams are conducting deactivation activities in other buildings to prepare them for demolition in the coming months. This involves removing hazards to ensure safe and efficient demolition of the structures: disconnection of electricity and utilities;

characterization and sampling; removal of asbestos and other wastes; and inspection, drainage and removal of piping and equipment as required. Deactivation of the entire complex was expected to be completed in early 2020 and demolition and waste removal expected to be completed in the summer of 2020 [WNN 19k].

Centrus Energy Corp. completed the decontamination and decommissioning (D&D) of the US Department of Energy (DoE) K-1600 facility in Oak Ridge, Tennessee. This work included the removal and disposal of all equipment and materials to render the facility radiologically uncontaminated and unclassified [WNN 19k].

5.4.1.3. *The Savannah River site*

The Savannah River site (SRS) is a US nuclear materials processing facility located in South Carolina in Aiken, Allendale and Barwnell counties on the Savannah River, 40 km from Augusta, Georgia. The site was contaminated from 1951 to 2001 by tritium and various actinides. A reprocessing plant was located on this site [AMI 18, AMI 19a].

Born amidst the tensions of the Cold War, the Department of Energy (DoE) Savannah River site (SRS) played a critical role in the US national security strategy for more than 40 years. Facilities at the site range in size and complexity from large nuclear reactors to dozens of small storage buildings. Many SRS facilities have underground structures.

Decontamination and decommissioning (D&D) is a major priority at the Savannah River site with clean-up initiatives by the Ministry of Energy, for whom reducing the site's footprint is a high priority. There are approximately 6,000 buildings on the site, covering approximately 930,000 m^2. In 2006, more than 250 buildings were demolished, but the D&D program does not necessarily mean demolition. For some buildings, a more reasonable and safer solution is *in situ* decommissioning, or sarcophagi.

As with other US nuclear sites, SRS is divided into zones. Historically, production at SRS began in zone M. Here, the facilities produced materials for the reactors installed on SRS. The only radiological contamination in the zone comes from the uranium used to make reactor fuel.

Zone T, also known as TNX, is a research and development area built near the Savannah River in 1950. It was the oldest operational area on the site. Historically, equipment was transported to the SRS by barge. Contamination in zone T comes from a small quantity of uranyl nitrate, which does not pose any problems to human health, as well as chlorinated volatile organic compounds, principally trichloroethylene (TCE), classified in Group 1 by the IARC, tetrachloroethylene

(PCE), classified in Group 2A by the IARC, and carbon tetrachloride, uranium and radium-226.

Zone D is located only a few hundred meters from zone T. The function of the zone D facilities was to extract heavy water from the Savannah River for use in the SRS reactors. Contamination in the zone includes tritium and mercury.

Zone	Planned in contractual period	Completed in 2004	Planned for D&D in 2005	Total expected by the end of 2005
A	40	15	5	20
C	9	0	4	4
D	35	29	4	33
E	3	0	3	3
F	61	22	4	26
G	6	0	6	6
H	2	0	0	0
K	8	0	1	1
L	10	0	5	5
M	23	6	13	19
N	8	0	1	1
P	12	0	4	4
R	7	0	1	1
T	28	28	0	0
Total	252	107	44	151

Table 5.9. *Nuclear facilities undergoing decommissioning in various areas of the Savannah River site (source: [POD 05])*

Zone F (F Canyon) is located near the center of SRS. It is one of the two chemical separation zones of the site. Historically, plutonium from irradiated nuclear fuel was extracted in this zone for national defense. This was carried out by FB Line using the PUREX process (simultaneous extraction of plutonium and uranium). PUREX operations were finished in both installations in 2002. Since the end of operations, work has been in progress to shut down obsolete systems that have not been used for decades, to empty and rinse the piping, to demolish surplus installations and to liquidate planned operations. One of the major projects in zone F is the proposed closure of the 247-F complex, which includes five buildings, covering more than 12,000 m^2, and in which the radiological situation is of concern. This complex is to be deactivated and demolished. Here, the stock of uranium has been converted into a form suitable for naval fuel. The installation operated from 1985 to 1989 and was closed because its services were no longer needed [POD 05]. The status of the decommissioning of the Savannah River BNI in 2005 is listed in Table 5.9.

SRS employees successfully decommissioned 264 facilities during the period 2005 to 2009, reducing the facility's footprint by nearly 1,600 m^2. More than 750 additional large facilities are expected to be decommissioned by 2035 [SRS 09].

Recently, the D&D programs have progressed well and the M and T zones have been completely decommissioned. Similarly, there has been *in situ* decommissioning of the Heavy Water Cooled Test Reactor (HWCTR), decommissioning of the K-Zone cooling tower and *in situ* decommissioning of the P and R plutonium production reactors. The future challenges of decommissioning are numerous on the F Canyon, with, notably, the FB line (integrated into the F Canyon structure), the plutonium-235-F fuel fabrication facility, the 221-1F A line (depleted uranium oxide), Canyon H (similar to Canyon F) and many others [BER 15].

5.4.1.4. *The Los Alamos site*

Los Alamos National Laboratory (LANL, also known as the Los Alamos Scientific Laboratory) is a laboratory of the US Department of Energy, managed by the University of California, located in Los Alamos, New Mexico. It was the main assembly site for atomic bombs [AMI 18, AMI 19a].

On this site, the decommissioning mainly concerns buildings 3 and 4, south of technical zone 21. The operations provide for decontamination, dismantling and demolition of the two buildings for treating enriched uranium, still containing this radioelement in retention in the conduits. The laboratory has adopted two successful management strategies to implement this project. First, investigators are using the observational approach rather than a total characterization of the site, supplemented by more in-depth characterization as the work progresses. Next, task hazard analyses are performed to fully define specific tasks and inform workers of the hazards. Advanced technical strategies also contribute to the success of the project with non-destructive analysis methods, such as a long-range alpha detector that detects and quantifies ions generated by alpha emitters [MAC 93, GAR 94, MAS 94, RAW 94]. The results obtained support the characterization work, an innovative duct-lining technique and techniques to treat perchlorate-contaminated items without delaying the schedule. Together, these strategies allowed the laboratory to accelerate the work schedule by one year and reduce fixed costs by $1.4 million [STO 95].

In the TA-41 technical area of the Los Alamos National Laboratory (LANL), 16 buildings are to be decontaminated and decommissioned. This proposed D&D action stems from an increased risk of severe flooding following the Cerro Grande fire and will result in the demolition, modification or abandonment of all TA-41 properties (TA-41-1, -2, -3, -4, -6, -7, -8, -9, -16, -30, -44, -47, -53, -54, -56 and -64). These facilities were built between 1948 and 1951, and are located in Los Alamos Canyon, immediately south of the city of Los Alamos, New Mexico. These historic LANL properties are owned by the Department of Energy (DoE) and were used during the Cold War for the development of weapons of war and long-term studies of weapons subsystems from the late 1940s to the 2000s [MCG 02].

The National Environmental Policy Act (NEPA) and the National Historic Preservation Act (NHPA) have classified several LANL buildings in the TA-16 zone as historic monuments [MCG 97].

5.4.1.5. *The Livermore site*

The Lawrence Livermore National Laboratory (LLNL) is a national laboratory of the US Department of Energy, located in Livermore, California. Its primary mission is to develop nuclear weapons [AMI 18, AMI 19a].

In the management of LLNL's environmental remediation (ER) and decontamination and decommissioning (D&D) programs, the company in charge is TPMC (TerranearPMC). These programs include *in situ* groundwater remediation, management and operation of treatment facilities, decontamination and decommissioning of radioactively clean and contaminated buildings and environmental remediation research and development projects [TER 17].

5.4.1.6. *The Idaho Falls site*

The Idaho National Laboratory (INL or INEEL, Idaho National Engineering and Environmental Laboratory), is located in the western United States, in the state of Idaho. It is located in typical desert country in eastern Idaho, with an average elevation of approximately 1,843 m (4,883 ft). It covers an area of 2,315 km^2. Over the years, the activity of this site has led to the construction of 52 reactors. In addition to the reactors, a full range of support facilities has also been built. These include nuclear fuel processing facilities, waste management facilities, laboratories, hot cells and other nuclear research facilities. All of these facilities have operated very successfully for many years, and each has played an important role in the development and use of nuclear energy as a new energy source. The D&D program began in 1970 and, at that time, identified more than 70 nuclear facilities to be decontaminated and decommissioned [MES 97].

Most nuclear reactors and fuel and waste treatment facilities have now been decommissioned [MES 12].

In May 2005, the prime contractor for the US Department of Energy (DoE), CH2M-WG Idaho, was awarded the contract to clean up, deactivate and decommission INL. The scope of this work included the removal of more than 200 facilities and three reactor complexes (ETR, Engineering Test Reactor, MTR, Materials Test Reactor and PBFR, Power Burst Facility Reactor). Two additional reactors were added to the scope of the contract during the execution period. The Zero Power Physics Reactor (ZPPR) provision was added under a separate sub-contract with the contractor for the INL laboratory and the Fast Reactor (EBR-II) [SZI 12].

In addition, the US Department of Energy (DoE) has awarded a multi-million-dollar contract to a consortium of companies to manage and operate the Advanced Mixed Waste Treatment Project (AMWTP) at its Idaho site [WNN 11a].

On November 8 2011, up to 17 workers were exposed to plutonium during an incident at the INL. They were performing work inside the zero power reactor that was shut down in 1992 (ZPPR) when a container was opened for normal work, resulting in their exposure to plutonium. Six of the 17 workers were externally contaminated [WNN 11a].

5.4.2. *The dismantling of Soviet and Russian military nuclear facilities*

Spent nuclear fuel and high-level nuclear waste temporarily stored on "floating bases" (ships) and on land in the Kola Peninsula are of concern to the authorities and the people of Norway. Indeed, this peninsula is surrounded to the west by Finland, to the north by the Barents Sea and to the east and south by the White Sea [GUS 96]. The amounts of cesium-137 and iodine-131 present in the Kola Peninsula are listed in Table 5.10.

	"Worst case" emission (Bq)	
	^{137}Cs	**^{131}I**
Kola Reactor	10^{16}	10^{17}
Ship nuclear reactors	10^{15}	10^{16}
Spent fuel from a reactor core	10^{15}	-(?)

Table 5.10. *Quantities of cesium-137 and iodine-131 present in various structures on the Kola Peninsula (source: [GUS 96])*

Research and development and assembly of nuclear weapons are carried out in Russia at many sites. The Scientific Research Institute of Experimental Physics in Sarov specializes in nuclear weapons R&D, small-scale production of bomb cores and storage of weapons-grade plutonium and uranium. The Scientific Research Institute of Technical Physics in Snezhinsk performs the same tasks as the first institute. The Electrochemical Instrument Combine in Lesnoy produces HEU-based weapon components, deals with the assembly and disassembly of nuclear warheads and the storage of nuclear weapon materials and components. The instrument construction plant in Trekhgorny is responsible for the assembly and disassembly of nuclear warheads. The Avangard plant in Sarov was, in the past, responsible for the production of HEU-based weapon components, assembly and disassembly of warheads and storage of weapon materials and components. Likewise, Start Production Association in Zarechny used to assemble and disassemble warheads [POD 09].

The Russian plants involved in manufacturing and research and development of nuclear fuel are listed in Table 5.11.

Plant	HEU fuel	HEU and Pu fuels
Machine building plant, Electrostal	WWER-440, RBMK, EGP-6	Civil and military fuels for naval reactors Oxidized fuels for SM-3, BOR-60, VVR PIK reactors HEU fuel for BN-600 reactor
Chemical concentrates plant, Novosibirsk	WWER-440, WWER-1000 HEU Cermet fuel for converted reactors	HEU Cermet fuel for MR, VVR, IRT, IVV research reactors Fuel for tritium producing reactors Advanced fuel for ADE reactors
Scientific Research Institute for Nuclear Reactors (NIIAR), Dimitrovgrad		Research and development Pilot production of MOX fuel for the BN-600 reactor
Bochvar Research Institute of Inorganic Materials (VNIINM), Moscow		Research and development
Luch Scientific Production Association, Podolsk		Research and development

Table 5.11. *Russian fuel fabrication and research and development (R&D) facilities in the military nuclear field (source: [POD 09]). CERMET: ceramic-metal composite materials*

5.4.3. *The dismantling of French military nuclear facilities*

The CEA has two divisions involved in nuclear decommissioning, the Nuclear Energy Division (DEN, *Direction de l'énergie nucléaire*) and the Military Applications Division (DAM, *Direction des applications militaires*). The DEN has set up a Nuclear Decommissioning Directorate (DADN, *Direction assainissement démantèlement nucléaires*). In general, for military dismantling, the project owner (MOA, *Maître d'ouvrage*) is the DAM, with the exception of the dismantling of UP1, where the MOA is the DEN.

The military dismantling in progress at Pierrelatte UDG, at Marcoule UP1 and Les Célestins, and at DIF (DAM Îles de France) concerns buildings G, D and ERA. In the future, the dismantling will be at Pierrelatte URE, at Marcoule G2 and G3, the Pu, U and T2 buildings at Valduc and the INBS PN at Cadarache. For the DAM, the priorities are Pierrelatte URE, Marcoule les Célestins and ATM, and to complete the clean-up of the DIF nuclear center [DES 13]. The estimate for the dismantling of the Besse no. 1 plant now amounts to €1.397 million, of which €1.318 million remained to be done by the end of 2018, according to the Court of Auditors [COU 20].

5.4.3.1. *THE INBS PN*

The INBS PN, located on the CEA Cadarache site, groups together the onshore testing facilities for onboard nuclear boilers for naval propulsion and the associated core fabrication facilities. Comprising of numerous operating premises, building 400 houses two major installations, the PAT (prototype on land, *prototype à terre*) and the RNG (new generation reactor, *réacteur nouvelle génération*), which have been used to test a type of boiler room capable of equipping nuclear submarines launching new generation vehicles (SNLE NG, *sous-marins nucléaires lanceurs d'engins nouvelle génération*) and a nuclear aircraft carrier (PAN, *porte-avions nucléaire*).

Building 400 is gradually being replaced by the new RES (test reactor, *réacteur d'essais*) facility and is currently in the phase of definitive abandonment.

With the objective of completely dismantling the installation excluding civil engineering, the CEA has entrusted Onet Technologies with all the studies allowing the physical and radiological characterization of the installation, the orientation of the project and the development of the dismantling strategy, the estimation of the associated costs and deadlines and the drafting of the new safety guidelines [ONE 12].

5.4.3.2. *French test sites in the Sahara*

Following its various nuclear tests in the Sahara (Reggane and In Ekker), France buried radioactively contaminated military equipment, including trucks, bulldozers and planes, as well as rocks and lava, at several sites. Collin and Bouveret [COL 20] provide various testimonies on this questionable practice. The locations of these sites have not yet been revealed, nor their exact contents.

5.4.3.3. *The Centre d'expérimentations du Pacifique*

The dismantling of the *Centre d'expérimentations du Pacifique* (CEP) was a complex operation involving the dismantling of infrastructure and buildings, including concrete and scrap metal buildings, the monitoring of underground test pits and the management of nuclear waste.

The dismantling of infrastructures and the destruction of buildings that were useless or likely to deteriorate rapidly in an oceanic environment were carried out. The decontamination station, the solid waste treatment laboratory, the measurement laboratory and the storage room for radioactive sources, facilities that presented residual risks, were cleaned up and dismantled. Non-recoverable equipment was cut up, put in metal drums and covered with concrete.

Underground test wells, as well as underwater wells that did not pose a current or future risk to the inhabitants of the atolls, were cleared of unnecessary equipment protruding from the ground. The test well heads in the coral reef were sealed with concrete and a steel slab. Heavily concreted facilities that were free of contamination were not dismantled. Their equipment was removed, and accesses were sealed off. The drums containing the waste from the clean-up and dismantling of the installations were stored in two shafts specifically drilled for this purpose, at a depth of 1,200 meters. The most radioactive waste was stored in drums in the lower volcanic part of the shafts. The less radioactive waste was placed in the upper sedimentary part, either in drums or in bulk. These shafts were finally capped with coral aggregates and cement [CEA 15].

5.4.3.4. *The cost*

The dismantling of military installations requires substantial investments. For the UP1 plant at Marcoule, the total effort was 6 billion euros. For the ARDEMU program at Pierrelatte, the total cost of the program completed at the end of 2010 was 676 million euros. As of January 1, 2009, the total cost of the remaining dismantling operations represented an expenditure of more than 7.7 billion euros [GRA 11]. For its part, the Court of Auditors [COU 20] underlines a regular drift in the cost of dismantling military (and also civilian) BNIs, with new expenditure and a year-on-year extension of the dismantling schedule.

5.4.4. *The decontamination of the British site of Maralinga*

The main British nuclear tests took place in Maralinga and, very occasionally, in Emu. In Australia's 3,200 km^2 Maralinga Desert, in addition to the 21 atmospheric tests qualified as major tests, the United Kingdom conducted 12 safety tests that included a slight release of nuclear energy, all of which took place in the same area. In addition to these 12 explosions, a total of 550 quite diverse experiments were carried out from 1953 to 1963, and in many places at the same site in Maralinga. Principally 24 kg of plutonium and 100 kg of beryllium were spread across the Maralinga site, especially in the Taranaki sector.

In 1967, the British government carried out a "final clean-up" operation, Operation Brumby, at the Maralinga site, which was limited to plowing the soil to quickly disperse the most dangerous elements by burying them at a very shallow depth. The result was a dilution of radioactivity but which left the soil incompatible with human occupation.

Following Australian protests, the United Kingdom had to carry out further clean-up operations, which were more complex and costly than the first disastrous campaign. The prospect was to make the site habitable for aboriginal Australians again. In 1987, airborne gamma-ray spectrometric examinations made it possible to complete the exercise and to specify, in particular, the contamination by cesium-137 and americium-241. Part of the contamination was vitrified *in situ*. The cost of this clean-up was 104.4 million Australian dollars (1994 value) or about 64 million euros [PAR 02].

5.5. The destruction of atomic weapons and their vectors

In the framework of disarmament agreements, or because of their obsolescence, a large number of nuclear weapons and nuclear vessels, in particular nuclear-powered submarines, must be dismantled. All countries face the same challenge [MAM 16b]. Three states have been involved in the destruction of some of their nuclear weapons: the United States, Russia and France. The United Kingdom does not have weapons itself since it rents them from the United States, and has not set up a destruction system. The other states possessing the atomic bomb have not destroyed their stockpiles at all and even tend to increase them.

5.5.1. *The deconstruction of atomic bombs*

In France, nuclear warheads are manufactured by the CEA center in Valduc. These warheads equip the ballistic missiles of the Strategic Oceanic Force

(FOST, *Force océanique stratégique*) and those of the Strategic Air Force (FAS, *Forces aériennes stratégiques*). They are, respectively, the TN 75, gradually replaced by oceanic nuclear warheads (TNO, *têtes nucléaires océaniques*), and the TN 81, gradually replaced by airborne nuclear warheads (TNA, *têtes nucléaires aéroportées*) [GRA 11].

When a weapon has reached its guaranteed life span, the CEA takes charge of the warhead and proceeds with its deconstruction, i.e. the separation of the different elements. This phase is carried out at Île-Longue for the FOST missiles and at the Valduc special military center for airborne component missiles. The cost of dismantling nuclear missiles remains in the order of less than 20 million euros per year [GRA 11].

Submarine dismantling expenses are also low, since it is only a matter of completing previous programs [GRA 11].

5.5.2. *American disarmament*

The dismantling process involves almost all sites of the US nuclear weapons complex, the National Nuclear Security Administration (NNSA) has declared. It has accelerated sharply since 2007 and has been applied to the design laboratories and production facilities.

For plutonium-based weapons, disassembly takes place at the NNSA's Pantex plant in Texas, where high explosives are removed from the special nuclear material and the plutonium core is removed from the weapon. The plutonium is placed in secure storage at Pantex, but will eventually be converted into mixed oxide (MOX) nuclear fuel at the MOX fuel fabrication facility currently under construction at the Savannah River site in South Carolina, for use in nuclear reactors.

In the case of uranium-based components, they are removed at the Y-12 National Security Complex in Oak Ridge, Tennessee. The last remaining W56-type nuclear weapon in the United States has been definitively dismantled and dismantling is currently continuing on the W62 and B61 types. This work is expected to continue for several years [WNN 07].

During the 2011 exercise, the Y-12 dismantling operations used approximately $2 million with high efficiency to dismantle more units than planned, continuing a trend that began in 2006. Materials from the components removed and dismantled have been recycled for use in life-extension programs, as fuel for the nuclear navy, or converted for other civilian uses such as fuel for nuclear power plants, scientific research reactors, and even for the production of medical isotopes that fight cancer [DOE 20].

An alliance has conducted an analysis of the capacity and facility requirements of the Y12 nuclear weapons complex in Oak Ridge in light of declining production needs and increasing demand for decommissioning [ORE 09]. The Y-12 complexes in Oak Ridge and Pantex are managed and operated under contract by Consolidated Nuclear Security (CNS) in association with various companies.

5.5.3. *Russian disarmament: international collaboration*

On November 12, 1992, the agreement to assist in the dismantling of nuclear weapons (AIDA I) was signed. This agreement aims to develop Franco-Russian cooperation in the field of the elimination of decommissioned Russian nuclear weapons and the civilian use of nuclear materials derived from these weapons. This has made it possible, in particular, to define the process for recycling military plutonium into MOX fuel. The second AIDA-MOX II agreement is tripartite (France, Germany and Russia) and was signed on June 2, 1998. It specifies the technical modalities of transforming plutonium into MOX from the dismantling of Russian nuclear weapons [IRS 19c].

At the Kananaskis Summit (June 26–27, 2002), G8 leaders launched the Global Partnership against the Spread of Weapons of Mass Destruction (WMD) in the Commonwealth of Independent States (CIS) states, including Russia and Ukraine. Within this framework, the G8 countries committed to raising up to $20 billion to promote non-proliferation, disarmament, counter-terrorism and nuclear safety over the years 2002–2012; this represented the largest non-proliferation agenda ever.

From this date, the Franco-Russian AIDA Agreements for technical cooperation in the field of nuclear non-proliferation (AIDA 1 and AIDA-MOX 2), signed in 1992, came to be included in the Global Partnership. Thus, France is committed to supporting Russia in order to reduce the risks presented by the large arsenal it inherited from the Cold War [FRA 19b].

The cooperative threat reduction program the "Nunn-Lugar Initiative", named after two US Senators, Sam Nunn and Richard Lugar, was responsible for the deactivation and destruction of many weapons. As of October 13, 2004, the number of weapons systems deactivated or destroyed by the United States under the Nunn-Lugar programs amounted to 6,462 nuclear warheads, 550 intercontinental ballistic missiles (ICBMs),

469 ICBM silos, 13 mobile ICBM launchers, 135 bombers, 733 air-to-ground nuclear missiles, 408 submarine missile launchers, 530 submarine-launched missiles, 27 nuclear submarines and 194 nuclear test tunnels. The Nunn-Lugar programs also include provisions aimed at avoiding proliferation [LUG 04].

Cooperation in the Arctic zone between the United States, Norway and Russia was aimed, in particular, at correcting environmental damage caused by military activities [DOD 10]. Similarly, numerous international cooperations have allowed the decommissioning and dismantling of Russian submarines (see section 6.4.2).

5.5.4. *Disarmament of submarines and other military nuclear vessels*

Among the means of delivery of atomic weapons, nuclear submarines are large structures that present additional risks due to their nuclear propulsion. Thus, the dismantling of the submarine is necessarily accompanied by operations to dismantle the nuclear reactor.

5.5.4.1. *The American example*

The methods used in the United States to inactivate and dismantle nuclear submarines are described by MacKinson and Burritt [MAC 96]. The current US Navy program for the inactivation, recycling and disposal of nuclear submarines and surface ships is simple and uses relatively light technology. Nevertheless, it is a very costly process (Table 5.12) due, in part, to environmental and safety concerns, particularly due to the presence of PCBs and radioactive materials in these vessels. As a result, some of the dismantling has to take place in the United States, because demilitarized warships cannot be sent abroad without first being cleaned of any material containing PCBs exceeding 50 ppm [MAC 96].

Type	Inactivation cost	Decommissioning cost
Nuclear ship	103.9	To be determined
Nuclear ballistic submarine	29.6	9.2
Nuclear attack submarine	29.6	6.9

Table 5.12. *Cost in millions of US dollars for the inactivation and decommissioning of various types of nuclear ships and submarines (source: [MAC 96])*

The "classic" method of decommissioning nuclear submarines consists of cutting out the compartment containing the reactor and the primary system, sealing it and eliminating it as a whole, in the form of radioactive waste. Normally, this requires about 800 to 1,000 m^3 of final storage space.

Studsvik RadWaste in Sweden has developed a one-step decontamination process, at room temperature, based on nitric acid, cerium nitrate and ozone. This cleaning technique was validated on the steam generators of an 80 MW PHWR, an Agesta submarine-sized reactor that operated outside Stockholm from 1964 to 1974. Another demonstration was made on two sections of 330 tubes of a steam generator removed from the Dampierre 900 MWe PWR in France, within the framework of a project supported by the EU [MEN 96].

During the dismantling of a nuclear submarine, workers are subjected to gamma radiation and contact with beta contamination from various wall surfaces (Table 5.13).

	Electrical compartment part				
Exhibition	**3rd level corridor**	**Pump wall**	**Equipment wall**	**Rear equipment wall**	**Bilge**
γ radiation (mR.h^{-1})	0.02–0.05	0.02–0.03	0.08–4.3	0.07–4.8	0.06–3.0
β contamination (part/min/cm^2)	10	10–100	10–18	10–80	15–100

Table 5.13. *Gamma radiation (γ) and beta radiation (β) contamination levels on the surface of nuclear submarine parts during dismantling operations (source: [NIK 96])*

5.5.4.2. *The Russian example*

Russia faces technical, economic and organizational difficulties in dismantling its oversized and dangerous fleet of nuclear-powered submarines. Russia's inability to effectively manage the submarine decommissioning crisis increases the risk of environmental disaster.

A huge fleet of submarines

In 1995, the USSR had built 245 submarines, equipped with 445 nuclear reactors, in addition to 4 cruisers, with 2 reactors each, and 9 civilian vessels with 15 reactors, for a total of 468 marine nuclear reactors [BUK 95].

A large number of naval reactors

The reactors of the submarines Hotel, Echo, November, Charlie, Victor II and III, Delta and Yankee are one or two PWRs of 70 to 90 MWth with fuel enriched to 20%. The reactors of the submarines Typhoon, Oscar, Akula, Sierra and Mike are 190 MWth PWRs with fuel enriched between 20% and 45%. The submarines November 645 (73 MWth) and Alfa (155 MWth) are equipped with two LMR (liquid metal reactor) reactors whose fuel is of military quality. The icebreaker Lenin is equipped with two or three PWRs (90 MWth) with fuel enriched to 5%. The icebreakers Arktika (2 reactors) and Taymyr (1 reactor), as well as the auxiliary vessel Seymorput (1 reactor) have PWR reactors of 135 MWth with fuel enriched to more than 90% [RAW 02].

Numerous installations

Shipyards for submarine construction were located in Severodvinsk, Murmansk, Rybachy and Vladivostok. Shipyards for submarine repair were located in Severodvinsk, Nizhny Novgorod, Komsomolsk-na-Amure and St Petersburg. Submarine parking bases are in Murmansk, Gremikha, Rybachy and Vladivostok. Nuclear fuel management installations are in Murmansk, Gremikha, Chelyabinsk-65, Electrosal, Rybachy and Vladivostok [BUK 95].

Dismantling is secondary

In the midst of the Cold War, the Soviet Union was more concerned with building up its fleet than with arranging for the disposal of aging nuclear-powered vessels. It was not until 1986 that the Central Committee of the Communist Party and the Supreme Soviet took the first regulatory steps. Progress was hampered by the multiplicity of government agencies involved, and the break-up of the Soviet Union in 1991 further complicated matters. Russia assumed responsibility for all nuclear-powered vessels of the Soviet Navy. The major problems were financial. The dismantling of nuclear submarines is a costly task, estimated at between $6 and $10 million per submarine. As a result, decommissioned ships are often left in floating storage until funds could be allocated for dismantling. As a result, they receive very little maintenance. In November 2001, about 40% of decommissioned submarines had been in floating storage without much maintenance for more than 10 years [ØLG 01].

The fleet to be decommissioned

Information on decommissioned vessels in various stages of decommissioning is compiled in Tables 5.14 and 5.15 [ØLG 01].

	Northern fleet	**Pacific fleet**	**Total**
Submarines with fuel	60	35	95
Submarines without fuel	15	25	40
Reactor sectors shut down	40	15	55
Total	115	75	190

Table 5.14. *Russian submarines shut down and being decommissioned in 2001 (source: [ØLG 01])*

Class	Operational period	With fuel in the core	Without fuel	Total
November	1958–1989	6	2	8
Victor-I	1967–1991	11	1	12
Victor-II	1967–	5	0	5
Alfa	1971–1995	3	4	7
Echo-II	1963–1992	12	2	14
Papa	1970–1988	1	0	1
Charlie-II	1975–1996	5	1	6
Hotel	1960–1989	3	3	6
Yankee	1967–	9	9	18
Delta-I	1972–	7	0	7
Delta-II	1975–	4	0	4
Delta-III	1976–	1	0	1
Total		67 (75%)	22 (25%)	89

Table 5.15. *Northern fleet submarines shut down and stored in 1997 (source: [ØLG 01])*

A threat of proliferation

The threat of proliferation posed by these vessels is global, because of the large quantities of highly enriched uranium (HEU) and plutonium contained in their spent fuel. The new nuclear fuel used in most Russian submarines is enriched to contain 21–45% uranium-235. Twenty-four submarines used fuel enriched to 90% in ^{235}U. A typical reactor core filled with spent fuel can thus contain tens of kilograms of uranium-235 and several kilograms of plutonium-239 [CHU 01].

International cooperation needed

The decommissioning of Russian nuclear-powered vessels was a major concern for the United States and Scandinavian countries neighboring Russia [ØLG 01]. These countries were concerned because they were aware of the security record of the Soviet and Russian navies and the budgetary constraints imposed on the Russian government. This raises concerns about possible environmental damage due to accidents or improper storage.

Foreign aid was therefore necessary to support Russia's efforts to dismantle its fleet. The largest funds came from the United States, while Japan, Norway, the European Union and a few other countries also provided aid.

Initially, the Arctic Military Environmental Cooperation (AMEC) was a Norwegian, Russian and US government consortium [DOD 10] (later joined by the British) set up to deal with military environmental issues, mainly the dismantling of Russia's nuclear submarine fleet in Europe. After the dissolution of AMEC, the Norwegian and British governments shared the £3.9 million cost of dismantling a November Class submarine.

Subsequently, the United States provided assistance with the Cooperative Threat Reduction (CTR) programs, administered by the Department of Defense (DoD), and the Nuclear Materials Security Task Force Program, administered by the Department of Energy (DoE). The CTR program was mandated to help eliminate 564 SLBM (Submarine Launched Ballistic Missile) launchers and dismantle 36 SSBNs (Sub-Surface Ballistic Nuclear), while the DoE continued programs to improve accounting and physical protection at selected sites, to provide additional physical protection measures for ship refueling and to consolidate fresh fuel storage with a central facility for each fleet [CHU 01]. The CTR has enabled the dismantling of two Oscar I submarines.

Similarly, there is a Russian–Japanese cooperation "Star of Hope" for the dismantling of Russian nuclear submarines. Six decommissioned submarines have already been dismantled. They are (a) Victor-III Class (hull no. 304); (b) Victor-I Class (hull no. 614); (c) Charlie-I Class (hull no. 714); (d) Victor-III Class (hull no. 333); (e) Victor-III Class (hull no. 271); and (f) Victor-III Class (hull no. 308). The program began in June 2003 and was successfully completed in December 2009. For decommissioning work, the Japanese government allocated JPY 4.92 billion (the Japanese Yen is worth approximately €0.0082, so approximately 40.03 million euros), while other participants in the G8 Pacific Group (Australia, Republic of Korea and New Zealand) provided financial assistance amounting to JPY 880 million (approximately 7.22 million euros) [TEC 15a].

Significant efforts have also been made on the Pacific submarine fleet, as noted in the *ex post* evaluation report of the cooperation program for the dismantlement of decommissioned nuclear submarines in the Russian Far East [TEC 15a]. The construction of a long-term storage facility for the units of the reactor compartment of dismantled nuclear submarines, called "Reactor Compartments", has been completed [TEC 15b], as well as the inauguration of the shot-blasting and painting facility for the units of the reactor compartment of dismantled nuclear submarines in the Russian Far East [TEC 14].

Reactor disposal procedure

The removal of fuel from a nuclear reactor requires a specially trained team. The coolant is drained first, but a reactor must be cooled for at least three years after its final shutdown before this is possible [ØLG 01]. The shell above the reactor is then removed, followed by the upper shield. The fuel elements are extracted and transported by ship and then by rail to a storage facility. The still highly radioactive reactor compartment can then be shut down. Most Russian submarines have two reactors that are in separate parts but in the same compartment or unit.

Due to the lack of land-based storage facilities, two adjacent empty compartments, one at the front and one at the back, are usually also cut off. This creates buoyancy for the submarine's hull and allows for underwater storage. In some cases, however, only the reactor compartment is removed and pontoons are attached to it to keep it afloat [ØLG 01]. A third method consists of filling the reactor compartment with polystyrene to achieve its buoyancy. Reactor compartments from various shipyards on the Kola peninsula, such as Polyarny, and from Severodvinsk in the county of Arkhangelsk, were towed to Sayda Bay.

The decommissioning, the unloading of fuel from reactor cores and the recycling of submarines were provided by Danilyan and Vysotsky annually from 1985 to 1995 [DAN 96].

Dismantling status

In 1995, 54 submarines were decommissioned, 22 cores unloaded and 9 recycled. In 2000, considerable progress was made in the general process of dismantling nuclear submarines in Russia. The dismantling of submarines is well advanced, and the storage of new nuclear fuel is also in much better condition than it was in 1991, but major efforts are still necessary [MOL 00].

Among the 14 submarines of the November class (1986–1990), several of them were scrapped or as carcasses in Russian naval bases (K-14, K-42, K-115 and K-133 of the Pacific fleet; K-11 and K-21 of the Northern fleet). The project of transforming the first submarine of the class (K-3) into a museum ship in Saint Petersburg had been mentioned. At least two submarines of this class were damaged (K-27 and K-159). The eight submarines of the Hotel class were taken out of service for demolition between 1987 and 1991, the 34 Echo class submarines were taken out of service for demolition between 1985 and 1995 and on June 4, 2010, the only member of the Papa class (K-222) was scrapped in Severodvinsk.

As a result of the SALT I and START I II treaties, 33 of the 34 Yankee Class vessels were decommissioned. Of this class, K-219 was lost on October 6, 1986, after an explosion and fire near Bermuda. The 18 Delta I vessels were decommissioned in 1998 and scrapped in 2005, while the four Delta II's were decommissioned in 1996. The 11 Charlie I and 7 Charlie II vessels were decommissioned for demolition between 1990 and 1994. The 16 Victor I's and 7 Victor II's were decommissioned prior to 1996.

The Alfa class consisted of seven units that joined the Soviet Navy in December 1971. The first of the class (K-64) suffered a major reactor problem and was dismantled. The other submarines were decommissioned for demolition, five in 1990 and one in 1996. By February 2003, K-463, K-316, K-432 and K-493 had been dismantled at Sevmash, the other two were at Bolshaya Lopatka and Zapadnaya Litsa, awaiting the same fate [RAW 02]. Alfa reactors use a liquid metal coolant that must be kept at a temperature above 123°C to prevent solidification. A 2002 document reports that the coolant of K-123 and K-373 solidified [RAW 02].

Both the Oscar I's were decommissioned in 1996 and their dismantling was financed by the United Kingdom under the Cooperative Threat Reduction program. Three of the six typhoons were removed. The TK-202 was deactivated in 1997 and was deposited from July 1999 at the Sevmash shipyard in Severodvinsk. Thanks to funding from the Nunn–Lugar Cooperative Threat Reduction program, the unloading of its reactors began in June 2002 at the Zvezdochka Shipyard. The K-284 of the Akula class, the leading ship of this class, was decommissioned in 1995 to avoid the cost of refueling a reactor.

Among the surface vessels, the *NS Lenin*, the world's first civilian nuclear-powered surface ship, was decommissioned in 1989 and was later converted into a museum ship. Of the six icebreakers of the Arktika class, two are no longer in service. *NS Arktika* was decommissioned and its reactor was removed in 2008. *NS Sibir* was decommissioned in 1992 due to a problem in the ship's steam generation system.

In the Bay of Andreyeva, nearly 250 nuclear submarines were taken out of service in the aftermath of the collapse of the Soviet Union, and installations were left in a perilous state [WAL 17]. Spent fuel from the nuclear submarines of Russia's Northern fleet has been stored in a pool-type storage facility at the former Andreeva Bay base on the Kola Peninsula since the 1960s. The base was closed in 1992. Following a leakage from the pools some 30 years ago, spent fuel was transferred to three dry storage tanks at a new facility on the same site. About 22,000 irradiated fuel assemblies remain, the equivalent of 100 reactor cores. The clean-up of the Andreeva Bay site is funded by grants from the Northern Dimension Environmental Partnership (NDEP). The NDEP was created in 2002 to provide technical assistance to environmental projects in the northwestern region of Russia. Its support fund is administered by the European Bank for Reconstruction and Development (EBRD) and has received support from the European Union and Belarus, Belgium, Canada, Denmark, Finland, France, Germany, the Netherlands, Norway, Russia, Sweden and the United Kingdom. Thirteen countries have provided 165 million euros in funding since 2003 [WAL 17]. Discharge of this spent fuel from nuclear submarines into Andreeva Bay began in May 2017. More than 20,000 assemblies of irradiated fuel had to be recovered, packaged and removed from the site [WNN 14a, WNN 17c,

WNN 17d]. Following the success of the operation, radiation levels in the installation have fallen by more than 40%. However, a “large volume” of radionuclides remained in the radioactive sludge at the bottom of the pools, according to the EBRD, and six assemblies of irradiated fuel remained at the bottom of one of the pools in building 5 [WNN 19n]. These are metallic oxides that are formed with the water of the primary circuit and that are activated by passing through the core of the reactor in operation.

The removal of spent nuclear fuel from the Lepse floating technical base can now begin, thanks to the installation of a special protective cover at the Nerpa Shipyard in Snezhnogorsk in the Murmansk region of Russia. The Lepse base was decommissioned in 1988. It still contains 639 irradiated fuel assemblies, some of which have been damaged. Its decommissioning is carried out with the support of the European Bank for Reconstruction and Development (EBRD) with a coverage of 23 million euros (approximately $27 million) [WNN 18d].

Nuclear waste

The decommissioning of Russian nuclear submarines has created three types of low-level liquid radioactive waste (LRW). The chemistry and concentration of contaminants vary considerably in the three waste streams, necessitating a wide range of treatment technologies (Table 5.16). Unfortunately, waste storage facilities are inadequate.

	Primary circuit	**Decontamination solutions**	**Contaminated sea water**
Total activity ($Ci.L^{-1}$)	10^{-2}–10^{-4}	10^{-5}–10^{-6}	10^{-5}–10^{-6}
^{137}Cs	60–80%	70–80%	15–60%
^{90}Sr	5–10%	10–15%	5–30%
Other radionuclides	10–20%	1–5%	1–8%

Table 5.16. *Composition of liquid radioactive wastes according to their location (source: [CHI 96])*

5.5.4.3. *The French example*

The option chosen by France to dismantle its nuclear propulsion vessels complies with the dismantling steps standardized by the IAEA. It is also compatible with France’s national waste treatment strategy, does not release any radioactive products into the environment and minimizes the doses received by workers carrying out the dismantling [MAS 96].

Decommissioning is an essential step in the life of a nuclear-powered vessel. In France, this operation has been entrusted to the French defense procurement agency (DGA, *Direction générale de l'armement*). It currently involves the first six nuclear-powered ballistic missile submarines (SNLE, *sous-marins nucléaires lanceurs d'engins*) and, in the 2020s, nuclear attack submarines (SNA, *sous-marins nucléaires d'attaque*). The dismantling of the *Redoutable* was completed in 2002 and it is on display at the Cité de la Mer in Cherbourg. It is appropriate to distinguish between two perfectly independent approaches: dismantling, which deals with operations affecting nuclear security, and deconstruction, which deals with the hull after the separation of the reactor unit [DGA 10].

Decommissioning is a sensitive operation from the point of view of nuclear safety, technically difficult to carry out, but controlled, particularly in terms of radiation protection and waste treatment. Dismantling comprises five stages: final shutdown, level I, II and III dismantling phases and final storage.

The final shutdown is an operation lasting a few months. It consists mainly of unloading the nuclear reactor core and unloading the various energy sources. The irradiated fuel elements constituting the core are stored in the pool of the reactor workshop at Homet (area of the military port of Cherbourg, near the pier of the same name) while awaiting their return to the CEA for storage and/or reprocessing. At this stage, the submarine is secured.

Level I dismantling is the stage where the reactor unit equipment is unloaded and additional monitoring facilities are set up.

During the level II dismantling phase, the confined area is reduced to a minimum and the sealing and biological protection is reinforced. Extensive decontamination is carried out, all fluids are drained and the high-capacity primary circuit is reduced by evaporation. The openings in all partitions are sealed and the tightness of the reactor compartment is checked. This operation lasts from 10 to 12 months depending on the radiological condition of the submarine. Then, the hull is put under water in order to cut out the reactor unit. The three-month operation consists of separating this 700-ton unit, 10 m in diameter and 8 m wide, from the rest of the vessel and transporting it on a rolling platform to its storage site. Then, the two parts are re-welded to make the hull watertight and to allow the vessel to be launched.

Long-term storage is carried out under supervision at Le Homet on a seismic-resistant slab, protected from the weather by a suitable structure. An annual penetration is carried out under ventilation in order to check the evolution of the materials over time and to guarantee adequate ambient conditions. This storage must last a few decades.

Level III dismantling, which takes place after this long period of time, consists of removing all nuclear materials and equipment. The installation is entirely cut up and packaged in waste drums for ANDRA. This work will have to be carried out on site in a specific infrastructure allowing the complete dismantling of the reactor unit. Feasibility studies for this phase are currently being carried out by the CEA.

The deconstruction, which finalizes the whole process, only starts after authorization by the safety authorities, and the hull can be recycled outside the nuclear field. Deconstruction is currently the subject of studies that take into account the treatment of hazardous materials and respect for the environment. The hull is more than 100 m long, 10 m in diameter and weighs 7,000 tons [DGA 10].

To date, four ex-SNLEs have reached the level II dismantling phase. The first was *Le Redoutable*, commissioned in 1967 and disarmed in 1991. Work was carried out on *Le Tonnant* (1980–1999), *Le Terrible* (1973–1996) and *Le Foudroyant* (1974–1998), the hulls of which were re-welded and put back into the water after extraction of the reactor unit. These ex-SNLEs are now waiting, afloat, in the military port of Cherbourg. Work on *L'Inflexible* (1985–2008) was to be completed in 2013. As for the last vessel in the series, *L'Indomptable* (1976–2005), it was still equipped with its reactor in 2013 and was subject to dockside monitoring. Its passage from level I to level II of dismantling was planned for 2015 [GRO 13]. It was completed in 2016 [HER 18]. Naval Group is the prime contractor, while Veolia and Neom (a subsidiary of Vinci Construction France) were its two main industrial partners [GAR 18].

The dismantling of the SNA is planned for between 2020 and 2030, the SNLE NG from 2030 to 2040, the aircraft carrier (PA, *porte-avions*) *Charles de Gaulle* from 2040 to 2045 and the Barracuda from 2050 to 2075 [GRO 13].

Only *Le Redoutable* has avoided deconstruction. It has been preserved since 2002 at the *Cité de la mer* in Cherbourg.

5.5.5. *Denuclearization of rocket bases*

France's deterrent force was originally based on three components (missile-launching submarines, aircraft and rockets). This last component was located on the plateau d'Albion in three departments (Vaucluse, Drôme and Alpes-de-Haute-Provence) and seven communes. From 1971 to 1996, this plateau hosted the launch site for the ground-to-ground ballistic nuclear missiles of the French nuclear deterrent force, mainly in the commune of Lagarde-d'Apt. It included 18 missile silos and two fire control stations.

From 1997 to 1999, the former intercontinental ballistic missile silos and the command and fire facilities on the plateau d'Albion were dismantled and reconverted for conventional civilian or military activities. One of the silos now houses an astronomical observatory and a low background noise underground laboratory that replaces a former command and firing post.

On Thursday June 18, 2015, representatives of some 30 member states of the Conference on Disarmament (CD) in Geneva visited the dismantled military nuclear facilities on the plateau d'Albion. This visit, organized as part of France's efforts to promote transparency in nuclear disarmament, allowed the visitors to see the transformations that have taken place [PIC 15].

In France, according to the Ministry of Defense, the total cost of the denuclearization of the plateau d'Albion and its 18 silos was close to 75 million euros, a detail in view of the 23.3 billion euros allocated to the French nuclear deterrent by the 2014–2019 military programming law (more than 10% of the defense budget) [MAM 16a].

The denuclearization of the plateau has not eliminated the social problems of the region. Twenty years after the demilitarization of the plateau d'Albion, some former military personnel continue to denounce nuclear missile-related activities as responsible for radioactive contamination and persistent diseases [MON 19]. These problems are probably mainly related to irradiation by neutrons emitted by the plutonium payload. Indeed, while alpha particles and X-rays are easily stopped by the various materials surrounding the nuclear warhead, neutrons originating from spontaneous fission, on the other hand, easily pass through all structures. The workers were not warned of this risk. The same problem was repeated at Île Longue with military and civilian personnel.

5.6. Conclusions

Huge efforts to clean up the former American nuclear weapons manufacturing centers have been undertaken by the United States. While some dismantling operations are complete and these sites may return to conventional activity, many other sites will remain nuclearized and controlled zones for centuries to come.

The decommissioning and dismantling of enrichment, conversion, manufacturing and reprocessing plants, as well as plutonium-producing and tritium-producing reactors for military purposes, are effective for the United States, the United Kingdom and France, and are much slower for Russia. The other nuclearized states (China, Israel, India, Pakistan and North Korea) maintain, and even increase, the activity of these various plants.

The disarmament of four states (United States, Russia, United Kingdom and France) is effective. However, this must be qualified by the continuous improvement of the new weapons handed over to the armies of these countries, and we must remember that, on the other hand, the armament of the five other nuclearized nations is steadily increasing.

The dismantling of nuclear submarines is progressing steadily in the United States, the United Kingdom and France, even though the process is extremely long. For Russia, on the other hand, the dismantling of its huge nuclear submarine fleet depends entirely on international financial aid.

France is the only nation to have dismantled its rocket launch base carrying atomic bombs which constituted its ground deterrent force.

6

The Dismantling of Electronuclear Reactors

6.1. Introduction

As of 2019, globally there were 417 operational nuclear reactors, 28 in extended shutdown and 173 had been permanently shut down; only 36 reactors were under construction and 82 sites had been permanently abandoned among the states that already have nuclear energy. [SCH 19].

Several states that have never had a nuclear industry are building new nuclear power reactors: Bangladesh (2), Belarus (2), Turkey (1) and the United Arab Emirates (4). In addition, three states have signed construction contracts, Lithuania (suspended), Poland (?) and Vietnam (suspended). Plans are underway for Egypt (1), Jordan (abandoned) and Turkey (2 plans, 1 of which has been abandoned). Some plans are successfully pledged for Indonesia, Kazakhstan, Saudi Arabia, Thailand and Uzbekistan [SCH 19].

More than 115 commercial power reactors – along with 48 experimental or prototype reactors – more than 250 research reactors and several fuel cycle facilities have been decommissioned. Of these shutdowns, 12 were due to damage during a severe nuclear accident and 37 reactors were shut down prematurely following political decisions or popular referendums, often as a result of the accidents at Chernobyl or Fukushima. However, a large number of reactors (122) were shut down after reaching their target or were no longer economical to operate. A small number of these reactors have been completely dismantled [WNN 19h].

Most parts of a nuclear power plant do not become radioactive or are contaminated only at very low levels. Also, most of the metal can be recycled.

Proven techniques and equipment are available to safely dismantle nuclear facilities and are now well handled in many countries. The costs of decommissioning nuclear power plants, including the disposal of associated waste, are high compared to other industrial plants, but they are decreasing and represent only a small fraction of the total cost of electricity generation [WNN 19h].

The situation at nuclear power plants is changing very quickly. Thus, in 2019, nine power reactors with a combined capacity of 5,976 MWe were officially shut down. These were Bilibino 1 in Russia, Chinshan 2 in Taiwan, Genkai 2 in Japan, Mühleberg in Switzerland, Philippsburg 2 in Germany, Pilgrim in the United States, Ringhals 2 in Sweden and Three Mile Island 1 in the United States. Unit 1 of the South Korean Wolsong plant, which had not operated since June 2018, was declared closed on December 24. At the end of 2019, 442 reactors were operational worldwide, for a total of 392.4 GWe net and 54 were under construction (59.9 GWe gross) [WNN 20b].

6.1.1. *The various types of electronuclear reactors*

The main characteristics of a nuclear reactor are defined by the nuclear fuel it uses, the type of moderator used to slow neutron velocities, the type of coolant used to transfer the heat created by the fissions that occur in the reactor core and the nature of the tubes that contain the fuel (Table 6.1).

Branch	Fuel	Neutrons	Moderator	Heat transfer fluid	Girdle
NUGG	Natural metallic U	Thermal	Graphite	Carbon dioxide	Mg
Magnox or AGR	Enriched U (2%)	Thermal	Graphite	Carbon dioxide (or helium)	Stainless steel
RBMK	Enriched U (1.8–2.4%)	Thermal	Graphite	Ordinary boiling water	Zr–Nb
PWR or REP and WWER	Enriched U (3–4%) (or MOX)	Thermal	Ordinary pressurized water	Ordinary pressurized water	Zr
BWR or REB	Enriched U (2–3%) (or MOX)	Thermal	Ordinary boiling water	Ordinary boiling water	Zr

Branch	Fuel	Neutrons	Moderator	Heat transfer fluid	Girdle
CANDU or PHWR	Natural U (or slightly enriched)	Thermal	Heavy water	Carbon dioxide (or heavy water or ordinary water)	Zr
FRB or RNR or Rapid	Enriched U and Pu (15–20%)	Rapid	No moderator	Sodium molten salts	Stainless steel
High temperature (HTR and VHTR)	Enriched U or thorium (5–10%)	Thermal	Graphite	Helium	Graphite

Table 6.1. *The main nuclear power reactor systems (modified from [NAU 08, AMI 13]). AGR: Advanced Gas-Cooled Reactor; BWR: Boiling Water Reactor; FRB: Fast Breeder Reactor; HTR: High Temperature Reactor; PHWR or CANDU: Pressurized Heavy Water Reactor; PWR: Pressurized Water Reactor; NUGG: Natural Uranium Graphite Gas; RBMK: Reaktor Bolshoy Moshchnosti Kanalniy; BWR: Boiling Water Reactors; RNR: Fast Neutron Reactor; VHTR: Very High Temperature Reactor*

6.2. The dismantling of graphite-moderated reactors

These systems have been developed mainly in France and the United Kingdom, with carbon dioxide (CO_2) or helium (He) as the heat transfer fluid for high-temperature types, and in Russia with boiling water as the heat transfer fluid. The French sector uses natural uranium as fuel, while the British and Russian sectors use slightly enriched uranium.

Fifty-six GCRs have operated commercially in the United Kingdom, France, Spain, Italy, Japan, Germany and the United States, and 38 are currently in various stages of decommissioning. GCRs are characterized by low energy density, large sizes and high volumes of graphite waste [BRO 12].

6.2.1. *Decommissioning of French nuclear power reactors (UNGG)*

The first generation of French electronuclear reactors consisted of graphite-moderated reactors with carbon dioxide as a coolant and natural uranium as nuclear fuel. The prototype was the Chooz A reactor (see Chapter 7, section 7.4.1.1), followed by seven other reactors built in Spain and primarily in France (Table 6.2). These eight reactors have been shut down and are in the process of being decommissioned.

Country	Reactor name	Power (MWe)	Operating life in years (shutdown date)	Decommissioning status
Spain	Vandellos-1	480	18 (1989)	SAFSTOR (30 years old)
France	Bugey-1	555	22 (1994)	Phase 3
	Chinon-A1	80	17 (1973)	Delayed
	Chinon-A2	230	20 (1985)	
	Chinon-A3	480	25 (1990)	
	Saint-Laurent-A1	500	21 (1990)	Delayed
	Saint-Laurent-A2	530	21 (1992)	

Table 6.2. *French nuclear power reactors (UNGG, Uranium Naturel Graphite Gaz) with graphite moderator (according to [NEA 16a])*

The first generation of French reactors included three UNGG reactors, Chinon A1, A2 and A3, that had been operating at the Chinon site since the early 1960s, and which were shut down in 1973, 1985 and 1990, respectively. Their partial dismantling was completed in 1984, 1992 and 2007. At Saint-Laurent, Units A1 and A2 were commissioned in 1969 and 1971 and shut down in 1990 and 1992. The removal of fuel from both reactors was completed in 1992 and 1994. Another UNGG reactor was also commissioned at the Bugey site in 1972 and shut down in 1994 [WNN 16a]. For more details on the various reactors, the operators and the performance of each nuclear sector, the CEA census [CEA 16] is very comprehensive.

6.2.1.1. *Decommissioning planning for French UNGG reactors*

These reactors were operated by EDF, and it is this company that is the prime contractor for complete decommissioning. Initially, EDF chose the strategy of immediate dismantling. The dismantling of the installations (excluding the reactor building) occurred quickly after the shutdown of the power plant (in practice, within 5 to 10 years) and the dismantling of the reactor building was deferred, leading to the total and unconditional release of the site for a period of 25 to 50 years after the shutdown of the power plant. This solution of maintaining containment for decades allowed EDF to benefit from the natural decay of the radioactivity. However, the ASN had expressed doubts about the likelihood for the operator to maintain the integrity of the installations, especially the civil engineering structures, during this period [ASN 19a].

In April 2001, EDF informed the ASN of a major change in its strategy for the dismantling of first-generation reactors, for it had opted for a complete and immediate dismantling program over 25 years (by 2025), with no waiting period. This new strategy, which calls for full completion by 2036, was reviewed by the

ASN's Standing Advisory Group of Experts (*Groupe Permenant d'Experts,* GPE) in March 2004. The ASN accepted it on the condition that EDF respected various remarks made on the feasibility of dismantling, safety, radiation protection and waste and effluent management [ASN 19a].

In July 2009, EDF sent the ASN an update on its dismantling strategy. The ASN considered the strategy to be generally appropriate in its principles but it lacked proposals for alternative solutions for graphite waste management. In June 2012, the ASN asked EDF not to make the dismantling of the UNGG reactor casings conditional on the commissioning of the storage center for long-life, low-level waste and to consider the possibility of intermediate storage. EDF sent an update on the strategy for dismantling its reactors in October 2013. This update was the subject of several reviews by the ASN and its permanent group of experts, with new questions and new answers from EDF in 2014 and 2015 [ASN 19a].

On March 29, 2016, the ASN interviewed EDF, who presented a new strategy for dismantling UNGG reactors. This new strategy led to postponing the dismantling of certain reactors by several decades, in light of the strategy announced by EDF in 2001 and updated in 2013. In July 2016, the ASN sent EDF a letter of additional requests. EDF submitted these responses in 2017. In view of this situation, the ASN planned to regulate the dismantling of UNGGs by decision. The draft decision required EDF to carry out the dismantling operations that could already be carried out and to submit the necessary dismantling modification files before 2022. The scenarios to be studied by EDF are imposed by the ASN [ASN 19a].

In EDF's new strategy, the dismantling of the six reactors will take longer than expected as the company wants to complete the decommissioning of one unit before working on the others. The company originally planned to dismantle the reactors using underwater techniques, but has now decided to dismantle them dry in the air. EDF has told ASN that it will complete the dismantling of the peripheral installations of all UNGGs within 15 years [WNN 16a].

6.2.1.2. *Technical challenges*

Graphite is an interesting material as a neutron moderator. The reactor grade of this material is produced at about 3,000°C. During reactor operations, this graphite accumulates some contamination by radionuclides – in particular carbon-14 – at levels that often mean that it must be classified as intermediate-level radioactive waste. This graphite contamination is produced by neutron activation of impurities. These are cobalt-60, iron-55 and tritium for short-lived activation products, and carbon-14 (activation of carbon-13, a stable carbon isotope), chlorine-36, calcium-41, nickel-59 and 63, niobium-94 for long-lived activation products. In addition, there are fission products (strontium-90, cesium-137 and 135, technetium-99, etc.) and actinides (americium, plutonium, etc.) that have been released by fuels during reactor operations. The total activity of graphite waste will decrease by two orders of magnitude over a period of 50,000 years.

This explains why the dismantling of graphite reactors is complex by design. Indeed, they are made up of a stack of graphite layers inside the reactor and the volume of this material to be removed is significant. Currently, there are about 60 reactors of this type in the world and, so far, only two small graphite power reactors have been dismantled. The market is large.

An original "underwater" dismantling method consists of filling the concrete chamber with water to protect the workers from radiation, in order to proceed with the cutting of the upper part of the chamber before reaching the graphite stack; the water being in permanent circulation, which allows it to be filtered. This underwater dismantling method was successfully used during the dismantling of the Fort Saint-Vrain reactor in Colorado, USA. However, EDF indicated to the fact-finding mission [AUB 17] that material differences between the American reactor and the French model did not allow the same core dismantling technique to be used.

Furthermore, a notable change of strategy has been decided very recently, and unilaterally, by EDF with the choice of dismantling "in the air" instead of "under water". The change in EDF's strategy came 20 years after the shutdown of the reactors and 15 years after the beginning of dismantling. This explains the questioning, not to mention the irritation of the ASN. EDF's explanation is that the extraction of the graphite piles will be carried out by subcontractors who, instead of taking 3 years, will require 15 years and the underwater method of dismantling the chamber over this long period would pose serious corrosion problems [AUB 17].

To this end, EDF and Veolia have created a joint venture to dismantle reactors using graphite technology. The first objective of this company, Graphitech, will be to provide EDF with an optimized scenario for the dismantling of the Chinon A2 reactor (shut down more than 35 years ago in 1985) in 2028 and to propose a test program to evaluate the technological solutions required to carry out the project. This program will begin in 2022. In addition, Graphitech will seek contracts to assist in the dismantling of these reactors in France, Italy, Japan, Lithuania, Spain and the United Kingdom [WNN 19q].

6.2.1.3. *Waste management*

Waste from the decommissioning of nuclear power plants is managed in the same way as that from operating plants. It is sorted, disposed of, compacted and conditioned, before being transported to storage centers adapted to its nature. Most (80%) of the waste generated by the deconstruction of a nuclear power plant is conventional waste, i.e. non-radioactive. The remaining (20%) is, for the most part, weakly or very weakly radioactive [EDF 19].

The dismantling of the first generation of UNGG reactors will generate large quantities of waste in the form of contaminated graphite. How will the 17,000 tons of radioactive graphite be stored? [REN 18].

Since the storage centers managed by ANDRA are expected to be saturated in the medium term, EDF has decided to build a temporary storage center for some of its waste. This project is called ICEDA (*Installation de conditionnement et d'entreposage de déchets activés*). Construction work began in 2010 but was halted by a court decision that canceled the building permit, confirmed on appeal in 2012. Construction work resumed in early April 2015, and the final finishing work and pre-commissioning tests continued in 2017. The facility is currently in the testing phase, and the ASN is experiencing a significant delay in the progress of this program. EDF planned to commission the facility in the second quarter of 2018 [ASN 19c].

ICEDA is a warehouse unit and not a storage unit. It is therefore a temporary solution with an expected operating life for ICEDA of 50 years. The ultimate goal of the stored waste is to be subsequently evacuated to ANDRA's future final storage center. For reasons of synergy and optimization of waste management, the site selected is the Bugey power plant. It will be the first UNGG reactor to be dismantled, and the site is connected to the rail network, which will facilitate the transport of waste from other sites undergoing deconstruction. The storage facility, owned by EDF, will occupy a surface area of 8,000 m^2.

ICEDA will make it possible to condition and store long-lived, intermediate-level radioactive waste (with a radioactive half-life of more than 30 years) from the nine EDF nuclear power plants undergoing decommissioning, in particular graphite waste. This waste represents less than 0.1% of total decommissioning waste, or about 500 metric tons. It will also receive metal waste from operating plants (about 1,500 tons). The company will ensure the containment of radioactivity and guarantee the absence of impact on the public and the environment [EDF 20b].

6.2.1.4. *The Spanish UNGG reactor*

Unit 1 of the 480 MWe Vandellòs Spanish gas–graphite reactor was supplied by France and operated between 1972 and 1989. Enresa began the dismantling and decommissioning of Vandellòs 1 in 1994. Phase 2 of the dismantling, which consisted of removing all buildings, systems and equipment outside the reactor pressure vessel, was successfully completed in 2003. The pressure vessel, emptied of its nuclear fuel, was sealed and remained so for 25 years. During this period of dormancy [NEA 16a], the activity of the internal structures naturally decreased. In about 2028, the pressure vessel and all the internal structures will be removed. The site will then be completely cleared. Engineering services for the dismantling of Unit 1 of the Vandellòs nuclear power plant will be provided by Westinghouse and

EDF, under a contract awarded by the Spanish company for dismantling and waste management (Enresa) [WNN 18c].

6.2.2. *The dismantling of British reactors*

British first-generation reactors were graphite-moderated with carbon dioxide as a coolant. The nuclear fuel was slightly enriched uranium. Twenty-seven reactors were built in the United Kingdom, as well as three outside the British Isles (North Korea, Italy and Japan) (Table 6.3). Their life and performance were very good.

Country	Reactor name	Number and power (MWe)	Operating life in years (shutdown date)	Decommissioning status
North Korea	Yongbyon	1 x 200	20 (2005)	SAFSTOR(?)
Italy	Latina (Foce Verde)	1 x 160	24 (1987)	SAFSTOR (30 years old)
Japan	Tokai-1	1 x 166	32 (1998)	SAFSTOR (10 years)
United Kingdom	Berkeley 1, 2	2 x 166	27 (1988–1989)	SAFSTOR (30 years old)
	Bradwell 1, 2	2 x 146	39 (2002)	SAFSTOR (30 years old)
	Calder hall 1 to 4	4 x 60	44–46 (2003)	SAFSTOR (30 years old)
	Chapelcross 1 to 4	4 x 60	44–45 (2004)	SAFSTOR (30 years old)
	Dungeness-A1, A2	2 x 230	41 (2006)	
	Hinckley Point-A1, A2	2 x 267	35 (2000)	
	Hunterston-A1, A2	2 x 173	25 (1989–1990)	
	Olbury-A1, A2	2 x 230	44 (2011–2012)	
	Sizewell-A1, A2	2 x 245	41 (2006)	
	Trawsfynydd-1, 2	2 x 235	26 (1993)	
	Windscale AGR	1 x 36	18 (1981)	
	Wylfa 2	2 x 540	44, 41 (2015)	

Table 6.3. *Graphite Moderated Gas-Cooled Reactors (GCR) in the British nuclear power industry (source: [NEA 16a])*

Table 6.4 details the actions taken by the NDA (Nuclear Decommissioning Authority) to decommission UK facilities for the years 2020 and 2021, including the areas involved, the estimated date of completion of decommissioning and the budget allocated to each site [NDA 19].

Site	Surface area (ha)	Remaining area under nuclear license (ha)	End of remediation	Cost in 2020/21 (millions of pounds)
Sellafield	265	265	2125	2.15
Magnox Ltd				515
Berkeley	27	16	2079	
Bradwell	20	20	2092	
Chapelcross	96	96	2095	
Dungeness A	20	20	2097	
Harwell	108	86	2064	
Hinkley point A	20	20	2090	
Hunterston A	15	15	2080	
Olbury	47	15	2103	
Sizewell A	14	14	2097	
Trawsfynydd	15	15	2083	
Winfrith	95	85	2023*	
Wylfa	21	21	2105	
Dounreay Site Restoration Ltd	60	60	?	200
LLWR Ltd	99	99	2135	82
RWM Ltd				
Springfields	81	81		18
Capenhurst	31	14		55
TOTAL	1034	927		3.02

Table 6.4. *Decommissioning of UK facilities (source: [NDA 19]) LLWR: Low-Level Waste Repository; RWM: Radioactive Waste Management (* not insured)*

6.2.3. *The dismantling of the RMBK sector*

One of the Soviet nuclear reactor systems used graphite as a moderator and boiling water as a coolant. The fuel was slightly enriched uranium.

At the time of the Chernobyl accident, the RBMK fleet had 17 reactors in operation. Eleven were in Russia, each with a power of 1,000 MWe, distributed over three power plants, Leningrad (4 reactors), Kursk (4 reactors) and Smolensk (3 reactors); four in Ukraine (Chernobyl power plant), 800 MWe for the first and 1,000 MWe each for the others (a fifth was under construction); and two in Lithuania (Ignalina power plant), 1,500 MWe each [IRS 16]. However, the second reactor at Ignalina never materialized.

The 11 Russian reactors are currently in operation after having been modernized. Their service life has been extended to 45 years, so the last reactor that was built (Smolensk-3 started up in 1990) is expected to operate until 2035.

The situation is different in Ukraine, since the Chernobyl reactor No. 4 was seriously damaged and must be dismantled in a particularly delicate manner (see section 6.6). The three other RBMK nuclear reactors with a power of 1,000 MWe are currently shut down permanently. Reactor number 1, commissioned in September 1977, was shut down in November 1996. Reactor number 2, which started up in December 1978, has been out of service since October 11, 1991, following a fire in the engine room. The Ukrainian authorities decided to shut it down permanently in March 1999 because the cost of refurbishment was too high. Reactor number 3 was put into service in 1981. After 1995, it underwent numerous shutdowns for maintenance, inspection and repair operations. In accordance with a decision taken in June 2000, the Ukrainian authorities shut it down definitively on December 15, 2000. The construction of reactors 5 and 6, which had started in 1981, was abandoned following the 1986 accident.

Lithuania agreed to close Ignalina 1 and 2 as a condition of its accession to the European Union. The European Commission is providing substantial funding for the project, which is expected to be completed in 2038.

The work that began in October 2011 to dismantle the turbine halls of Unit 1 of the Ignalina nuclear power plant has been completed. The dismantled equipment weighs more than 18,000 tons and consists mainly of turbines, generators, heat exchangers, pumps, valves and pipes. In April 2016, the IAEA stated that Lithuania should plan for potential project risks in the decommissioning of the Ignalina power plant. A five-day expert mission to the site, led by the IAEA, examined the project risks and uncertainties associated with the decommissioning of the two graphite-moderated light-water RBMK-1300 reactors, which were shut down in late 2004 and late 2009, respectively [WNN 16b].

Construction of the second unit (Ignalina 3?) had begun in 1985 but was suspended by the government in 1988 when the unit was 50% complete and some equipment had already been delivered to the site. Work on the dismantling of the buildings of the unfinished third reactor of the Ignalina nuclear power plant in Lithuania has been completed [WNN 08].

6.3. The dismantling of the pressurized water system (PWR)

The pressurized water reactor (PWR) industry has been very successful. It uses nuclear fuel slightly enriched in uranium and pressurized water as both a moderator and a coolant. This system constitutes the first two generations of reactors for several countries. France has opted for this sector for its second generation of reactors. Twenty-eight reactors in this sector have been shut down and are in the process of being decommissioned (Table 6.5).

Country	Reactor name	Power (MWe)	Operating life in years (shutdown date)	Decommissioning status
Germany	Biblis-A	1,225	36 (2011)	
	Biblis-B	1,300	34 (2011)	
	Muelheim-Kaerlich	1,302	2 (1988)	
	Neckarwestheim-1	106	35 (2011)	
	Obrigheim	357	36 (2005)	
	Stadium	672	31 (2003)	
	Unterweser	1,410	32 (2011)	
Austria	Zwentendorf	723	Never	
Belgium	SCK-CEN-BR3 (Mol)	12	25 (1987)	Completed (2011)
Spain	José Cabrera	160	38 (2006)	Completed (2018)
United States	Crystal River-3	890	33 (2009)	SAFSTOR 2019–2067
	Haddam Neck	603	29 (1996)	ISFSI
	Indian Point-1	277	12 (1974)	Deferred
	Kewaunee	595	39 (2013)	Deferred

Country	Reactor name	Power (MWe)	Operating life in years (shutdown date)	Decommissioning status
United States	Maine Yankee	900	24 (1996)	ISFSI
	NS Savannah*	75	9 (1970)	Deferred
	Rancho Seco-1	917	15 (1989)	ISFSI
	San Onofre-1	456	25 (1992)	Deferred
	San Onofre-2 and 3	2 x 127	31, 30 (2013)	Deferred
	Saxton	3	6 (1972	Completed
	TMI-2**	959	(1978–1979)	Deferred
	Trojan	1,155	16 (1993)	ISFSI
	Yankee NPS	180	31 (1991)	ISFSI
	Zion-1 and 2	2 x 1,085	25 (1998)	Deferred 15 years
France	Chooz-A	320	24 (1991)	Complete
Italy	Enrico Fermi	270	6 (1972)	SAFSTOR (30 years old)

Table 6.5. *Pressurized Water Reactors (PWRs) in shutdown and in the process of being decommissioned (source: [NEA 16a, WNN 19s]). *NS: Nuclear Ship; **TMI-2: Three Mile Island-2*

6.3.1. *The dismantling of German reactors*

Due to its new nuclear policy, Germany has a large number of nuclear reactors to decommission and has gained experience in this field.

On May 6, 2013, the German power utility EnBW officially applied to the State for permission to decommission and demolish Neckarwestheim 1 (GKN I). It received approval from the Baden-Württemberg Ministry of the Environment to decommission and dismantle this unit in February 2017. The company anticipated that dismantling work would begin in March 2017 and be completed within 10 to 15 years. The unit was one of the eight oldest German reactors decommissioned by order of Chancellor Angela Merkel a few days after a tsunami hit Japan's Fukushima Daiichi nuclear power plant in March 2011. Unit 2 is scheduled to close in 2022 [WNN 17a].

PreussenElektra, a German public utility, operates three of the eight nuclear units currently in operation in Germany – Brokdorf, Grohnde and Isar 2 – which are scheduled to close by 2022. It also operates Grafenrheinfeld, which was closed in

2015, and Isar 1 and Unterweser, which, along with seven other nuclear power plants, were closed in March 2011 [WNN 18a].

The department announced the decommissioning and dismantling of the Unterweser nuclear power plant (KKU), which was the second PreussenElektra reactor shut down in 2011 to receive such a license, after Isar 1 in January 2017. KKU was a 1,410 MWe PWR reactor operated between 1978 and 2011. Dismantling work inside the controlled area is expected to take about 13 years,

The dismantling of the company's Stade reactor is "well advanced" and should be completed around 2021, the company said, while the dismantling of Würgassen was completed in 2014 [WNN 18b].

In Philippsburg, the Baden-Württemberg Ministry of the Environment issued the German utility EnBW with a permit for the decommissioning and dismantling of Unit 1 of its nuclear power plant in 2017 [WNN 17b]. EnBW announced its intention to permanently close Unit 2 in 2020. This unit began commercial operation in April 1985. The permit from the Ministry of the Environment of the Land of Baden-Württemberg for the decommissioning and dismantling of the 1,392 MWe PWR reactor was issued in December 2019. Since its commissioning, Unit 2 in Philippsburg generated about 10 terawatt hours of electricity per year and more than 355 TWh over its lifetime [WNN 20a]. EnBW expects the dismantling of the entire unit to take about 10 to 15 years [WNN 19s].

6.3.2. *The dismantling of American reactors*

As with Germany, the United States has a large number of PWRs (14 and 17 for Germany and the United States, respectively) and BWRs (10 and 13 for Germany and the United States, respectively) that have been shut down, are undergoing decommissioning and some of them are fully decommissioned. La Guardia [LAG 12b] presents the results of the oldest decommissionings of US reactors. He details the removal of the largest parts, including internal components and buildings, as well as site remediation and reuse.

At Crystal River in Florida, the 860 MWe pressurized water reactor was to be modernized and its capacity increased in September 2009, but this was never completed after complications with the containment of the reactor building. The plant was declared permanently closed in February 2013. At that time, the US utility Duke Energy stated that it expected the decommissioning to be completed by 2074. Since then, Duke Energy has decided to gather pace and has announced its intention to decommission the Crystal River nuclear power plant by 2027. The utility has contracted Accelerated Decommissioning Partners, a joint venture between

NorthStar Group Services and Orano USA, to carry out the decommissioning work [WNN 19w].

The Connecticut Yankee reactor was operational for a few months in 1996 and was in the transition phase until 2007. Due to its early shutdown, immediate dismantling was delayed. The final decommissioning provides for a return to greenfield – except for one hectare which is for small dry fuel storage – and dismantling began in 2007 [NEA 18a].

EnergySolutions is currently carrying out decommissioning projects at the Zion Nuclear Power Plant in Illinois. This project is expected to be completed in early 2020. In May 2019, the company also decommissioned the San Onofre power plant in California (three reactors) and the Fort Calhoun power plant in Nebraska [WNN 19j].

Unit 1 at Three Mile Island was closed in September 2018 for economic reasons, after more than 45 years of production. Exelson Generation chose the SAFSTOR deferred dismantling option for the unit. The spent fuel will be placed in the site's pool and then stored in dry containers by the end of 2022 [WNN 19j].

The current status of PWR reactor decommissioning in the United States shows 10 reactors under deferred decommissioning with Crystal River-3 (2073), Indian Point (2026), Kewaunee (2073), Nuclear Ship Savannah (2031), San Onofre-1, shut down in November 1992 (2030), San Onofre-2 and 3, shut down in June 2013 (2031), Three Mile Island-2, damaged in March 1979 with half of its core in the form of corium (2036) and Zion-1 and 2 (2026). Five reactors were placed under sarcophagi without complete dismantling (ISFSI): Haddam Neck, Maine Yankee, Rancho Seco, Trojan and Yankee NPS (Rowe). Two decommissionings have been completed for the Saxton and Shippingport reactors [NEA 16a].

6.3.3. *The dismantling of French reactors*

The oldest French PWR reactor (Fessenheim no. 1, 1977, 900 MWe) was shut down on February 22, 2020, and the Fessenheim no. 2 reactor was shut down in June 2020. Between now and 2023, EDF must evacuate the irradiated fuel constituted by 157 assemblies of 264 rods (4.12 m long, 665 kg, including 513.9 kg of UO_2-PuO_2, thousands of enriched uranium pellets) to the storage pool of the power station. Then, this fuel will be sent to La Hague (Manche) to the ORANO reprocessing center. Five years after the final shutdown, the reactor will begin to be dismantled (around 2025), and this deconstruction should take about 15 years. The cost of dismantling the two reactors at Fessenheim should be 1 billion euros. The site should be entirely decommissioned around 2041 and retain industrial use.

The construction of a technocenter is planned to melt down steel and metal parts with very low radioactivity [ANO 20].

6.3.4. *Reactor decommissioning in other countries*

The José Cabrera plant, also known as Zorita, is the first to be decommissioned in Spain. The single-loop pressurized water reactor is located in the municipality of Guadalajara. It operated from 1968 to 2006 when it was shut down by ministerial order. It is a small 141 MWenet PWR reactor (gross power of 150 MWe) that supplied more than 75% of the electricity needs of the province of Guadalajara [WNN 19m]. The José Cabrera plant was in a transition phase from 2006 to 2010 and has been in the dismantling phase since 2010. Part of the site will be reused for other purposes [NEA 18a]. The strategy chosen for the José Cabrera PWR reactor is immediate dismantling [NEA 16a]. The Spanish decommissioning and waste management company Enresa has begun to dismantle the dome of the containment building for this reactor. The underwater segmentation of the reactor vessel was complex. It was undertaken in the former spent fuel pool, using remote-controlled cutting tools. Dismantling should generate about 104,000 tons of material, and about 4% of them will be classified as radioactive waste [WNN 19m].

In Sweden, the Rhinghals power plant has one BWR and three PWR reactors, operated by Ringhals AB which is 70% owned by Vattenhall. Reactors 1 and 2 were scheduled to be shut down in 2019 and 2020, and reactors 3 and 4 in 2040. Unit 2 (PWR of 907 MWe net) of the Ringhals nuclear power plant in Sweden was permanently shut down in December 2019, ending 44 years of operation and 215 TWh of production. Ringhals 1 (BWR of 878 MWe net), which started industrial operation in 1976, was expected to close in 2020. Two newer and larger PWR units at the Ringhals site, Ringhals 3 and 4, are expected to remain in service until 2041 and 2043 according to current plans [WNN 19u].

6.3.5. *The dismantling of WWER reactors*

The Soviet WWER (Water-Water Energy Reactor) was of a design close to PWRs with less safety. In addition to the two Russian prototypes, 17 reactors were built in Armenia, East Germany and Slovakia (Table 6.6). For the WWER reactors, the "impromptu" decommissioning project of Greifswald in Germany is used as a basic reference.

With the reunification of West and East Germany, East German reactors were shut down and decommissioned. Upon the accession of Bulgaria, Lithuania and Slovakia to the European Union, the latter made the shutdown of Soviet nuclear reactors a condition. European credits were allocated to these three countries to carry out complete decommissioning. These concern four Bulgarian WWER reactors in Kozloduy, two Slovak WWER reactors in Bohunice and a Lithuanian RMBK reactor in Ignalina.

The key dates for the decommissioning of the Kozloduy WWER reactors are as follows. The shutdown of reactors 1 and 2 occurred on December 31, 2002, and they were in a transition phase from 2002 to 2012. Reactors 3 and 4 were shut down at the end of 2006 and were in the transition phase from 2007 to 2012. Spent fuel was removed from all four reactors on December 31, 2012. Entry into the final phase of dismantling began in 2013–2014 and safety from 2015 to 2049. The complete dismantling and the release of the site are planned for 2049, 2059 and 2061 [THI 17j]. Similarly, Slovakia has its two WWER reactors, Bohunice VI-1 and VI-2, in the process of being dismantled [NEA 16a].

Country	Reactor name	Operation dates	Reactor type	Power (MWe)
Armenia	Medzamor-1	(1976–1988)	WWER-440W-270	408
Bulgaria	Kozloduy-1	(1974–2002)	WWER-440W-230	440
Bulgaria	Kozloduy-2	(1975–2002)	WWER-440W-230	440
Bulgaria	Kozloduy-3	(1980–2006)	WWER-440W-230	440
Bulgaria	Kozloduy-4	(1982–2006)	WWER-440W-230	440
Germany	Rheinberg-1	(1966–1990)	WWER-70	70
Germany	Greifswald-1	(1973–1990)	WWER-440W-230	440
Germany	Greifswald-2	(1974–1990)	WWER-440W-230	440
Germany	Greifswald-3	(1977–1990)	WWER-440W-230	440
Germany	Greifswald-4	(1979–1990)	WWER-440W-230	440
Germany	Greifswald-5	(1989–1989)	WWER-440W-213	440
Germany	Stendal-1 to 4	Never	WWER-1000W	1000
Russia	Novovoronezh-1	(1964–1988)	WWER-440W-210 prototype	210
Russia	Novovoronezh-2	(1969–1990)	WWER-440W-365 prototype3	65
Slovakia	Bohunice-V1-1	(1978–2006)	WWER-440W-230	440
Slovakia	Bohunice-V1-2	(1980–2008)	WWER-440W-230	440

Table 6.6. *Shutdown nuclear power reactors of the WWER-type (source: [BAE 12, NEA 16a, WNN 19s])*

The European Court of Auditors regularly audits the use of these credits and the state of progress of the decommissioning of the seven Soviet-built reactors. Since 2011, progress has been made on the decommissioning of the nuclear power plants of Ignalina in Lithuania, Kozloduy in Bulgaria and Bohunice in Slovakia. Essential components in the uncontrolled areas of the plants have been dismantled, but the three member states have yet to address the critical challenges of working in the controlled areas, including the reactor buildings. Although the authorities in these member states ensure that the plants are irreversibly shut down, not all of the expected accomplishments, against which the Commission assesses progress towards irreversible closure, have been fully realized. The estimated cost of decommissioning at the three plants will amount to at least €5.7 billion in total, or even double if the cost of final disposal is included [COU 16j].

The cost of decommissioning the five-unit Greifswald nuclear power plant (KGR) is estimated at 1,653 million euros, including 833 million euros for operations still to be carried out [BAE 12].

The mass of concrete and the size of the buildings of the WWER reactors are large (Table 6.7), generating a lot of nuclear waste.

Building	WWER type	Mass of reinforced concrete (Mg) Tons	Interior area of the building (m^2)	Building volume (m^3)
Building I (reactors 1 and 2)	230	134,100	95,000	264,119
Building III (reactors 5 and 6)	213	311,379	125,000	465,585

Table 6.7. *Concrete masses of WWER reactor buildings (source: [BAE 12])*

The Greifswald nuclear power plant (KGR) is operated by EWN (Energiewerke Nord GmbH). It is located 20 km north-east of Greifswald. Its construction started in 1970. Units 1 to 5 began operation in 1973, 1974, 1977, 1979 and 1989, respectively. All units were shut down in 1990, and the decommissioning license was obtained in 1995. EWN is carrying out the world's largest decommissioning project of eight Russian WWER-440 pressurized water reactors.

In 1972, construction of the Bohunice V1 plant began, with two WWER-440 version V-230 reactors (adapted to Finnish safety requirements with a gross power of 440 MWe) supplied by the Russian company Atomenergoexport and Skoda. The first was connected to the grid in December 1978, and the second 15 months later (March 1980). Despite extensive upgrading work on the two V1 units, these were shut down at the end of 2006 (unit 1) and 2008 (unit 2) as a condition for Slovakia's accession to the European Union.

In 1976, construction began on two V-213 reactors (the V2 plant) built by Skoda. The V2 units began operating in 1984 and 1985. All of them were designed by Atomenergoproekt. From 2005 to 2008, the operator Slovenské Elektrárne (*SE*) carried out a major modernization program of the two Bohunice V2 units to improve seismic resistance, cooling systems and instrumentation and control systems, with a view to extending the operational life to 40 years (2025). This was followed by a progressive upgrade program for both units, eventually increasing the capacity of each unit from 440 to 505 MWe gross (472 MWe net) in November 2010. SE plans to extend the licenses of the V2 units to 2045 after the upgrade.

Westinghouse Electric Company has signed a contract with Jadrová a vyraďovacia spoločnosť (JAVY) for dismantling of the cooling systems of the two WWR-440 units of the Bohunice V1 nuclear power plant in Slovakia. The project is financed by the European Bank for Reconstruction and Development (EBRD). The contract, announced in September 2017, covers the decontamination, dismantling and cutting prior to dismantling of the power reactor pressure vessels, as well as various other internal reactor components [WNN 17e].

6.4. Dismantling the heavy water sector

A Pressurized Heavy Water Reactor (PHWR) is a nuclear reactor that uses natural uranium as fuel and heavy water (D_2O) both as a coolant and as a moderator. Reactors that are currently shut down and decommissioned are listed in Table 6.8.

Today in Canada, five nuclear power plants in three provinces house 22 nuclear reactors. These are the Bruce (Ontario), Pickering (Ontario), Darlington (Ontario), Gentilly-2 (Quebec) (recently shut down) and Point Lepreau (New Brunswick) nuclear power plants. Approximately 15% of the country's energy comes from nuclear power. All nuclear reactors in Canada are CANDU reactors.

In Canada, decommissioning activities are regulated by the CNSC (Canadian Nuclear Safety Commission), which regulates the entire lifecycle of nuclear power plants. It is the licensee who decides to shut down the reactor and is responsible for decommissioning activities. The CNSC ensures that these decommissioning activities are carried out in accordance with its regulatory requirements in order to protect workers, the public and the environment and to respect Canada's international commitments on the peaceful use of nuclear energy.

Country	Reactor name	Power (MWe)	Number of years of operation and shutdown date	Decommissioning status
Germany	MZFR	57	18 (1984)	
Canada	Douglas Point	218	17 (1984)	
	Gentilly-2	675	30 (21012)	
	Pickering-A2	542	30 (2004)	Early 2020
	Pickering-A3	542	36 (2008)	
	Rolphton NPD	25	25 (1987)	
United States	Carolina-Virginia Tube	19	4 (1967)	License terminated
India	Rajasthan-1	100	44 (2014)	
Sweden	Agesta	12	10 (1974)	

Table 6.8. *Pressurized Heavy Water Reactors (PHWRs) shut down and decommissioned (source: [ASP 16a])*

On average, decommissioning plans for nuclear power plants span 50 years. At the same time, the operator files for a nuclear power plant operating license and a decommissioning plan that outlines how it intends to manage the decommissioning of its plant. The CNSC verifies the financial guarantee based on the plan. The operator may place the nuclear facility in a guaranteed shutdown state as an initial step in decommissioning.

The Gentilly-2 nuclear power plant in Bécancour, Quebec, whose operation recently ended, will soon be placed in a guaranteed shutdown state. Units 2 and 3 of the Pickering Nuclear Generating Station in Pickering, Ontario, are already in a guaranteed shutdown state.

Three prototype nuclear reactors, built and owned by Atomic Energy of Canada Limited, were shut down and prepared for decommissioning in the 1980s at Rolphton and Kincardine (Ontario) and at Bécancour (Gentilly-1) [SAB 19].

A CANDU reactor is responsible for radioactive contamination by four main radionuclides: ^{124}Sb (6%), ^{95}Zr (14%), ^{95}Nb (29%) and especially ^{65}Zn (51%). Decontamination tests were carried out on a heat exchanger at the Bruce nuclear power plant that had been contaminated by neutron-activated corrosion products, as well as fission products. Decontamination factors between 8 and 12 and between 10 and 50 were achieved on the tube bundle and tube sheet, respectively [LAC 80].

Among the heavy water reactors, the 106 MWe (gross) Niederaichbac reactor was the first power reactor to be dismantled in the world. Its dismantling was completed in 1995 [AEN 10].

In the United States, the decommissioning of the PHWR Carcinas Virginia Tube Reactor has been completed. On the other hand, in Sweden, dismantling of the Agesta reactor (PHWR) has been deferred [NEA 16a].

6.5. Dismantling of the boiling water reactor sector

Boiling Water Reactors (BWRs) that have been shut down and are being decommissioned worldwide are listed in Table 6.9.

Country	Reactor name	Net and gross power (MWe)	Number of years in operation (year of shutdown)	Decommissioning status
West Germany	Brunsbueffel	771/808	35 (2011)	
	Gundremmingen A	237/250	11 (1977)	
	HDR Grosswelzheim (proto)	25/27	1 (1971)	Finished
	ISAR-1	21 878/912	34 (2011)	
	Kruemmel	1,346/1,402	28 (2011)	
	Lingen	183/268	10 (1979)	
	Philippsburg	890/926	31 (2011)	
	VAK Kahl	15/16	24 (1985)	Finished
	Wuergassen	640/670	22 (1994)	Finished
Spain	Santa Maria de Garola	466	(1966–2013)	Fuel removed
United States	Big Rock Point	67/71	35 (1997)	ISFSI
	BONUS	17/18	4 (1968)	Entombed

Country	Reactor name	Net and gross power (MWe)	Number of years in operation (year of shutdown)	Decommissioning status
United States	Dresden-1	197/207	18 (1978)	Deferred
	Elk River	22/24	5 (1968)	Finished
	Humboldt Bay	63/65	13 (1976)	Deferred 30 years
	Lacrosse	48/55	19 (1987)	Deferred
	Millstone-1	641/684	28 (1998)	Deferred
	Pathfinder	59/63	1 (1967)	Finished
	Pilgrim	677/711	47 (2019)	
	Shoreham	820/849	3 (1989)	Finished
	Shippingport	60/68	25 (1989)	Completed (5 years)
	Vallecitos	24	6 (1963)	Deferred
	Vermont Yankee	605/635	42 (2014)	Deferred
Iraq	Osirak/Tammuz	40	Destroyed by Israel (1981)	
Italy	Caorso	860/882	3 (1987)	SAFSTOR
	Garigliano	150/160	(1982)	SAFSTOR (30 years)
Japan	Fukushima Daiichi-5	760/784	(1977–2011)	
	Fukushima Daiichi-6	1,067/1,100	(1979–2011)	
	Fukushima Daini-1	1,067	(1981–2011)	
	Hamaoka-1 and 2	515/540 and 806/840	26, 25 (2001–2004)	
	Japan Power Demonstration Reactor	12/13	13 (1976)	Finished
The Netherlands	Dodewaard	55/60	28 (1997)	SAFSTOR (40 years old)
Sweden	Barsebäck-1, 2	2 x 600/615	24, 28 (1999–2005)	SAFSTOR (2020)

Table 6.9. *Boiling Water Reactors (BWRs) shut down and decommissioned (source: [NEA 16a])*

In the United States, the BWR reactors under deferred decommissioning are Dresden-1 (2036), Humbolt Bays-3 (2017), La Crosse (2026), Millstone-1 (2056), Vallectos (2023) and Vermont Yankee (2020). The Big Rock Point reactor is only decommissioned under ISFSI. Decommissioning is complete for the Elk River, Shoreham and Superheat BWR Pathfinder reactors. Burial was carried out for the BONUS reactor [NEA 16a].

The Vermont Yankee reactor was shut down in December 2014. It was in a transition period until 2018 and has since been placed in a safe storage phase for 10 years. The end of decommissioning is planned for 2074 [NEA 18a].

Pilgrim, a 690 MW BWR located in Plymouth, Massachusetts, was closed in May 2019 after 47 years of operation. In June 2019, GE Hitachi Nuclear Energy (GEH) was scheduled to decommission the reactor's internal components and reactor pressure vessel. The dismantling contract was awarded to Holtec International and the joint venture, SNC-Lavalin Comprehensive Decommissioning International [WNN 19r].

In July 2019, GEH was awarded a contract to decommission the internal components and pressure vessel of the Oyster Creek reactor in New Jersey [WNN 19p].

EnergySolutions is currently carrying out decommissioning projects at the La Crosse Nuclear Power Plant (BWR) in Wisconsin. This project was expected to be completed in early 2020 [WNN 19j].

In Germany, the 250 MWe Gundremmingen A (KRB-A) BWR reactor operated from 1966 until its shutdown in 1977, following an incident. Dismantling began in 1983 and was well advanced in 2010 [AEN 10]. In 1989, the dismantling produced more than 6,200 tons of material. The European Commission had chosen this KRB-A reactor as a pilot project for dismantling for its 1989–1993 Research and Technological Development (RTD) program on the decommissioning of nuclear installations. The Commission, by sharing the costs in specific parts of the project, intended to promote the use of advanced techniques, and the conduct of collateral investigations, in order to improve the generation of knowledge and experience useful in subsequent decommissioning tasks. In particular, the generation of specific data on costs, working hours, dose received during the work and the amount of secondary waste generated was considered an important objective of this pilot project [EU 05b].

In Switzerland, the Mühleberg nuclear power plant ceased production in December 2019 after 47 years of operation. The Swiss BWR reactor had a power of 373/390 MWe and had started operating in 1972. It had generated about 130 TWh of electricity, supplying about 5% of the country's electricity demand. It will be the first Swiss nuclear power plant to be decommissioned [WNN 19s].

In Sweden, the two boiling water reactors, designed by ABB in Barsebäck, are located about 30 km from Malmö in southern Sweden. They were shut down in November 1999 and May 2005, respectively. The two 600/615 MWe reactors, which began operating in 1975 and 1977, were shut down earlier than planned due to political pressure from neighboring Denmark. All nuclear fuel, which contained most of the radioactive substances, has now been removed from the Barsebäck reactors.

Under the terms of the contract signed in 2015 with the operator of the Barsebäck Kraft AB plant, Westinghouse is to dismantle, segment and package the internal components of the reactor pressure vessel for final disposal. Westinghouse said it would implement its proven, remotely controlled underwater mechanical cutting techniques. The project began immediately in 2016 and was expected to take approximately four years [WNN 15b].

EDF subsidiary Cyclife Sweden AB signed an eight-year contract in 2019 with Uniper for the treatment and recycling of contaminated metals generated by the dismantling of Oskarshamn 1 and 2 and Barsebäck 1 and 2. Uniper has a common strategy for dismantling and demolition of the reactors, which were shut down between 1999 and 2017. According to Uniper, parts of the reactors – roughly 6%, which represents about 67,000 tons of material – are radioactive and must be removed first during decommissioning. Under the contract with Cyclife, OKG plans to treat between 1,500 and 2,500 tons of waste each year until 2027. The total cost of decommissioning the Barsebäck and Oskarshamn units is estimated at US$1.1 billion [WNN 19i].

The three BWR reactors at Oskarshamn 1, 2 and 3, with respective powers of 473, 638 and 1,400 MWe (net power), started up in 1972, 1974 and 1985, respectively. GE Hitachi Nuclear Energy (GEH) has completed segmentation of the internal components of the reactors of Units 1 and 2 of the Oskarshamn Nuclear Power Plant in Sweden. The pressure vessels of the units' reactors will also be segmented. This work, scheduled to begin in 2020, will mark the beginning of the large-scale decommissioning and demolition of Oskarshamn 1 and 2 [WNN 19r].

In Japan, the Hamaoka 1 and 2 reactors are both boiling water reactors, which started up in 1973 and 1978, respectively. Unit 1 had a capacity of 515/540 MWe, while Unit 2 had a capacity of 806/840 MWe. Chubu Electric Power Company decided to shut down both units in January 2009 because they required costly modifications to meet the seismic standards imposed after the Niigata Chuetsu Sea earthquake in July 2007. Japan's nuclear regulator gave Chubu the green light to begin removing peripheral equipment from Units 1 and 2 of its Hamaoka nuclear power plant in the Shizuoka prefecture. The overall plan divided the decommissioning of the two units into four stages. The first stage was devoted to preparation of the decommissioning work of the units, while the second stage concerns the dismantling and removal of the equipment surrounding the reactors. The third stage concerns the dismantling of the reactors themselves, while the fourth covers the demolition of the buildings of the two units. Stage 2 began in 2016. During this phase, the company began dismantling the equipment surrounding the reactors – including the exhaust chimneys, steam pipes, turbines and main condensers – and continued to conduct contamination and decontamination investigations of the reactor pressure vessels. This phase is expected to be completed in 2022. Chubu plans to complete the decommissioning of Hamaoka 1 and 2 by 2036.

After the 2007 earthquake, then Prime Minister Naoto Kan decided to close Units 4 and 5 in Hamaoka and not restart Unit 3, which was already offline for regular inspections, until the sea defenses had been reinforced. The enhanced safety measures were completed in September 2016 for Unit 4 and September 2017 for Unit 3. Chubu was studying the necessary measures for Unit 5 [WNN 16c].

Units 1 and 2 of the Ohi Nuclear Power Plant in Fukui prefecture began operation in March 1979 and December 1979, respectively. Kansai Electric Power Company announced in December 2017 that it would not seek permission to restart the two 1,120/1,175 MWe pressurized water reactors. Japan's Nuclear Regulatory Authority approved the decommissioning plan for the two reactors in December 2019. The decommissioning plan for Ohi 1 and 2 will last approximately 30 years and will be carried out in four stages. The first stage, lasting approximately eight years, will consist of preparing the reactor for dismantling (including the removal of all fuel and monitoring for radioactive contamination), while the second stage, lasting 10 years, will consist of dismantling the major and peripheral equipment. The third stage, which will take about six years, will involve the demolition of the reactor itself, while the fourth stage, which will take about three years, will see the demolition of all the remaining buildings and the release of the land for other uses. The disposal of fuel in the first stage includes 76 unused fuel assemblies and 323 assemblies used in Unit 1, and 140 unused assemblies and 306 assemblies used in Unit 2. A total of 400 tons of high-level radioactive waste is expected to result from the decommissioning of Ohi 1 and 2, as well as 2,850 tons of low-level waste and 20,240 tons of very low-level waste; an additional 13,200 tons of non-radioactive waste will also be generated by the clean-up of the site [WNN 19x].

In Italy, the Italian nuclear power plant of Caorso was the largest and most advanced in the 1970s and 1980s. It was an 860/882 MW BWR reactor, which was definitively shut down on July 26, 1990. The transition from passive protection to deactivation was accelerated. Spent fuel was stored dry on a dedicated site. The Marzano decree of December 2, 2004, maintains the reactor at the SAFSTOR stage [GUE 06a]. At the Garigliano power plant (BWR, 150/160 MWe), dismantling produced 187,000 t of conventional materials, 76,000 t of radioactive materials that can be decontaminated and reused and 3,100 t of non-reusable (2%) radioactive materials [GUE 06b].

In Taiwan, the Democratic Progressive Party (DPP) was elected to the government in January 2016, with a policy of creating a "nuclear-free homeland" by 2025. Under this policy, Taiwan's six power reactors will be decommissioned when their 40-year operating licenses expire. Construction of two units at Lungmen began in 1999 but the project has faced political, legal and regulatory delays. The dismantling of Unit 1 has been completed, and it was mothballed in July 2015. However, the deconstruction of Unit 2 was suspended in April 2014 [WNN 18g].

Taipower announced that the operating license for Chinshan 1, a 604/636 MWe boiling water reactor commissioned in December 1978, had expired in December 2018. The unit has generated approximately 162.5 TWh of electricity since its start-up. Under Taiwanese regulations, a decommissioning plan for a power reactor must be submitted by the operator three years prior to the planned final shutdown. Decommissioning must be completed within 25 years of obtaining a decommissioning license from the regulator. Taipower's decommissioning plan for Chinshan 1 was approved by the Atomic Energy Commission (AEC) on June 28, 2017. AEC submitted an Environmental Impact Assessment (EIA) in January 2016 covering the disposal of spent fuel and other materials to the Environmental Protection Administration for review and approval. The decommissioning license for Chinshan 1 will be granted by the CEA once the EIA is approved. The decommissioning plan for Chinshan 1, encompassing Unit 2 whose operating license expired on July 15, 2019, includes the construction of a dry storage facility for used fuel. Construction of this facility was completed in 2013. However, the New Taipei City municipal government has not yet issued a license for its use. Decommissioning of the unit cannot begin until all 816 fuel rods have been removed from the reactor core and placed in a nuclear reactor's pool. The final phase with the recovery of the site should be completed by the end of 2043 [WNN 18g].

6.6. Dismantling following a nuclear accident

Decommissioning when a nuclear installation has suffered a serious accident is a very special case. This is due to the fact that the event is sudden – by definition unforeseen and unprepared for – where the situation is difficult to assess. The site is inaccessible and urgent measures must be taken to protect personnel, the neighboring population and the environment. In addition, the technical means available are generally inadequate to manage the vast area affected by the accident.

In a serious accident, the nuclear installation is disrupted, the known information about the internal structures is no longer valid. In the case of a nuclear reactor, when dismantling of the core began, not only was the core not evacuated as it would be in a normal situation but it was also totally or partially melted down in the form of a highly radioactive corium. Its actual volume may have expanded considerably into the environment. The soil around the damaged facility may be heavily contaminated and may require decontamination over large areas. Two previous volumes of this collection have detailed the environmental and health consequences of military accidents [AMI 18] and civil accidents [AMI 19a].

Three nuclear accidents (Three Mile Island, Chernobyl and Fukushima) provide partial feedback to assist in future decommissioning. On the other hand, feedback on short-term safety measures is obvious [CEA 17, AMI 19b].

At Three Mile Island, because of the high radioactivity at the site, no dismantling is envisaged for another four to five years. The first inspection of the interior of the tank did not take place until 1982 and its opening under water until 1984 (i.e. five years after the accident with an irradiation dose of 35 $mSv.h^{-1}$ in the hall). The evacuation of the 20 tons of debris and the 342 fuel assemblies was carried out from the end of 1985 to 1986. The information collected during this partial dismantling should be used for subsequent dismantling, in particular following the shutdown of TMI-1 at the end of 2019.

In Chernobyl, corium is a shapeless magma of complex composition and no action has yet been taken. However, the risk of criticality is low [CEA 17]. Thirty years after the accident, preparations for dismantling began under the shelter of a new enclosure (arch) but no timetable is available.

In Fukushima, the accident simultaneously devastated four of the site's six nuclear reactors. The first actions were, after their automatic shutdown, to ensure the safety of the reactors and access to the devastated areas. It was also the evacuation of the debris from the explosions, the emptying of the storage pool of reactor No. 4 with the construction of an overhanging hall. In a second stage, there was the construction of an underground frozen wall, offering protection to a depth of 27 m, surrounding the four reactors. The purpose of this wall was to prevent the water that cools the corium from escaping into the sea. Finally, considerable volumes of liquid effluents, contaminated by a multitude of radionuclides, had to be purified. The decontamination of soils over a large area created a huge volume of nuclear waste, the management of which is gradually being put in place [AMI 19a].

The scale of dismantling in Fukushima is considerable. To prepare for all these dismantling operations, a dedicated institute has been created, the IRID (International Research Institute for Nuclear Decommissioning). The IRID has focused on three projects. The first is the removal of fuel from the pools of the four reactors. The second is preparation for the extraction of corium debris, which is probably at the bottom of the containments and therefore difficult to access. Finally, the third is the treatment, storage and dispatch of liquid and solid radioactive waste to dedicated storage centers to be built. IRID's budget for the last few years has been about 120–127 million euros.

The removal of spent nuclear fuel from the pool of reactor No. 4 was very urgent. It began on November 18, 2013, and its completion was planned by TEPCO for the end of 2014 [ZER 14]. Initially planned to begin in 2023, the removal of spent fuel from the storage pool of Reactor 1 will not begin until 2027 or 2028, and that of Reactor 2 has also been postponed to a window from April 2024 to March 2026. Each of these tasks must last two years. The Japanese authorities foresee the duration of the complete dismantling of the plant to be about 40 years, which seems very optimistic.

The number of accidents in nuclear reactors is relatively large (Table 6.10); six reactors have extensive damage (Three Mile Island, Chernobyl and four Fukushima reactors). This list is not exhaustive, and some authors include other accident-affected reactors [NEG 17, AMI 19a].

Country	Reactor	Type	MWe	Operational	Shutdown
Germany	Greifswald-5	WWER-440	408	0.5	1989
	Gundremmingen A	BWR	237	10	1977
Spain	Vandellos-1	GCR	480	18	1990
United States	SREc	FBR	6.5	7	1959
	Three Mile Island 2	PWR	880	1	1979
France	Saint-Laurent-A2	UNGG	465	9 + 3	1980d
Japan	Fukushima Daiichi-1	BWR	439	40	2011
	Fukushima Daiichi-2	BWR	760	37	2011
	Fukushima Daiichi-3	BWR	760	35	2011
	Fukushima Daiichi-4	BWR	760	32	2011
	Monju	FNRa	246	1	2016
Switzerland	St-Lucens	GCHWRb	6	3	1966
Slovakia	Bohunice-A1	GCHWRa	93	4	1977
Ukraine	Chernobyl 4	RBMK	925	2	1986

Table 6.10. *List of damaged reactors (source: [LAR 12a, WNN 19l, AMI 19a]) a: prototype; b: experimental; c: SRE Sodium Reactor Experiment; d: temporary shutdown*

6.6.1. *IAEA and NEA recommendations*

The IAEA has conducted studies and analyzed the dismantling of accident-affected reactors in order to make recommendations. It appeared that the IAEA should strengthen its clean-up program after a nuclear accident to help Member States facilitate the return of affected areas to normal conditions. The accident at the Fukushima Daiichi power plant highlighted people's concern for their safety. The international community should strive to develop a working definition of the term "safe" to facilitate communication with the public. The IAEA should assist Member States in developing desired end states and decommissioning strategies for accident facilities. Many volumes of radioactive waste and materials containing residual amounts of radionuclides are present in many countries. The IAEA should review its guidelines on the management of such waste and materials, with a view to ensuring their management in the event of a nuclear accident [LAR 13].

The IAEA [IAE 19d] analyzed the results of the decommissioning of eight accident-affected reactors: Windscale Reactor No. 1 (Pile), Bohunice-A1, TMI-2, Chernobyl-4 and Fukushima Daiichi-1 to 4. Each accident is unique with specific post-accident decommissioning issues and challenges, and requires a specific research and development (R&D) program.

The IAEA notes that no final state of decommissioning has yet been reached in any of the eight cases. Safe storage (care and maintenance) has been established for the Windscale Pile-1 and TMI-2 reactors. The Chernobyl nuclear power plant is moving towards an environmentally safe system, and the Bohunice-A1 nuclear power plant at the Jaslovské site is moving towards decommissioning to a brownfield site. These provide useful experience with many technical variants that could be implemented and completed in the future. Valuable and unique examples of R&D results, such as innovative characterization techniques, special robots and other remote handling systems or advanced decontamination and dismantling procedures, can inspire and help future nuclear decommissioning worldwide [IAE 19d].

The final decommissioning states of Bohunice and TMI-2 are defined. The final decisions for the Windscale Pile-1, Chernobyl and Fukushima Daiichi reactors can be flexible and should be left to future generations.

The importance of effective management of the large quantities of radioactive waste produced during decommissioning activities should be stressed [IAE 19d].

Within the OECD NEA, members of the EGFWMD (Expert Group on Fukushima Waste Management and Decommissioning) presented basic and contextual information on decommissioning and waste management at five sites, the Chernobyl Nuclear Power Plant (Ukraine), the Three Mile Island nuclear power plant (United States), the temporary storage site for spent fuel and radioactive waste in Andreeva Bay (Russia) and the Windscale Pile fire site (United Kingdom), as well as the Fukushima Daiichi Nuclear Power Plant (Japan). The main problems concern the health risks for workers and the local population and the management of nuclear waste. Each accident is different. The details of any post-accident scenario (after an emergency) are unpredictable and specific to the existing circumstances. The management of decommissioning and radioactive waste after a major accident may also require a different approach than that used after planned normal operations.

The NEA notes that the dismantling of an accident-damaged reactor requires centralized authority and stakeholder involvement. An implementation strategy must be proposed. Dismantling must be optimized. This optimization is an important aspect of radiation protection, and it is preferable to do it taking into account social and economic factors, and not only radiological factors, for example, to reach reference levels. It is essential that the storage and disposal of nuclear waste are planned and assured. Safety analyses and radiological and environmental impact assessments are required to support the identification of priorities, identify feasible management options and select preferred options from among feasible alternatives. In any decommissioning of an accident-affected reactor, international cooperation is rewarding [NEA 16b].

6.6.2. *The dismantling of Three Mile Island*

All licenses and assets of Unit 2 of the Three Mile Island Nuclear Power Plant (TMI-2) in Pennsylvania will be transferred to a subsidiary of EnergySolutions under an agreement signed in 2019 by US utility FirstEnergy and EnergySolutions. The reactor, which underwent a partial meltdown in 1979 and never reopened, will then be decommissioned. In the early 1980s, 99% of the nuclear fuel was removed from TMI-2 (this figure does not take into account the fact that 45% of the total fuel has melted and is currently in the form of corium without having pierced the vessel), packaged and transported to a storage facility at the Idaho National Laboratory site. The unit has remained in a safe and stable storage condition for the past 26 years. To perform the decommissioning work at TMI-2, EnergySolutions and the New Jersey-based construction company Jingoli formed a joint venture called ES/Jingoli Decommissioning LLC [WNN 19j].

6.6.3. *The dismantling of Chernobyl*

The dismantling of Chernobyl has not really started yet. Very soon after the accident, a sarcophagus (Shelter Object) was hastily built above reactor No. 4. Despite consolidations in 2004 and 2008, this structure threatened to collapse. Thus, a giant arch structure (NSC, New Safe Confinement) was built to enclose the damaged reactor and the sarcophagus in order to reduce the consequences of a collapse while allowing the dismantling of the sarcophagus under controlled conditions. Built at a cost of 1.5 billion euros ($1.7 billion) and financed by 45 donor countries and institutions, the arch was built in two halves near the site of the accident between 2010 and 2016, and was finally placed over the damaged reactor in November 2016. It is the largest mobile land structure ever built, with a span of 257 m, a length of 162 m, a height of 108 m and a total weight of more than 36,000 tons, equipped with a crane system for future dismantling work. The keys to the arch were symbolically handed over to the Ukrainian authorities in June 2019 [WNN 19b]. The project manager for the new sarcophagus was the French joint venture NOVARKA, equally owned by the French companies VINCI Construction Grands Projets and Bouygues Travaux Publics. It is important to note that the irradiated fuel from the three other Chernobyl reactors had not been evacuated. The last reactor was shut down in 2000.

Beforehand it was necessary to carry out dismantling work on the old ventilation chimney shared by units 3 and 4 of the Chernobyl nuclear power plant in Ukraine. This was indispensable to allow the installation of the giant protective structure (arch) on the damaged reactor. This ventilation chimney was 75.5 m high and had a diameter of 9 m. It comprised seven sections weighing a total of 330 tons. The top of the chimney was 150 m above ground level. The dismantling work was carried out by Ukrtransbud Corporation using a giant crane with a carrying capacity of 1,600 tons. The removal of the entire pile was completed on December 10, 2013 [WNN 13].

To dismantle the old sarcophagus, a contract was signed on July 29, 2019 between the Chernobyl nuclear power plant and the construction company Ukrbudmontazh (UBM). The cost of the operation is 70 million euros. This sarcophagus still houses the molten core of the reactor and about 200 tons of highly radioactive material. UBM will carry out the work in three stages. The first two stages include inspection of the sarcophagus and design work. The third stage of the work includes the acquisition and installation of equipment for the dismantling and handling of the dismantled fragments. This last stage of the work will be the most complicated [IAE 19b].

At Chernobyl, the dismantling will be carried out under extremely difficult conditions because the irradiation levels are measured in sieverts per hour.

6.6.4. *The decommissioning of Fukushima*

At Fukushima, the cores of slabs 1, 2 and 3 melted and formed a material called corium. It is believed that this corium is currently distributed between the reactor vessel and the base of the vessel. The Japanese government has decided to dismantle these damaged plants, which implies in particular the cutting and recovery of the corium. A French team has been in charge of R&D on corium laser cutting technology, and its principal results are presented by Journeau and Roulet [JOU 18].

The test withdrawal of fuel debris from the damaged reactors of the Japanese nuclear power plant of Fukushima Daiichi will begin at Unit 2 in 2021. The removal of fuel from Unit 1 has been postponed from four to five years and from one to three years for Unit 2. The removal of all fuel assemblies from Units 1 to 6 is expected to be completed in 2031. The installation of a large cover on Unit 1 to suppress dust diffusion is also expected to be completed by that time. The volume of contaminated water generated at the plant has already been significantly reduced from approximately 540 $m^3.d^{-1}$ in May 2014 to an average of 170 $m^3.d^{-1}$ in 2018. The schedule has been revised and maintains the current target of reducing production of contaminated water to 150 $m^3 \cdot d^{-1}$ in 2020, but sets a new reduction target of 100 $m^3 \cdot d^{-1}$ in 2025. The Ministry noted that the overall timeline for decommissioning the Fukushima Daiichi power plant is unchanged and will be completed within 30 to 40 years [WNN 19t].

The French group ORANO recently signed a cooperation agreement with the Japanese electricity company Tokyo Electric Power Company (TEPCO) to strengthen their cooperation in the framework of the decommissioning of nuclear facilities at the Fukushima Daiichi site. ORANO will share its experience in providing safe and innovative technological solutions for the design, manufacture and operation of nuclear waste recovery and treatment facilities [WNN 19f]. The waste generated by the Fukushima accidents is very large (Table 6.11).

Type of waste	Dose rate ($mSv.h^{-1}$)	Volume stored (m^3)
Debris (metals, concrete)	< 0.1	115,600
Debris (metals, concrete)	0.1–1.0	31,400
Debris (metals, concrete)	1.0–30	19,700
Debris (metals, concrete)	> 30	6,200
Debris (metals, concrete)	Total	172,900
Felled trees	Total	85,100
Protective clothing	Total	66,000

Table 6.11. *Volume of nuclear waste produced at the end of 2015 in Fukushima, after the accident (source: [NEA 16b])*

The NEA's Radioactive Waste Management Committee (RWMC) established the Fukushima Waste Management and Decommissioning R&D Expert Group (EGFWMD) in 2014. The EGFWMD was established with the primary purpose of advising on the management of large quantities of waste at the Fukushima Daiichi site that have complex properties and sharing experiences with the international community and NEA member countries [NEA 16b].

6.6.5. *Decommissioning of other damaged nuclear facilities*

At Vandellòs, phase 1 of the dismantling was carried out from 1991 to 1997 and phase 2 was finished in June 2003. Since then, the reactor has been in a latency phase for 25 years. Of the waste, more than 15,000 t were reused, 78,000 t of concrete were stored on the site and 1,763 t of low and medium activity were stored in the El Cabril storage center [NEA 11].

At Sellafield, the cesium extraction plant suffered a serious accident (INES, level 3). To decontaminate and dismantle this plant, a 900-ton mobile module and a remote-controlled robot were used. This was the first remote decommissioning activity undertaken in areas with extremely high irradiation doses [SLA 13].

Also at Sellafield (Windscale), the dismantling of the Pile One Chimney, building B6, was undertaken. The 1957 fire caused significant contamination of the chimney. It is a concrete structure 120 m high and 14 m in diameter. It is located at a very constrained site location. Decommissioning began in 1988 with the removal of the upper concentrator, filters and support structures, insulation (bricks) and liner boxes and metal support structures. Decommissioning stopped in 2003 following a fatal accident and resumed in 2007. The main difficulty was due to the fact that the work was carried out at a height of 120 m. The scope of the work can be understood with the following figures: 5,000 tons of concrete (main duct), 100 tons of reinforcing bars (main duct), 600 tons of concrete (diffuser), 15 tons of reinforcing bar (diffuser), 750 tons of concrete (filter) and 20 tons of reinforcing bar (filter). This dismantling presented a radiological challenge. Indeed, the average dose at the base of the chimney was 433 $\mu Sv.h^{-1}$ and in the chimney 250 $\mu Sv.h^{-1}$. Although the doses were not significant, the duration of the work resulted in significant individual doses. However, the problem was essentially one of waste due to the large quantities, in particular tritium whose external levels (mass contamination) exceeded the current UK exemption limits. Therefore, all materials needed to be disposed of as radioactive waste. This resulted in the very large flooding of the low-level waste repository [SLA 13].

6.7. Future reactor shutdowns

Following the accident at Fukushima, safety regulations were tightened in all countries. However, the oldest reactors required too much investment to be upgraded. The states decided to shut them down and then decommission them.

France wants to reduce the share of nuclear energy (currently 72%) in the country's electricity production. The French government, in its Multi-Year Energy Program, states its intention to close 14 nuclear reactors by 2035. EDF agrees to close reactors, but not nuclear power plants. Also, it proposes that the reactors be shut down in pairs on sites with at least four production units. Each of these plants should therefore close two reactors. The plants selected are Bugey, Chinon, Cruas, Dampierre, Gravelines, Le Blayais and Tricastin. These 14 closures of nuclear reactors are thus in addition to the closure of the Fessenheim power plant, already scheduled for September 2020 at the latest for reactor number 1 and August 2022 for reactor number 2 [GOT 20]. These last two reactors are in fact currently closed; the first was shut down on February 22 and the second on June 29, 2020.

Scheduled nuclear reactor closures for Belgium (Table 6.12), Japan (Table 6.13), South Korea (Table 6.14), Taiwan (Table 6.15) and the United States (Table 6.16) extend from 2012 to 2029.

Reactor	Type	Power (MWe)	Start of operations	Scheduled closing
Doel 1	PWR	433	1974	2025
Doel 2	PWR	433	1975	2025
Doel 3	PWR	1,006	1982	2022
Doel 4	PWR	1,038	1985	2025
Tihange 1	PWR	962	1975	2025
Tihange 2	PWR	1,008	1982	2023
Tihange 3	PWR	1,038	1985	2025

Table 6.12. *The official and planned shutdown of Belgian nuclear reactors (source: [SCH 19])*

In Germany, the nuclear reactors Biblis-A and B (PWR, 1,167 and 1,240 MW, 1974 and 1976, respectively), Brunsbüttel (BWR, 771 MW, 1977), Isar-1 (BWR, 878 MW, 1976), Krümmel (BWR, 1,346 MW, 1983), Neckawestheim-1 (PWR, 785 MW, 1976), Philippsburg-1 (BWR, 890 MW, 1979) and Unterweser (BWR, 1,345 MW, 1978) lost their licenses on August 6, 2011. The Grafenrheinfeld reactor (PWR, 1,275 MW, 1981) lost its license on December 31, 2015, Gündremmingen-B (BWR, 1,284 MW, 1984) lost its license on December 31, 2017 and Philippsburg-2

(PWR, 1,402 MW, 1984) on December 31, 2019. The Brokdorf PWR (1,410 MW, 1986), Grohnde (PWR, 1,360 MW, 1984) and Gündremmingen-C (BWR, 1,288 MW, 1984) reactors will follow on December 31, 2021, followed on December 31, 2022 by the Isar-2 (PWR, 1,410 MW, 1988), Emsland (PWR, 1,329 MW, 1988) and Neckarwestheim-2 (PWR, 1,310 MW, 1989) reactors [SCH19].

Reactor	Type	Power (MWe)	Start of operations	Scheduled closing
Fukushima Daiichi 1 to 4	BWR	439, 760, 760, 760,	1970, 1973, 1974, 1978	2012
Fukushima Daiichi 5, 6	BWR	760, 1,067	1977, 1979	2014
Mihama-1 and 2	PWR	320, 470	1970, 1972	2015
Ohi-1, 2	PWR	2 x 1,120	1979, 1979	2018
Genkai-1, 2	PWR	2 x 529	1975, 1980	2015 and 2013
Ikata-1, 2	PWR	2 x 538	1977, 1982	2016, 2018
Monju	FBR	246	1995	2017
Tsuruga-1	BWR	340	1970	2015
Shimane-1	PWR	439	1974	2015
Onagawa-1	BWR	498	1984	2018

Table 6.13. *The official and scheduled shutdown of Japanese nuclear reactors (source: [SCH 19])*

Reactor	Type	Power (MWe)	Start of operations	Scheduled closing
Kori-1	PWR	576	1978	2017
Kori 2	PWR	640	1983	2023
Kori 3	PWR	1,011	1985	2024
Kori 4	PWR	1,012	1985	2025
Hanbit 1	PWR	955	1986	2025
Hanbit 2	PWR	988	1986	2026
Wolsong 2	PHWR	611	1997	2026
Hanul 1	PWR	966	1988	2027
Wolsong 3	PHWR	641	1998	2027
Hanul 2	PWR	2,967	1989	2028
Wolsong 4	PHWR	622	1999	2029

Table 6.14. *The official and scheduled shutdown of South Korean nuclear reactors (source: [CEA 19, SCH 19])*

Reactor	Type	Power (MWe)	Start of operations	Scheduled closing
Chinshan-1, 2	BWR	2 x 604	1978, 1978	2018, 2019
Kuosheng-1, 2	BWR	2 x 951	1981, 1982	2021, 2023
Maanshan-1, 2	BWR	2 x 890	1984, 1985	2024, 2025

Table 6.15. *The official and planned shutdown of Taiwan's nuclear reactors (source: [SCH 19])*

Reactor	Closing date	Duration of operation (years)
Oyster Creek	2018	49
Pilgrim	2019	47
Diablo Canyon-1, 2	2025	40
Fort Calhoun	2016	43
Palisades	2022	51
Indian Point-2, 3	2021	47/44
Three Mile Island-1	2019	45
Beaver Valley-1, 2	2021	45/34
Davis Besse-1	2020	43
Perry	2021	35.5

Table 6.16. *Official and scheduled shutdown of US nuclear reactors (source: [SCH 19])*

6.8. Conclusions

The dismantling of nuclear power reactors represents a very important market worldwide for various reasons. The first is that many power reactors are reaching the end of their lives. In the next 30 years, there will be a large number of nuclear power plant closures. The second reason is that, following the serious nuclear accidents of Chernobyl and Fukushima, a number of states have decided to give up the use of nuclear power to produce electricity. Two states have gained some experience in decommissioning and dismantling. They are the United States and Germany.

The decommissioning and dismantling of power reactors are very long and very costly operations that give rise to large volumes of radioactive waste of various categories.

Most states choose the strategy of immediate decommissioning with the exception of the United States, which in some cases has opted for deferred decommissioning with safety of the nuclear facility.

Decommissioning and dismantling are different depending on the nuclear industry. Some sectors generate significantly greater quantities of nuclear waste than others. This is particularly the case for those processes whose moderator is graphite, such as the first generations of French (UNGG), British (Magnox) and Soviet (RMBK) reactors. Solutions for the storage of graphite waste do not yet exist. Similarly, heavy water-cooled reactors generate large quantities of tritiated waste that no nation knows how to control. These difficulties lead to serious delays in decommissioning time. In some cases, it is future generations who will have to complete these operations, with the resulting technical difficulties and financial costs.

On the other hand, the dismantling of PWR reactors is easier with more reasonable volumes of radioactive waste. The boiling water reactor produces more radioactive waste than the PWR reactor.

The greatest challenge in the field of decommissioning and dismantling is, without doubt, the case of severely damaged nuclear reactors such as those at Chernobyl and Fukushima. These have indeed become veritable chaos, and the mapping of radioactivity zoning will be difficult to achieve. In particular, the spent fuel has melted, mixing with the reactor structures and the concrete to form a corium that will be difficult to split and evacuate to storage centers.

7

The Decommissioning of Research Reactors and Other Basic Nuclear Facilities

7.1. Introduction

In addition to power or nuclear reactors (Chapter 5), many other reactors have been built throughout the world. It is customary to distinguish between prototype reactors (Brennilis, Phénix), lead reactors, fourth generation reactors (FNRs) and experimental or research reactors.

Nuclear reactors are not the only nuclear facilities to be decommissioned and dismantled. There is also a multitude of other facilities. These are mainly facilities located upstream in the nuclear fuel cycle such as uranium and thorium mines, conversion, enrichment and nuclear fuel fabrication plants. Some are gigantic plants containing long-life radionuclides of the uranium family.

For some nations, there are also the plants at the end of the fuel cycle, i.e. the spent fuel reprocessing plants. These are, by far, the most polluting plants for the environment. While all the states possessing nuclear weapons have developed military reprocessing plants, only two nations, France and Russia, continue to operate this type of facility for civilian purposes. This is due to two main facts. Reprocessing is expensive, dangerous and polluting, and uranium ore is of low value. This makes the recovery of uranium and plutonium economically unprofitable and without military value because of the stocks already acquired by many states possessing the atomic bomb. Secondly, reprocessing was originally envisioned to supply fuel for the fast neutron reactor industry. However, in the great majority of projects, this sector has mainly encountered safety difficulties, but also technical and economic difficulties.

There are a large number of nuclear facilities around the world. The example provided in Table 7.1 illustrates the case of France which had, at the end of 2013, 140 BNIs, operated mainly by three manufacturers.

Location	EDF	CEA	AREVA	ANDRA	Others	Total
Operation	58	22	10	2	10	102
Decommissioned	9	21	7	0	1	38
Total	67	43	17	2	11	140

Table 7.1. *Number of French nuclear facilities at the end of 2013 (based on [NEA 16a])*

According to the ASN's 2019 annual report, as of December 31, 2019, the number of BNIs (as defined as legal entities) in France was 124 and 55 had been decommissioned [ASN 19d].

7.2. The dismantling of experimental reactors around the world

Experimental reactors are nuclear facilities in which experimenters create and maintain a chain reaction to obtain a flow of neutrons. This neutron flux will support the various applications being experimented with. The energy produced by fission reactions is generally not recovered. The first experimental atomic pile, which came into existence on December 2, 1942, was done so under the direction of Enrico Fermi. It was installed in the basement of a stadium in Chicago. The first French atomic pile, ZOÉ, was created in 1948 in Fontenay-aux-Roses.

7.2.1. *The main roles of experimental reactors*

Experimental or research reactors are mainly used as a source of neutrons for research and development in the nuclear power industry. They have three main functions: to validate neutron calculations, to test materials under irradiation and to study accident situations.

A research reactor can also be used to train personnel in the nuclear power industry, in nuclear medicine for the production of medical radioisotopes or in the military nuclear industry. In France, for example, these are the ULYSSE, MINERVE, ISIS and AZUR reactors. Some reactors produce radionuclides for medical and industrial use. This is the case of OSIRIS in Saclay, which has supplied 95% of the ^{99}Mo in the world. Other reactors serve as tests and prototypes, such as the RES (*réacteur d'essais*, test reactor) at Cadarache to develop compact reactors for naval nuclear propulsion [CEA 12]. This project has experienced many delays. It was supposed to start at the end of 2008, then October 2010 and finally on October 10, 2018.

The validation of neutronics calculations at the CEA is currently being carried out on three so-called critical models using very low power (EOLE, MINERVE and MASURCA) located at Cadarache. This is also the case in Belgium for BR-1.

Experimental reactors are used to study the behavior of materials and nuclear fuels in the face of neutronic, thermal-hydraulic or chemical stresses representative of the full-scale operation of an industrial power reactor. The structural materials or components of the fuel assembly must meet demanding specifications. This type of study is carried out, for example, at the CEA with the OSIRIS reactor (closed at the end of 2015), ORPHEE and the Jules Horowitz reactor (RJH) at the Institut Laue-Langevin (ILL), which is under construction. Other examples of neutron beam reactors include HFIR (United States), MURR (United States), BER-II (Germany) and JRR3M (Japan).

Experimental reactors are the preferred equipment for studying the behavior of nuclear fuels in the event of a nuclear accident. The CEA have used three reactors for this type of study (CABRI, PHEBUS and SILENE) [CEA 12b]. CABRI has been shut down since 2000, renovated and restarted in October 2015. PHEBUS at Cadarache has been shut down since 2009. There is little information on the SILENE reactor which was commissioned in 1974 at CEA-DAM in Valduc.

7.2.2. *The global overview of experimental reactors*

Since the early 1950s, more than 670 experimental reactors of all types have been built worldwide. In 2012, 234 of them were in operation in 58 countries, the others being permanently shut down, dismantled or awaiting dismantling [CEA 12]. In 2005, 287 research reactors were operational and nearly 600 were either shut down or decommissioned (Table 7.2). The vast majority consist of small reactors with a power of less than 1 MW (Table 7.3). Most are less than 40 years old, but the number of reactors over 40 years old is far from negligible (Table 7.4). Of the 434 research reactors being decommissioned in 2005, 207 decommissionings were completed with unrestricted site use status (Table 7.5).

Around the world, 100 experimental reactors are in Europe, 66 are in North America and 59 are in Asia. Russia (48) and the United States (41) have the most research reactors [CEA 12].

At the end of 2019, the IAEA listed 841 research reactors, of which 444 were decommissioned and 66 in the process of being decommissioned [AIE 20]. In France, 40 research reactors have been built. Currently, seven are still in operation [COU 19].

State	Developed country	Developing countries	All countries
Planned	2	11	13
Under construction	4	7	11
Operational	133	87	220
Temporarily shutdown	11	4	15
Extended shutdown	5	8	13
Definitive shutdown	46	13	59
In decommissioning stage	63	3	66
Decommissioned	413	31	444

Table 7.2. *Status of research reactors (according to [AIE 20])*

Power	Number of reactors
1 kW	306
1 kW–1 MW	245
1–5 MW	86
5–10 MW	52
> 10 MW	123
Unknown	20
Total	**832**

Table 7.3. *Power of research reactors (from [IAE 06])*

Age (year)	Number of reactors
< 40	205
40	76
Unknown	6
Total	**287**

Table 7.4. *Age of research reactors in operation (based on [IAE 06])*

Final status	Number of reactors
Planned decommissioning	
Unlimited use	4
Safe enclosure	2
Unknown or undecided	18
Decommissioning in progress	
Unlimited use	55
Safe enclosure	29
Unknown or undecided	44
Full decommissioning	
Unlimited use	207
Safe enclosure	26
Unknown or undecided	4
Decommissioning status unknown	45
Total	**434**

Table 7.5. *Final status at the end of the decommissioning of research reactors (from [IAE 06])*

7.2.3. The main types of experimental reactors

Several types of research reactors have been in operation. They can be classified according to the following six main types: pool or open basin reactors, tank reactors, argonaut reactors, homogeneous liquid reactors, fast reactors and the last category includes graphite reactors and all other reactors [IAE 17b]. The IAEA provided a breakdown of the various types of research reactors in the world in 2005 (Table 7.6).

For its part, the IRSN [COU 19] distinguishes five classes of research reactors: irradiation reactors or so-called MTR (Material Testing Reactor) reactors, reactors equipped with "neutron channels" (often called "neutron exit beams"), critical models (or assemblies), safety study reactors and teaching reactors.

The CEA [CEA 17] and the *Club des exploitants des réacteurs de recherche français* (CER) (French Research Reactor Operators' Club) use another classification distinguishing between neutron beam reactors (ORPHEE, RHF-ILL), technological irradiation reactors (OSIRIS, Phénix), teaching reactors (ISIS, AZUR), research reactors for safety (CABRI, PHEBUS), neutronics study reactors (ÉOLE, MINERVE, MASURCA) and new research reactors (RES, RJH, ZEPHYR).

Reactor type	Number of reactors	Subtotal
Pool type		***253***
TRIGA	74	
SLOWPOKE/MNSR	19	
Other	160	
Tank type		***189***
Heavy water	48	
ARGONAUT	29	
Pressurized	22	
Other	90	
Homogeneous type		***170***
Homogeneous liquid	45	
Homogeneous solid	44	
Fast	37	
Graphite	44	
Other		***220***
Zero power	185	
Miscellaneous	28	
Unknown	7	
Total		**832**

Table 7.6. *Classification of research reactors (based on [IAE 06])*

Among the open basin reactors (pool or swimming), two types are in the majority: TRIGA reactors (Training, Research, Isotopes, General Atomics, supplied by General Atomics, USA) and SLOWPOKE reactors (Safe LOW-Power Critical Experiment). TRIGA is a category of small nuclear research reactors, designed and built by the American company General Atomics, as its name indicates, which are used for training and research purposes as well as for the manufacture of radioisotopes. The fuel used by this type of reactor is based on uranium–zirconium hydride (UZrH). SLOWPOKE is a low-energy pool-type nuclear research reactor designed by Atomic Energy of Canada Limited (AECL, now Canadian Nuclear Laboratories) at the end of the 1960s. The reactor, which uses a very small quantity of uranium fuel, is surrounded by a beryllium reflector that reflects neutrons back into the core, prolonging the chain fission reaction as much as possible. In France, the 14 MW experimental reactor ORPHEE is of the type of a graphite-moderated pool. It has been shut down since October 29, 2019.

In tank reactors, the arrangement of the core and reflector and the position of these components inside the tank are similar to those of the pool type. However, the solid concrete shield has increased robustness around the sides of tank type reactors, whereas the biological shield relies primarily on water for pool type reactors. Some reactors combine a tank within a pool such as the Soviet WWR (Water-Water Reactor or pool-in-tank) reactors.

Argonaut reactors (ARGOnne Nuclear Assembly for University Training) were used mainly for training. France had two Argonaut Ulysse reactors at Saclay and RUS at the University of Strasbourg. Ulysse operated with uranium fuel enriched between 20% and 90%, moderated by ordinary water and graphite reflectors. It has been shut down since August 2014 and its dismantling was completed in June 2020.

Homogeneous liquid reactors are reactors in which the coolant and moderator are mixed. One example is the molten salt reactor (MSR), a nuclear reactor concept in which the nuclear fuel is in liquid form, dissolved in molten salt (600–900°C) which acts as both a coolant and a containment barrier. The reactor can be moderated by graphite (thermal neutrons) or a function without a moderator (fast neutrons).

Homogeneous reactors use a solution of nitrate or sulfate of uranium (or even plutonium) as a moderator. Thus, the French reactor PROSERPINE at Saclay used plutonium in solution as a fissile material and was moderated with light water [COU 19].

Experimental fast reactors do not have a moderator that slows neutrons. An example is provided by the CEA's CALIBAN reactor.

Many other experimental reactors exist, such as the zero power reactors (ZPRs), or in French, the *maquettes critiques*, which are based on the concept of the first Enrico Fermi reactor, with a stack of graphite blocks or "piles".

7.2.4. *Major incidents and accidents involving research reactors*

There have been seven major incidents and serious accidents in research reactors causing partial core meltdown. In chronological order [COU 19]:

– December 12, 1952, the 42 MW NRX reactor at Chalk River Laboratories (Ontario, Canada);

– May 24, 1958, the 135 MW NRU reactor at Chalk River Laboratories (Ontario, Canada);

– October 15, 1958, the Boris Kidrich Institute zero-power research reactor (Vinca, Yugoslavia);

– January 3, 1961, the SL-1 reactor (3 MW) at the National Reactor Testing Area (Idaho, USA);

– December 30, 1965, the VENUS reactor (0.5 MW) (Mol, Belgium);

– November 7, 1967, the SILOE reactor (15 MW) (Grenoble, France); and

– September 23, 1983, the critical model RA-2 (Constituyentes, Argentina).

7.2.5. *Cost*

The cost of decommissioning a research reactor depends on many different factors, including the type of reactor, its age and operating history, the end-state of the facility, the reactor owner's level of decommissioning experience, the overall costs of labor and equipment in the country, its location and the national spent fuel and radioactive waste management infrastructure [IAE 17b].

The IAEA analyzed the cost of decommissioning 14 research reactors. For the five TRIGA reactors analyzed, decommissioning costs ranged from US$3.18 million (2017 value) (IPR-R1, Belo Horizonte, Brazil, 0.1 MWth) to US$12.4 million (KRR-2, Seoul, South Korea, 2 MWth). Two WWR reactors were compared, WWR-SM10 in Budapest, Hungary, and WWR-M in Kiev, Ukraine, both with a power of 10 MWth. Decommissioning costs were estimated at more than $2 million and $5 million, respectively, for WWR-M and WWR-SM10. Seven pool reactors were analyzed by the IAEA, and the cost varied from $3.4 million (Apsara, Bombay, India, 1 MWth) to $23.5 million (Astra, Austria, 10 MWth) [IAE 17b].

The financial consequences of delayed decommissioning are related to the ongoing monitoring and maintenance of the facility, possible changes in decommissioning technology in the event of dose reduction and changes in waste management processes [IAE 17b].

7.2.6. *Some examples of the decommissioning of experimental reactors*

Decommissioning activities for small facilities have some specific features. According to Griffiths [GRI 12], the main challenges are due to factors such as inadequate post-operational management, a lack of resources and funding, a lack of required knowledge and a lack of ownership of the problem.

The variety of experimental reactors being high, only a few examples will be discussed below.

7.2.6.1. *TRIGA reactors*

The FiR 1 TRIGA Mark II research reactor owned by VTT and located in Espoo, Finland, is being decommissioned. An estimate of the radioactive contamination of the structures is difficult to obtain, and a complete characterization of the site in terms of activity measurements must be carried out. This is why an inventory of activities has been determined by calculation using two software packages (MCNP and ORIGEN-S). However, their results require validation by physical measurements. These physical measurements are carried out using γ spectroscopy for three different types of samples, concrete, a fluent neutron moderator (Al+AlF3+Li mixture) and aluminum using the In Situ Object Counting System (ISOCSTM). For the concrete sample, the measured and predicted activities for ^{60}Co and ^{152}Eu are in agreement with an uncertainty of 11% and 2%, respectively. This new prospective is promising [ACK 17].

7.2.6.2. *Pool reactors*

One of the examples of decommissioned pool reactors is the first reactor installed at the Grenoble nuclear center, the Mélusine reactor. This small reactor of 680 m^2 (34 x 20 x 19 m) had a power of 1 MW which was increased to 8 MW. It was shut down in 1993 and clean-up began in 2002, followed by dismantling in 2004. Decommissioning was completed in 2009 with the reuse of the building without radiological constraints. 11% of the waste is low-level waste (LLW, about 100 Bq.g^{-1}) and 89% is very low-level waste (VLLW, < 100 Bq.g^{-1}). Debris and rubble account for the vast majority (99%) of demolitions [NEA 11].

The MINERVE research reactor (INB 95) at Cadarache was commissioned in 1977. It was of the pool type and had low power (100 watts). It was used for studies on neutronic phenomena that occur in nuclear fuels and was also used to train future reactor operators. It was shut down in December 2017, and currently clean-up and dismantling operations are being prepared [CLI 19].

7.2.6.3. *The WWER-S reactors*

The experimental reactor built between 1955 and 1957 in Romania, owned by Horia Hulubei at the National Institute for Physics and Nuclear Engineering (IFIN-HH), was of the WWER-S type with thermal neutrons, cooled by water, whose fuel was enriched to 10% and then 36%. It operated until 1987 and stopped definitively in 2002. The 30 m^3 buffer tank of liquid effluents belonging to the nuclear reactor was contaminated. For this research reactor in accordance with the approved decommissioning plan and radiological characterization, the radionuclides important for the underground structures were ^{241}Am, ^{137}Cs, ^{60}Co, ^{90}Sr-^{90}Y and ^{152}Eu. The total level of contamination (^{137}Cs and ^{60}Co) measured during the cutting operations was between 11 and 75 Bq.cm^{-2} [IOR 19].

7.2.7. *Heavy water research reactors (HWRR)*

Many experimental reactors have used heavy water as a coolant. This is the case for the research reactor ÉOLE (INB 42) at Cadarache, which was commissioned in 1965. It had a power of 100 MW. It was shut down in 2017 [CLI 19]. The unloading of the fuel and the removal of the moderator were carried out in 2018. The dismantling file was sent to ASN in July 2018. In 2019, according to ASN (29/05/3020), the removal began with radioactive or hazardous materials[1].

7.2.7.1. *The dismantling of Ispra-1*

Ispra-1 is a 5 MW heterogeneous thermal reactor located in Ispra, in the province of Varese, Italy. It was the latest version of the Chicago Pile series of research reactors developed by Enrico Fermi. The reactor was the first research reactor built in Italy by the National Council for Nuclear Research between 1957 and 1958. It operated between 1959 and 1973. Ownership of this research reactor was officially transferred to *Societa Gestione Impianti Nucleari SpA* (Sogin), the Italian state-owned company responsible for dismantling the country's nuclear power plants. Decommissioning of the reactor, which ceased operating in 1973, will begin once Sogin's request is approved. The decommissioning of Ispra-1 will be carried out in three phases with preliminary activities, dismantling of the reactors and final clean-up of the site [WNN 19d].

7.2.7.2. *The dismantling of the Chinese HWRR reactor*

China has approved the decommissioning plan for its first nuclear reactor, a heavy water research reactor (HWRR). Construction of the HWRR began in 1956 in a suburb of Beijing. It reached its first criticality in June 1958 and was definitively shut down at the end of 2007. The HWRR was a tank type reactor that used heavy water as its primary coolant and moderator. Its nominal power was 7 MW, while its enhanced thermal power was 10 MW. As a replacement, a new research reactor, the China Advanced Research Reactor, with a thermal power of 60 MW, was built at the CIAE and reached criticality in 2010. The dismantling of the first reactor will be undertaken by the China Institute of Atomic Energy (CIAE). The decommissioning of the HWRR will be done in three phases [WNN 19o].

HWRRs are the first type of nuclear reactor to be decommissioned in China. Immediate decommissioning was chosen as the decommissioning strategy. The reactor was unloaded and all spent nuclear fuel was moved to the storage pool for interim storage. The primary cooling system and the secondary cooling system were drained after the reactor was unloaded. The operational waste on the site was cleaned up [LI 19].

---

1 https://www.asn.fr/L-ASN/L-ASN-en-region/Provence-Alpes-Cote-d-Azur/Installations-nucleaires/Eole.

However, the decommissioning project is made difficult by the lack of design and engineering experience in China and by the long duration of the project (more than 10 years) with the associated uncertainties and financial risks. In addition, key techniques such as the dismantling of internal reactor components, the recovery and conditioning of graphite blocks, and the decontamination and decommissioning of high dose rate hot cells have not yet been identified [LI 19].

After the adoption of a precise schedule, the dismantling will be carried out in three periods. During the first period, known as the transition period (2008–2010) (the removal of fuel, the draining of coolant, the transport of spent fuel, the cleaning of operational waste, the reconstruction of special decommissioning facilities (e.g. ventilation and radiation protection systems)), a characterization survey and the application for a decommissioning license were carried out.

Then, implementation period I (2011–2015) included the dismantling and removal of reactor systems and equipment, the evacuation of water in the spent fuel pool and the decontamination of radioactive structures. The third period is implementation II (2016–2020) with the dismantling and removal of the reactor core, internal components, experimental tubes, graphite reflector and water tank, decontamination of the reactor concrete body, decontamination and restoration of the site [YAN 06].

7.2.7.3. *The decommissioning of the Indian Cirus reactor*

Cirus, located in the Indian center of Bhabha (Mumbai), was a 40 MW (thermal) reactor of the tank type using heavy water as a moderator, graphite as a reflector, demineralized light water as a primary coolant, and natural uranium metal as fuel. It had been in service since 1960. After about three decades of operation, the reactor showed signs of aging. The reactor was shut down from 1997 to 2003 for renovation on the basis of a complete evaluation of aging. It was restarted in 2003 with a significant increase in availability and safety margins and was shut down definitively on December 31, 2010. After the final shutdown of the Cirus reactor, the core was relieved of all irradiated fuel assemblies, isotope production assemblies, experimental assemblies and shut-off rods. For this reactor, the strategy of deferred decommissioning and dismantling was considered the best option. The dismantling of the reactor structure and core components is envisaged after about 30–35 years, when the dominant radionuclide ^{60}Co will have decayed significantly to facilitate the dismantling and handling of the radioactive components. The reactor has moved to a state of “safe storage” [RAN 18].

The main radionuclides present in the various systems vary. They are ^{60}Co, ^{65}Zn, ^{54}Mn, ^{109}Cd and ^{137}Cs in the reactor structures, ^{152}Eu, ^{144}Ce, 125Sb, ^{137}Cs, ^{106}Ru, ^{54}Mn and ^{60}Co in the primary cooling water system, traces of ^{60}Co and ^{3}H in D_2O and helium in the heavy water system, ^{60}Co, ^{110m}Ag and traces of ^{137}Cs in the pressurized water loop, ^{60}Co and ^{152}Eu in traces in the ventilation system and ^{60}Co, ^{137}Cs and ^{125}Sb in the waste disposal system [RAN 18].

7.2.8. *Fast neutron reactors*

The Los Alamos Molten Plutonium Experimental Reactor (LAMPRE I) was decommissioned at the Los Alamos National Laboratory, Los Alamos, New Mexico, in 1980. LAMPRE I was sodium-cooled and this reactor was designed to develop various plutonium fuels with applications for fast neutrons. It was shut down in the mid-1960s. A report precisely describes the procedures for decommissioning, health physics programs, waste management and the costs of the operation [HAR 81].

The HARMONIE reactor (formerly INB 41) was a fast neutron reactor located at Cadarache, commissioned in 1965 with a power of 1,000 MW. It ceased its activities in 1997 and was completely dismantled in 2007 [CLI 19]. It was decommissioned by ASN after a decision from March 31, 2009 [ASN 09].

7.2.9. *Other research reactors*

In 2006, the IAEA [IAE 06] produced a detailed report on the problems encountered during the decommissioning of research reactors. Numerous examples are provided for BR-3 (Belgium), JANUS and BGRR (USA), JASON (UK), PSI (Switzerland), MELUSINE and SILOE (France), ETRR-1 (Egypt) and FRF1/FRF2 (Germany).

As an example, a few experimental reactors from the Idaho laboratory (INL) were reported. At the INL, more than 52 research reactors were in operation. Among them, the decommissioning of the REE, MTR, PBF, ZPPR and EBR-II reactors was mentioned [SZI 17].

The engineering test reactor (ETR) was the largest and most advanced material test reactor in the world when it started up in September 1957. The 175 MW thermal test reactor specialized in the evaluation of fuel, coolant and moderator materials. In 1972, the ETR was modified by the addition of a sodium loop safety facility in the reactor core to support the DoE's breeder reactor safety program. The ETR was deactivated almost immediately after its shutdown in December 1981. The reactor was removed and disposed of in September 2007.

The materials test reactor (MTR) was, in 1952, the second reactor operated at the INL. This 40 MW reactor contributed to the design of pressurized water reactors, moderated by organic liquids and cooled by liquid metal. The materials testing reactor was shut down in April 1970 and was removed and disposed of in November 2010.

The power burst facility (PBF) reactor was started up in September 1972. It was part of the reactor safety test program and was designed to simulate various types of imaginary accident scenarios caused by rapid power changes (a few milliseconds). It simulated severe fuel rod burst tests and loss of coolant accidents. The data from its tests were used to develop and validate fuel behavior codes for the Nuclear Regulator Commission. The reactor was shut down in 1985 and then removed and disposed of in 2008.

The zero power physics reactor (ZPPR) was started up in April 1969. The ZPPR was a low-power physics reactor that provided physics data on the spectrum of all fast neutrons. The reactor consisted of an array of two separate honeycomb core assemblies. The fuel was loaded into the two lattice halves and moved together to reach a critical low-power condition. The data were then easily extrapolated to full power conditions. The reactor was put on standby in April 1992 and was removed and disposed of in 2009. Because of its unique design, the ZPPR reactor had relatively low contamination and irradiation. Furthermore, it was easily accessible and relatively uncomplicated to dispose of.

The Experimental Breeder Reactor-II (EBR-II) was built and began operations in September 1961. It was a liquid metal-cooled breeder reactor and its main purpose was to demonstrate the feasibility of on-site fuel reprocessing as a complement to a power plant composed of a fast reactor. These objectives were achieved during the first years of operation. The reactor operated at 62.5 megawatts and supplied the site's electrical network until its shutdown in 1994. The reactor is currently undergoing D&D and the first phases of sodium treatment have been completed. The planned completion of the reactor layout was scheduled for March 2012.

For all INL reactors, prior to their decommissioning, a risk-based shutdown model was applied. This model was implemented under the Comprehensive Environmental Response, Compensation and Liability Act (CERCLA) and the Non-Time Critical Removal Action (NTCRA) process, which evaluated several options. Three options are being considered: no action and therefore the maintenance of the reactor as is with long-term stewardship and monitoring (mothballing), the burying in place and the complete dismantling (removal) of the reactor.

Before starting large-scale D&D, hazardous components were removed, including cadmium, beryllium, sodium (passivated and elemental), PCBs, electrical components, lead, asbestos and mercury.

Each reactor had to be isolated in order to be decommissioned. Due to the activated materials in the reactor vessels, the dose rates above the cores varied from 50 to 1,200 $R.h^{-1}$. Subsequent dose rates outside the vessels ranged from 60 $mR.h^{-1}$ to more than 50 $R.h^{-1}$. Due to the high dose rates, the project team decided to fill the ETR and MTR vessels with grout at a level above the core area to reduce the dose.

To remove the ETR reactor, access to the base (feet) of the support was necessary. These feet were enclosed in a high-density concrete biological shield about 20 cm below ground level. The project team used explosives to remove the biological shield. The demolition had to be controlled to avoid damaging the reactor vessel and to limit the seismic impact on a nearby operating reactor. At the end of the blast, the concrete was removed, exposing the support legs of the vessel.

Two reactor buildings (ETR and PBF) had to be removed to accommodate the lifting systems for the reactor vessels. Two reactors (PBF and MTR) were removed using mobile cranes, two reactors were removed in pieces (ZPPR and MTR) and the ETR, because of its weight, was removed using a double gantry lifting system [SZI 12].

7.3. Decommissioning and dismantling of fourth-generation reactors

The Generation IV International Forum, launched in 2000 by the US Department of Energy (DoE), was born out of the desire to create an international R&D framework capable of catalyzing the research efforts of different countries to bring about the most efficient technologies more rapidly. Eleven countries (South Africa, Argentina, Brazil, Canada, China, France, Japan, Russia, South Korea, Switzerland and the United Kingdom) and the European Union have chosen to join this forum and pool their efforts to develop a new generation of nuclear systems.

Many possibilities were considered. A selection using eight top-ranked objectives with 17 criteria and 26 metrics was used to evaluate the Generation IV systems. At the end of this evaluation exercise, six systems were selected as the most promising and were called upon to federate the cooperative effort within the framework of the forum for joint development, starting in 2004.

The first three systems are the fast neutron reactors with three different types: the GFR (gas-cooled fast reactor system), the SFR (sodium-cooled fast reactor system) and the LFR (lead-cooled fast reactor system).

The three other systems are the helium-cooled very high-temperature reactors (VHTR), dedicated to hydrogen production or hydrogen/electricity cogeneration, the supercritical water-cooled reactors (SCWR) and the molten salt reactors (MSR). Other potential models exist and have been tested [CEA 06].

7.3.1. *The dismantling of the fast neutron reactor (FNR) industry*

It is this sector that has been the subject of more tests in the world. The experimental and research phase on FNRs began in 1949 with the CLEMENTINE reactor (25 MW) at Los Alamos and was followed by many other reactors. The coolant is sodium for the majority of the reactors, and NaK for EBR-I and DFR. Testing of this type of reactor continued with BR5/BR10 in 1958 in the USSR and with the British DFR reactor at Dounreay in 1959.

The first power reactors were BN 350 in the USSR in 1972, PHENIX in France in 1973 and PFR in the United Kingdom in 1974. However, in as early as 1990, many of these FNR reactors were shut down because economic competition with PWR and BWR reactors was strong and FNRs experienced many incidents. Thus, this sector was abandoned by the Americans and the British [CEA 14].

In 2012, out of all the fast sodium power reactors in the world, four prototype reactors were still in operation (BOR-60 in Russia, JOYO in Japan, FBTR in India and CEFR in China). Fifteen prototype and power fast reactors are permanently shut down (Table 7.7).

Three FNR power reactors experienced significant operating difficulties: BN 350 in Russia, PFR at Dounreay in the United Kingdom and MONJU in Japan. French reactors have also had their share of incidents [CEA 14]. The shutdown of Monju was due to a sodium fire and that of Fermi was due to a partial core meltdown. The SNR-300 reactor never emerged because of safety problems, and Clinch River was never built.

Currently, three fast neutron reactors supply an electrical grid. These are the Russian Beloyarsk-3 (BN-600, 560 MWe since 1981) and Beloyarsk-4 (BN-800, 820 MWe since 2015) reactors and the Chinese CEFR reactor (20 MWe). The BN-600 operates particularly well with an efficiency of 74%. Two reactors should soon be in operation; one is approaching the operational phase (the PFBR (Prototype Fast Breeder Reactor, 472 MWe) in Kalpakkam, India) and another is under construction in China, the CFR-600.

A significant number of fast reactors are being decommissioned and to date, all commercial fast reactors being decommissioned, built or planned are sodium-cooled.

Country	Reactor	Capacity MW	Operational	Yield (%)
Germany	*SNR-300*	0	None	-
	KNK-II	21	1977–1991	
United States	*RBA-I*	0.2	1951–1963	
	RBA-II	20	1961–1991	
	EFFBR (Fermi-I)	65	1963–1972	0.6
	SEFOR	0	1969–1972	
	Clinch River	0	1972 (none)	-
France	*Rhapsody*	40	1967–1983	
	Phénix	250	1973–2010	41
	Superphénix	1240	1986–1996	8
Japan	*Monju*	280	1995 (3 months)	-
United Kingdom	*DFR*	15	1959–1977	
	LIC	250	1974–1994	27
Russia	*BR-5BR-10*	0	1958–2002	
	BN-350	130	1972–1999	

Table 7.7. *Prototype of fast and power reactors permanently shut down around the world (from [FEI 14, CEA 14, NEA 16])*

7.3.1.1. *The French example*

France has built three sodium-cooled fast neutron reactors – Rapsodie, Phénix and Superphénix.

The demonstration reactor Rapsodie (RAPide SODIum) located at Cadarache (INB 25) was commissioned in 1967 and shut down in 1983 due to a leak in the reactor vessel. It is currently in the dismantling phase. Its nominal power was 20 MW, then 24 MW and 40 MW in 1970 then, following an incident in 1980, at 22.4 MW [CEA 14]. It was dedicated to research on "fast reactor" fuel and component technology. The installation was definitively shut down in 1983 because of a leak in the reactor vessel.

Preparatory work for dismantling was started after the final shutdown and was interrupted in 1994 following an explosion in a tank that contained sodium. Since then, the reactor has been shut down and is under supervision. Twenty-three tons of sodium contained in the installation were evacuated to Marcoule (Gard) for treatment. The dismantling file was submitted to a public inquiry in 2018 and the decree authorizing dismantling is in preparation [CLI 19]. Dismantling is expected to take several decades. The treatment of sodium that can ignite is one of the specificities of the project [IRS 18b].

The Fuel Assembly Cutting Laboratory (*Laboratoire de découpage d'assemblages combustibles*, LDAC), located within RAPSODIE, was responsible for carrying out controls and examinations on irradiated reactor fuel. This laboratory has been shut down since 1997 and partially cleaned up. The final clean-up is planned in the project for dismantling the entire Rapsodie BNI [CLI 19].

The Phénix reactor, with a nominal power of 250 MW, was operational for two campaigns (1973–2000 and 2004–2009). During its operation, there were numerous leaks of sodium (32 incidents), as well as five reactions between sodium and water. Its dismantling, which began in 2009, will be carried out on the basis of feedback from the Superphénix's dismantling [CEA 14].

The Superphénix (SPX) reactor was built at Creys-Malville (Isère) and was operational from 1985 to 1997. With a power of 1240 MW, it was to be operational for 30 years and was to be the leading reactor of the French fast neutron reactors (FBR). It is known to have been the largest fast neutron reactor in the world.

Three incidents resulted in a shutdown of the reactor. The first was a sodium leak from the cylinder in 1987 (18-month shutdown). The second was a sodium pollution incident by an air entry into the argon in 1990 (shutdown until 1994). The last incident was an argon leak on the feed tube of an exchanger bell in 1995. Also, the operating balance was mixed with 53 months of operation and 55 months of administrative shutdown [CEA 14].

These various technical and economic problems led to its premature termination in 1998. On site, the number of EDF staff and subcontractors was, respectively, 650 and 450 in 1999 and 85 and 3,500 in 2016 [NEA 18a].

The dismantling of SPX1 began in 1999. Phase 1 was the removal of the fuel from the core and all neutron protection. Phase 2 consisted of evacuating all the components removable from the primary vessel. Phase 3 neutralized the sodium (NOAH process) and transformed it into soda which was incorporated in concrete. These emptying operations were carried out from 2010 to 2017. They allowed the extraction and treatment of a total of more than 3,500 tons of sodium, and required the use of innovative treatment techniques [ANO 17b].

During phase 4, the sodium retention areas in the tank will be identified and cleaned. Phase 5 corresponds to a washing of the whole unit, followed by a watering of the tank. The dismantling of the internal structures can then begin. Its shutdown is planned for 2025–2030 [CEA 14].

The EDF Group began on Thursday, June 1, 2017, a major step in its dismantling process: the watering of the reactor vessel, a major technical milestone essential for the continuation of cutting operations. This operation is a real technological challenge, due in particular to the size of its components. Indeed, its vessel, measuring nearly 24 meters in diameter, is still to this day the largest nuclear reactor vessel ever built in the world [ANO 17b].

Among the fast neutron reactors, the CEA has a project for a new generation of FNR-G, where gas replaces sodium. This CEA project is the Reactor for Testing and Technological Development (REDT) [CEA 06]. The demonstration reactor FNR-G of the CEA is called Allegro (75 MW) [CEA 12]. This project remains the only FNR reactor project since the project for the fourth-generation FNR reactor ASTRID was abandoned in August 2019 for economic reasons (too low a cost for uranium).

7.3.1.2. *The American example*

The oldest of the American fast reactors have been decommissioned. Such is the case of the RBE-II (62 MWe). The reactor began operations in 1961 and was shut down in 1994. This reactor was the basis of the American program of integrated fast reactors (IFRs). Its fuel was quickly removed and transported to another facility on site for safe storage. The main clean-up contractor (CWI, CH2M-WG, Idaho, LLC) at the Idaho site of the US Department of Energy (DoE) placed this historic nuclear reactor in an *in situ* sarcophagus and processed the sodium coolant remaining in the reactor six months ahead of schedule and budget [WNN 15a].

The sodium reactor experiment (SRE) was built in Chatsworth, California, at the Santa Susana Field Laboratory near Simi Valley. It was operational from 1957 to 1964. It was the world's first commercial reactor. In July 1959, the reactor underwent a partial meltdown in which 13 of the 43 reactor fuel elements partially melted and a controlled release of radioactive gas into the atmosphere occurred. The reactor was repaired and restarted in September 1960. In February 1964, the SRE was in operation for the last time. The removal of the deactivated reactor was completed in 1981. Technical analyses of the 1959 incident produced contrasting conclusions regarding the types and quantities of radioactive materials released. Some researchers consider this nuclear accident to be the most serious on US soil [AMI 19a]. Members of neighboring communities expressed concern about the possible impacts of the incident on their health and the environment. In August 2009, 50 years after the event, the Department of Energy organized a community workshop to discuss the 1959 incident.

The decommissioning of the SRE near Los Angeles, California, was nearly completed in 1980. The reactor vessel and all contaminated support systems were removed in order to return the building and site to unrestricted use, the fuel was removed and all piping removed. Many of the cuts were made under water and remotely controlled [KIT 80].

The Enrico Fermi-1 fast neutron reactor is under deferred decommissioning in the United States until 2032 [PAR 16a]. However, EnergySolutions has completed the decommissioning of the Southwest Experimental Fast Oxide Reactor (SEFOR) in Arkansas, cooled by sodium with MOX as fuel [WNN 19j].

7.3.2. *High-temperature nuclear reactors*

One of the most represented sectors in the nuclear domain is that with graphite as a moderator, a gas as a coolant and nuclear fuel in the metallic form of natural uranium. The first generation of this type of reactor, where the coolant is CO_2, was developed mainly by France (UNGG) and the United Kingdom with the Magnox reactors. Only the United Kingdom has developed a second generation with AGR (Advanced Gas-cooled Reactors), where the nuclear fuel is in the form of uranium oxide enriched to between 2.1 and 2.6%. Fourteen reactors of this type exist in the United Kingdom, a fifteenth being in the dismantling process at Sellafield.

The third generation is known as high-temperature gas reactors (HTGR), where the gas is helium under high pressure. Seven reactors of this type are in operation around the world. In the United Kingdom, the Dragon reactor at Winfrith, operational from 1964 to 1975; in Germany, two reactors, AVR at Jülich (1966–1988) and the THTR300 reactor at Schmehausen (1983–1989); in China, the HTR10 reactor, which emerged in 2001. In the United States, two reactors, Peach Bottom in Pennsylvania (1966–1974) and Fort St Vrain in Colorado (1974–1989), and in Japan HTTR in Oarai (1998–present). All have been shut down except the Japanese and Chinese reactors (Table 7.8) [CEA 06].

Several projects were aborted in Germany, the THTR Modul reactor (200 MWth), in South Africa the PBMR, in the United States the MHTGR (350 MWth) and MHTGR/NPR reactors, and in Russia the GT-MHR reactor (600 MWth). In the latter, the fuel was plutonium or uranium.

In this sector, the largest volumes of waste are mainly due to graphite.

In this sector, the Peach Bottom-1 reactor is, in the United States, under deferred decommissioning until 2034 and the Fort St. Vrain reactor (330 MWe) has only been in ISFSI since 1992 [AEN 10, NEA 16a].

Country	Reactor name	Power MWth (MWe)	Operational	Decommissioning status
Germany	AVR Juelich	46 (15)	1966–1988	
	THTR-300	750 (308)	1983–1989	
China	HTR10	10	2001–present	
United States	Peach Bottom-1	115 (42)	1966–1974	Deferred
	Fort St. Vrain	840 (342)	1974–1989	ISFSI
Japan	HTTR Oarai	30	1998–present	
United Kingdom	Dragon Winfrith	20 (100)	1964–1975	Completed

Table 7.8. *Main prototype and power reactors of the high-temperature reactor type (from [CEA 06, NEA 16a]). ISFSI: Independent Spent Fuel Storage Installations*

7.3.3. *The other fourth-generation sectors*

The other fourth-generation sectors are numerous, and a multitude of projects have been tried. Some examples of decommissioning are provided in Table 7.9.

Country	Reactor name	Type and power (MWe)	Operational	Decommissioning Status
Germany	Niederaichbach	HWGCR (106)	1 (1974)	Completed
United States	Hallam	EGSR (84)	1 (1964)	Enseveli
	Piqua	EOMR (12)	2 (1966)	Enseveli
Switzerland	DIORIT	Experimental graphite-gas	17 (1960)	SAFSTOR
	Sapphire	Swimming pool (0.01–0.1)	39 (1994)	

Table 7.9. *Major prototype reactors and power reactors from various sectors (from [AEN 16a]). EGSR: experimental graphite-sodium reactor; EOMR: experimental organic moderated reactor; HWGCR: heavy water gas cooled reactor; SAFSTOR: SAFe STORage*

The heavy water reactor at Niederaichbach in Germany (100 MWe) was completed in mid-1995 [NEA 05].

The experience gained in the decommissioning projects of two American reactors, OMRE and Hallam (EGSR), moderated with graphite and whose coolant was sodium, gives people more confidence regarding the realization of future more difficult decommissioning projects. In particular, safe dismantling and handling procedures have been developed. The Hallam reactor was buried [MES 80]. Similarly, the Piqua reactor (EOMR), cooled and moderated by an organic liquid, was buried with the conversion of the building into a warehouse. The reactor vessel is buried in sand, and no continuous monitoring is planned [DEL 85].

The DIORIT reactor at the Paul Scherrer Institute in Würenlingen was moderated and cooled with heavy water and had a power of 30 MW. It was started up on August 15, 1960 and operated until 1977. It was the first reactor of Swiss design and construction. This reactor was probably involved in Swiss military research for the nuclear weapons project of this state. It was secured.

Müller Geri, representative of the Swiss Ecological Party in the Swiss National Council, filed an interpellation (06.3162) on March 24, 2006 concerning the incomplete documentation of the decommissioning of Swiss nuclear power plants [MÜL 06]. In his intervention, he pointed out that on March 1, 2006, during the work to dismantle the DIORIT reactor, workers working on the dismantling of this federal research reactor were unexpectedly and repeatedly exposed to asbestos, while the necessary protective measures had not been taken.

The pool-type SAPHIR reactor of the Federal Institute for Reactor Research started up in 1957 was acquired from the Americans who had exhibited it at the "Atoms for Peace" Conference in Geneva in 1955. It was operational until 1994. According to Müller [MÜL 06], in the expert report (HSK-Bericht 2/313, July 19, 1999) on the decommissioning of this experimental reactor, the Main Division for the Safety of Nuclear Installations attested that the Paul Scherrer Institute had extensive technical and methodological experience in the dismantling of nuclear reactors, experience acquired at the site of the DIORIT reactor (Chapter 1.3, page 4). The political representative was surprised to learn that the information on the dismantling of the DIORIT experimental reactor was apparently so incomplete that the necessary documents were no longer available, and was also shocked at the unexpected discovery of asbestos. Given that the operating life of the five large Swiss nuclear power plants can reach 50 years and more according to the current legal interpretation, the constructors and specialists present at the time of the construction may well no longer be available at the time of the decommissioning, if only for reasons of human longevity, as well as for reasons of the life expectancy of the companies.

7.4. The dismantling of first-generation prototype reactors

Usually, before building several reactors of the same fuel cycle, the authorities carry out tests with a prototype reactor of this fuel cycle, often of a reduced size, to acquire the maximum amount of information on the qualities and defects of this new fuel cycle. Once this mission has been accomplished, the prototype reactor is decommissioned, again often the first to test the best methods and techniques for dismantling.

7.4.1. *PWR reactors*

7.4.1.1. *Chooz A reactor in France*

Chooz A, in the Ardennes, on the banks of the Meuse, was the first pressurized water reactor in France to be decommissioned. With a power of 305 MWe, it has four primary loops. The particularity of Chooz A was its reactor and its main nuclear auxiliaries installed in rocky caverns on the hillside [RON 15].

The reactor was a scale model of the second-generation power plants and was therefore a prototype. After 24 years of operation, the reactor was shut down in 1991. A long process, both administrative and technical, then began for its dismantling [IRS 18a]. It should be noted that 16 years separated the shutdown of the reactor from the publication of the decree authorizing its dismantling.

The first stages of dismantling mainly involved the evacuation of the fuel, the emptying of the circuits and their decalorization. Following the choice made by EDF in 2001 to decommission as quickly as possible and the publication of the decree for the complete dismantling of the installation in 2007, the operations were followed by the demolition of the engine room and the installations located at the top of the hill; the preparation of the main support functions for dismantling (ventilation, waste storage area); the dismantling of the least active nuclear installations (fuel cavern and nuclear auxiliaries) towards the most active (dismantling of the primary circuit – outside the reactor vessel) and preparation for the underwater dismantling of the reactor vessel is underway [RON 15].

One of the most important worksites at Chooz A was the project to dismantle the Chooz A tank. After a call for tenders was issued in 2007, the contract was awarded in December 2009, the detailed design was completed at the end of 2011 and the execution studies were completed in mid-2016. The company moved to the site in December 2014 and the end of the construction site withdrawal was scheduled for the end of 2020 [BOU 15].

On March 8, 2017, the dismantling site entered its final phase, intended to cut the reactor vessel, the most sensitive part of the installation. It is in fact the key part of the installation; this steel tank weighs more than 220 tons and is immersed underground, at the bottom of a pool 8 meters deep and about 1,000 m^3. This particularity requires the operator to take maximum precautions to limit the exposure of the teams of engineers and technicians in charge of the cutting operations. To this end, this operation is carried out in total immersion. Remote-controlled robots equipped with circular saws allow each operation to be carried out remotely in conditions of optimal safety, security and radioprotection [ANO 17a]. The dismantling of the tank should be completed in 2021 [IRS 20].

The main lessons learned from Chooz A are the demonstration of the technical feasibility of dismantling a PWR plant, including in an unfavorable environment (cavern) and after a 16-year waiting and shutdown period. There is also the ability to decontaminate large components and circuits and the importance of preserving the memory of equipment and operating documents for the purpose of dismantling. Confirmation that dismantling must be conducted by a strategy based on nuclear waste management with the assurance of the existence of the channels and the approvals must be obtained before dismantling begins [RON 15].

7.4.1.2. *BR3 reactor in Belgium*

The Belgian Reactor 3 or BR3, which was a prototype pressurized water reactor, was operational between 1962 and 1987. It was the first reactor of this type in Western Europe and also the first to be decommissioned. It was located at the nuclear center of SCK CEN in Mol in Belgium.

It has been selected by the European Union as a complete dismantling demonstrator. The objective of the project was to illustrate its technical and economic feasibility. The dismantling operations produced a wealth of data concerning costs, waste production and radiation protection. This knowledge now makes it possible to make realistic calculations of the technical feasibility and financial impact of other dismantling projects. A sampling guide for the initial characterization of nuclear sites in constrained environments has been published. This project also made it possible to develop a methodology for characterizing and quantifying the radiological characterization of the biological shield of the BR3 reactor [BOD 17].

The dismantling of the BR3 reactor was carried out by Belgoprocess. The first task during the dismantling of the BR3 reactor was to remove and dispose of the nuclear fuels, and then the highly radioactive components such as the reactor vessel and the contaminated circuits, including the primary cooling circuit. Innovative cutting techniques were developed for this purpose. Special robots and remote-controlled equipment made it possible to keep the radiation doses to which employees were exposed to at an absolute minimum. Specific software was developed to produce a 3D image of the radiation environment.

The planning of the decommissioning was decided in advance and methods to minimize the amount of radioactive waste were applied, such as the decontamination of metal components.

The final part of a decommissioning project involves the decontamination and demolition of all buildings on the site. The goal here was to restore the land and return it to a greenfield site.

7.4.2. *The dismantling of the boiling water reactor (BWR) process*

At the Los Alamos site, several BWR prototypes were built in the TA-2 building. The first, called LOPO (Low Power), was built in 1943, then a second, HYPO (High Power), in 1944 and a third, SUPO (Super Power), in 1951. The fuel was highly enriched uranium (14% and then 88.7%). SUPO was stopped in 1974 and dismantling began in 1989. The external structures and the underground pipes associated with the gaseous effluent pipe (chimney) of the BWR reactor, located in Technical Zone 2 (TA-2), were removed in 1985–1986 during phase I of the decommissioning of the reactor. Phase II of the BWR decommissioning involved the biological shield, components within the biological shield and piping located in the floor of the reactor building. The cost of Phase II was approximately US$623,000. The decommissioning operation produced 173 m^3 of solid low-level radioactive waste and 35 m^3 of mixed waste [MON 91].

In Japan, in regard to the 10 MWe boiling water demonstration reactor, decommissioning was completed in 1996 [AEN 05a].

7.4.3. *The dismantling of the gas reactor sector (AGR)*

The prototype of the industrial scale development reactor for the British AGR nuclear power plants was the Windscale advanced gas reactor (WAGR). It was gas-cooled and built and operated by UKAEA (United Kingdom Atomic Energy Authority); it was operational from 1963 to 1981. The 100 MWth reactor stopped operating in 1981. From 1982 to 1983, spent fuel was removed from the reactor. The decommissioning plan (EC R&D Decommissioning Program) was obtained in 1989. The European Commission selected this reactor as a pilot decommissioning project for its R&D program on the decommissioning of nuclear installations. It was decided to decommission the WAGR to stage III, thus returning it to greenfield status.

The dismantling of the WAGR took place in 10 campaigns. The last four had not yet started in 2003. This operation required the use of a machine with two extendable arms (RDM, Remote Dismantling Machine) [BAY 03].

In 2005, the decommissioning of the Windscale reactor in the United Kingdom was still in progress [AEN 05a]. In February 2019, the demolition of the second Windscale chimney, 125 m high and the site of the most serious nuclear accident in Great Britain in 1957, began [AEN 19].

7.4.4. *Dismantling the heavy water industry*

This technology has been developed mainly in Canada with CANDU reactors. In France, a low-power industrial prototype has been built by the CEA in Brittany. Construction of the 70 MWe heavy water reactor began in 1962 in the Monts d'Arrée at Brennilis (Finistère). It was commissioned in 1967 and shut down in 1985, after only 18 years of operation. EDF was responsible for controlling operation and then dismantling. Its official name is EL 4 (INB 126) but is better known as the Brennilis reactor. Its operation produced the equivalent of the annual electricity consumption of the city of Paris.

The EDF group promised an exemplary construction site from an energy point of view and "greenfield" status as soon as possible. At the beginning of the dismantling, the clearance date of the Brennilis site was 2018. EDF now proposes 2022.

The first phases of decommissioning are proceeding as normal. Until 1992, the installations were first shut down with the emptying of circuits and the transfer of 100 tons of fuel to the Cadarache nuclear site (Bouches-du-Rhône). The second phase, planned until 2007, involved partial, non-nuclear dismantling. Some buildings were demolished, others were transformed [BEU 17].

In 2007, the Council of State came to a halt when it annulled the decree of total dismantling signed the previous year by then Prime Minister, Dominique de Villepin. The reason was a failure to communicate the impact study. In July, authorization for partial dismantling was obtained. From 2011 to 2012, EDF cleaned up the old effluent discharge channel [EDF 14].

In October 2012, a new request for complete dismantling was rejected by the Nuclear Safety Authority. The latter required a new dossier, not filed until July 2018, with the additions requested by the ASN in December 2019. This file is currently being examined by the ASN.

The storage of waste was not anticipated. Finally, EDF decided to build a specific center for dismantling waste from various nuclear sectors. This will be the conditioning and storage facility for long- and medium-level waste (ICEDA) near the Bugey power plant [COT 18]. But the ICEDA facility planned near the Bugey power station (Ain) has been delayed due to legal challenges [BEU 17]. The site produced 86 tons of TFA waste and evacuated 76 t TFA to Morvilliers and 46 t FA to Soulaines in 2019 [EDF 20b].

To date, the reactor block, which concentrates 90% of the radioactivity, is still standing. In 2019 and 2020, EDF carried out the preparatory phases for the complete dismantling of the core (securing, removal of asbestos, sampling from the reactor block) [EDF 20a]. The operations of taking samples from the reactor block were authorized by the ASN by a decision of September 20, 2019 and the public inquiry is planned for 2021.

For all heavy water-cooled reactors, the major problem is tritium contamination during normal operations [GAZ 10]. During dismantling, this problem is accentuated. At Brennilis, gaseous releases of tritium are between $1.2.10^{-3}$ and $1.7.10^{-3}$ TBq per month, or an annual release of $1.8–2.4.10^{-2}$ TBq, on the site [EDF 20a].

Not only is the dismantling delayed, but the cost of this operation is increasing significantly. The first estimate in 2001 was 254 million euros, and it rose to 377 million in 2007, according to the Court of Auditors. The total cost of decommissioning is now estimated at 482 million euros [BEU 17].

The Prototype MZFR (> 20 MWe) located in Karlsruhe was a 200 MWth (57 MWe) PWR, moderated and heavy water cooled, which was operational from 1965 to 1984. Its functions were the development and testing of nuclear materials, tests of heavy water systems, tests on fuel elements and personnel training. Decommissioning took place from 1987 to 2013 in eight stages with "greenfield" as the final state. The activity inventory was approximately $5.0.10^{16}$ Bq. The concrete structures were contaminated with tritium and required treatment. The cost of the project was approximately 320 million euros and the total expected collective dose was less than 1 man.Sv. The approximate waste masses were for 403Mg of primary waste (disposal), 95Mg of secondary waste (treatment), (including about 80 Mg of contaminated water) and 250 Mg of tertiary waste (new facilities, decontamination or reuse).

The dismantling benefited from a model used to carry out underwater cutting tests, with 3D simulations. The chosen cutting method was highly flexible, thanks to the separate use of the tool holder and the handling system.

The dismantling of the MZFR reactor was relatively exemplary. Decommissioning technologies are generally based on methods and processes applied in conventional fields. However, special developments and optimizations were necessary for boundary conditions and cost reduction. The remote dismantling of the pressure vessel with highly activated and contaminated components was the most technically difficult step. Early planning and demonstration of the interaction of handling systems and dismantling technologies using test benches reduced dismantling times and thus costs. Advanced planning tools (3D simulation, modern control and robotic techniques) have increased the overall availability of the system. The experience gained from the dismantling of prototype installations could well be used for the dismantling of industrial nuclear power plants [PRE 09].

7.4.5. *The dismantling of prototype reactors from various sectors*

The 165 MWe Japanese Fugen reactor (ATR, Advanced Thermal Reactor) was cooled with boiling water, moderated with heavy water and the fuel was in pressure tubes. It was fed only with the fuel of mixed oxides (MOX) of uranium and plutonium. The reactor was located at Myōjin-chō, in the town of Tsuruga, Fukui prefecture. It began commercial operation in March 1979 and was stopped in March 2003 after 25 years of operation. The decommissioning program for the reactor was approved in February 2008. The decommissioning project will last until at least 2034. Planning for decommissioning is done by computer simulation [IGU 17]. After preparation from 2003 to 2008, the actual dismantling has begun [JAE 20]. During the dismantling operations, it was found that the walls were not thick enough to ensure resistance for 25 out of 34 points. ORANO is in charge of reprocessing the 731 used MOX fuel assemblies from Fugen in France [WNN 18f].

The graphite-moderated and water-cooled reactor system has been tested by several Russian prototypes. The dismantling of the equipment of shutdown units 1 and 2 of the Beloyarsk nuclear power plant in Russia started in 2014. The decommissioning of the two prototype reactors is expected to take 18 years. Beloyarsk 1 and 2 were water-cooled, graphite-moderated slow neutron prototype reactors with a capacity of 100 MWe and 200 MWe, respectively. They are called ABM-100 and ABM-200 units. Unit 1 began operation in April 1964 and was shut down in 1981, while Unit 2 started up in 1967 and was shut down in 1989. The fuel has since been removed from both units and stored in an on-site storage pool. Demolition of reactors 1 and 2 in the main building will then begin and should be completed by 2032. The site of the reactors will then be cleared for other industrial uses [WNN 14b].

7.5. The dismantling of basic nuclear fuel cycle facilities

All nuclear fuel cycle facilities are shut down and dismantled at the end of their life. These facilities are many and varied (Table 7.10). The management of radiological risks during the decommissioning of nuclear fuel cycle facilities (NFC, Nuclear Fuel Complex) and the typical waste generated during decommissioning are sometimes significant [FOU 12].

Stage of the fuel cycle	Operation	Construction License pending Planned	Shutdown	Decommissioned	Expected	Other
U recovery	0	1	2	2	6	4
Conversion	27	3	9	11	1	1
Enrichment	17	3	4	11	2	2
U manufacturing	52	1	7	27	3	1
MOX manufacturing	7	2	5	11	2	1
Wet fuel storage	33	0	1	5	0	0
Dry fuel storage	66	21	0	0	0	2
Fuel reprocessing	9	2	0	27	3	5
Zirconium alloy	9	1	3	1	0	0
Zirconium tube	16	0	4	1	0	0
Heavy water production	7	0	2	5	1	2
Fuel conditioning	0	0	0	0	1	0

Table 7.10. *Number of nuclear fuel cycle facilities at various stages of their life (source: [FOU 12])*

7.5.1. *The dismantling of extraction mines*

Uranium extraction and concentration mines are abandoned when they are exhausted, economically at least. Many countries take care to renovate the site. This is the case, for example, in Canada at the Beaverlodge mine (Saskatchewan) where the operation ended in 1985. The same is true at the Elliot Lake Uranium Mine Site Reclamation, Serpent River Watershed, Ontario. It operated from 1950 to 1996. It was decommissioned from 1985 to 2000 [AEN 10].

7.5.2. *The dismantling of enrichment plants*

The main objective of remediation programs for uranium production facilities, underway or planned around the world, is to create stable and long-term conditions to ensure the safe use of the site by present and future generations. The remediation objectives are to return the affected areas to existing environmental conditions or to a land use that will be sustainable over the long term.

For these facilities, the main problems concern radon and radioactive dust emissions and water, both routine and underground, which may have been contaminated. The operations to be carried out are the same as for other facilities: site characterization; decontamination, dismantling and decommissioning; waste management; water remediation; long-term stewardship and monitoring; policies and regulations; and cost assurance [NEA 02].

Country	U_{nat} (mg.L^{-1})	^{226}Ra (Bq.L^{-1})	As (mg.L^{-1})	pH
Czech Republic	0.1	0.3	0.1	6–9
France	1.8	0.37	0.1	6.5–9
Germany	0.2–0.5	0.2–0.4	0.1–0.3	6.5–9
Spain	-	-	0.5	5.5–9.5
United States	-	0.18	0.05	6.5–8.5

Table 7.11. *Examples of effluent concentration limits (source: [NEA 02])*

In France, uranium enrichment took place in three plants, for military uranium at Pierrelatte from 1958 to 1996 (Chapter 5) and for civil uranium at Georges Besse I (EURODIF, 1977–2012) and Georges Besse II (2004–). Only the first two, based on gaseous diffusion, have been dismantled.

The Georges Besse I plant, called the EURODIF (European Gaseous Diffusion Uranium Enrichment Consortium) plant from 1978 to 1988, was operational from 1997 to 2012 and located on the Tricastin site. It definitively ceased production in 2012.

Since this shutdown, a specific program has been carried out to prepare for the final shutdown, prior to dismantling the installation. The aim of this program is to reduce the quantity of nuclear and chemical materials in the facility in order to reduce the risks during the dismantling phase. At the same time, a study was carried out on the management of some 900,000 m^3 of low-level waste (LLW) resulting from this operation and several scenarios for dismantling equipment and cleaning up buildings were defined [DUG 15].

The dismantling concerns the 1,400 floors of the diffusion cascade, in particular, which represents 160,000 tons of steel, 30,000 tons of equipment in various metals and more than 1,300 km of piping. The operations will be carried out by dismantling (unblocking, unbolting) at the level of the connections and by cutting. The waste generated during dismantling operations is estimated to be 205,000 tons of radioactive waste and 106,000 tons of conventional waste (excluding plant buildings). AREVA chose not to deconstruct the buildings of the four EURODIF Production plants so that they can be reused for conventional or nuclear industrial activities. It is true that the deconstruction of these 190,000 m^2 of buildings would generate more than 1,400,000 tons of concrete and cladding waste (i.e. the dismantling of six 900 MW nuclear power plants).

The Public Inquiry Commission issued a favorable opinion on the project for the permanent shutdown, the dismantling of the basic nuclear installation (BNI 93) and the requested modification of the perimeter [BLA 17].

However, the Commission has some reservations. First of all, it considers that the storage of waste in one or more ANDRA Centers constitutes a real test for this project. Indeed, the current storage capacities are limited given the other dismantling projects in the national territory. Therefore, the operator will have to obtain an exemption from the legislation in force for the implementation of an adapted melting device that would allow a significant reduction in the waste to be stored. It considers that the maintenance of the plant buildings, as recommended by the operator, deserves a more in-depth examination and it advises the operator, from the start of the operation, to dismantle the two cooling towers and to specify the method of deconstruction. The Commission recommends that the operator take full account of the skills of the personnel responsible for the dismantling operations [BLA 17].

Other uranium enrichment plants are being dismantled throughout the world. In the United Kingdom, the Capenhurst enrichment plant is operated by UCL (Urenco Capenhurst Ltd.). The closure of this plant is envisaged for 2035. During the review, the NII (HM Nuclear Installations Inspectorate, part of HSE, the Health and Safety

Executive) checked that all the tasks required to completely decommission its site were taken into account. In general, this was found to be the case. UCL's current strategy assumes that the end point of decommissioning will be the return of the site to non-nuclear commercial activities. The current estimate of decommissioning and tailing liability is £54 million [HSE 04].

In the United States, the East Tennessee Technology Park (ETTP) plant in Oak Ridge was under the jurisdiction of the Department of Energy. Enrichment at this plant was stopped in 1985. This plant consisted of numerous buildings. The U-shaped K-25 building, erected in 1943, covered about 1.5 km and 18 ha under one roof. Building K-27, built in 1945, was rectangular and had a surface area of about 35,000 m^2. Demolitions were almost complete in 2010 despite the high level of radioactive contamination [AEN 10].

7.5.3. *The dismantling of conversion and manufacturing plants*

The experience of dismantling conversion plants is based on a few cases. The conversion of 100 tons per year of UO_2 was carried out at the Korean Nuclear Energy Research Institute (KAERI). The decommissioning study began in 2001 and deconstruction was carried out from 2004 to 2007 [AEN 05a]. Similarly, the Ningyo-Toge concentration and conversion facility in Japan operated from 1981 to 1999 with a production of 700 t of U and 1,500 t of radioactive waste (neutralization sludge, ore residues). Dismantling began in 2008 and was completed in 2012 [AEN 05].

The manufacturing of fuel based on enriched uranium was carried out in Hanau in Germany in four installations, one of which manufactured fuel for light-water reactors (1,350 $t.yr^{-1}$). Decommissioning was recorded in 2006 and the waste was stored temporarily on the site [AEN 05a, AEN 10].

Several plutonium fuel manufacturing plants have been decommissioned. This was the case for the Plutonium Fuel Fabrication Facility (PFFF) at Tokai-Mura in Japan, erected in 1972 to provide fuel for the experimental reactor of the Joyo fast neutron reactor. It was shut down in 2003 and is in the process of being dismantled, in particular the glove boxes (for the manufacture of MOX pellets). Dismantling is carried out on contact (workers touching the structures) and with remotely controlled tools [AEN 10].

The Plutonium Fuel Form (PuFF) production facility is located in building 235-F of the Savannah River site. It was used to manufacture fuel pellets from plutonium 238. These have been used to fuel deep space missions, such as the Galileo space probe's journey to Jupiter, launched from the Atlantis space shuttle in October 1989.

The PuFF facility contains nine cells with thick concrete walls and shielded windows. It is estimated that about 1,500 grams of ^{238}Pu would remain in these shielded cells. This plutonium was removed from building 235-F in 2018 [WNN 18e].

7.5.4. *The dismantling of reprocessing plants*

A small number of countries have developed a civil industry for the reprocessing of spent nuclear fuel. These are Germany, the United States, France, India, the United Kingdom and Russia (Israel, Pakistan and North Korea only have military reprocessing plants). With the exception of China, France and Russia, all countries have shut down their reprocessing plants and are in the process of decommissioning (Table 7.12).

In Russia, the decommissioning of the RT-1 reprocessing plant is scheduled for 2030. But this country has new construction projects such as in 2020–2025, the Construction pilot demonstration center with a capacity of 250 t per year and the RT-2 plant with an annual capacity of 700 t for WWER assemblies, whose construction will start in 2025 [GOR 17].

Country	Plant	Operational	Security	Capacity
Germany	Karlsruhe WAK	Shutdown	Yes	?
China	Lanzhou pilot plant	Start-up	No	50–100
European Union	Eurochemic	Decommissioned	Yes	60
France	La Hague UP2-800	Yes	Yes	1,000
	La Hague UP3	Yes	Yes	1,000
India	Tarapur*	Yes	No	100
	Kalpakkam*	Yes	No	100
Japan	Rokkasho	Start-up**	Yes	800
	Tokai-mura	Shutdown	Yes	200
Russia	Mayak RT-1	Yes	No	200–400
United Kingdom	B205	Shutdown	Yes	1,500
	THORP	Shutdown	Yes	1,200
United States	Savannah River H-Canyon	Operations***	No	15

Table 7.12. *Civilian reprocessing plants worldwide (source: [FEI 14]). *Civil and military; **delayed; ***special operations*

7.5.4.1. *The European EUROCHEMIC plant*

The EUROCHEMIC experimental reprocessing plant in Dessel, Belgium operated from 1966 to 1974. This plant has made industrial history in many ways. It was undoubtedly unique that scientists from 13 European countries shared their knowledge as part of a vast and ambitious project in the field of nuclear science. This undertaking was innovative and allowed a decisive advance in the process of recovering spent fuel.

Very soon after the commissioning of the plant, the main partners, France and Germany, abandoned the partnership with the intention of taking over the industrial reprocessing of nuclear fuel. With a maximum capacity of 60 tons of uranium per year, EUROCHEMIC could not compete on an industrial scale with these large national reprocessing projects. In 1975, it was decided, for the first time, to close the plant temporarily. For 10 years, the plant faced an uncertain future. Despite reindustrialization projects with the new owner, Belgoprocess, responsible for the reprocessing of spent fuel in Belgium, the plant never restarted. In 1985, it was decided that EUROCHEMIC would be closed permanently.

EUROCHEMIC then became a pilot project for the development of nuclear decommissioning techniques at the European Union level and was the first civil nuclear reprocessing plant to be decommissioned. Belgoprocess started its decommissioning activities in 1987 with a pilot project for the complete dismantling using the "greenfield" process of two small storage buildings for the finished products from reprocessing. The dismantling of the main treatment building of the former EUROCHEMIC reprocessing plant was started with a small team that was expanded to 24 operators in 1992. The demolition of EUROCHEMIC took place in three stages with the subdivision of the plant in 2004 into an eastern, a western and a central part [BEL 11].

The building had a surface area of 55,000 m^2, with a concrete volume of 12,500 m^3 and 1,500 t of metal components. Half of the concrete was reused, as well as 70% of the metals. The building was demolished in 2008, and completion of the decommissioning was planned for 2012 [AEN 10].

7.5.4.2. *The German WAK plant*

The Karlsruhe (WAK) reprocessing plant was built as a pilot plant for subsequent commercial reprocessing in the 1960s. During the reprocessing phase of the WAK from 1971 to 1990, about 60 m^3 of the concentrated solution of highly active waste remained in the storage building of the plant. In order to condition the concentrated solution of highly active waste from reprocessing into waste for final disposal, vitrification was used. For this purpose, the Vitrification Facility Karlsruhe (VEK) was built next to the high-level waste storage building. VEK operated from

December 1996 to November 2010, producing 140 containers with a total of 56 tons of glass. Five Castor containers were filled with 28 containers each and transported to the northern intermediate storage site. In this way, all of the high-level liquid waste was vitrified [SHA 14].

7.5.4.3. *The French La Hague plants*

Nuclear cycle facilities around the world, such as reprocessing plants, differ in their design and operation. The transferability of lessons learned, best practices and decommissioning tools and techniques may not seem as relevant as for a fleet of reactors. In France, the UP1 and UP2-400 reprocessing plants have been shut down and are being decommissioned, while the UK reprocessing plants at Sellafield Thorp and Magnox were shut down in 2018 and 2020. The management of AREVA and Sellafield Ltd. (SL) have decided to pool their experience to prepare for the Post-Operation Clean Out (POCO) [YTO 16].

The two companies carried out various benchmarks and technical reviews to take into account feedback from AREVA's post-operation clean-up and decommissioning projects (such as UP2-400 at the La Hague site) and those carried out for clients (such as the CEA's UP1 at the Marcoule site). Several lessons have emerged from this review. The definition of the end state has a strong impact for each stage of the program on the POCO and decommissioning scenarios. A waste-focused strategy is essential for the overall management of the program's cost and schedule. The safety issues associated with POCO and decommissioning programs are different from the commercial operation environment. The characterization plan must be thorough and include physical and radiological studies and active sampling. The transition from operations to decommissioning requires a major cultural change [YTO 16].

On the La Hague site, a reprocessing plant has been decommissioned, the *Atelier du combustible de la filière à neutrons rapides* (AT1). This workshop, operated by the COGEMA establishment at La Hague, was a pilot installation designed to reprocess the fuel of the fast neutron channel (Rapsodie and Phénix). In operation from 1969 to 1979, it reprocessed more than 1,000 kg of fuel. After its production was shut down in June 1979, a rinse phase and a decontamination phase were carried out with the objective of recovering the maximum amount of uranium and plutonium, and of greatly reducing irradiation. In 1981, the preliminary studies for dismantling began, followed from 1982 to 1985 by cleaning operations and the development of specific apparatus. The dismantling of process equipment for VHA (Very High Activity) cells was carried out in particular by remote operation in a highly irradiating environment, thanks to the use of the carrier ATENA and the remote manipulator arm MA23 (1990–1993). From 1993 to 1997, the dismantling of the container walls was carried out. In the end, the dismantling and sanitation lasted

20 years, from 1981 to 2001. However, there are still the shells and end caps in the pool. This storage has been renamed as SOC (*Stockage Organisé des Coques*). The ASN recently took new decisions on this subject (decision no. 2020-DC-0690 of July 28, 2020) which impose new deadlines for taking back and conditioning the waste contained in the "high oxide activity" (HOA) silo and the pools of the organized storage of the hulls of the INB 80 operated by Orano Cycle on the La Hague site (Manche).

The European Commission had selected AT1 as a decommissioning pilot project for its R&D program on the decommissioning of nuclear facilities [CE 05].

7.5.4.4. *The British Sellafield plant*

The THORP (Thermal Oxide Reprocessing Plant) has been in operation since 1997, was definitively shut down in 2018 and has been in a transition phase since then [NEA 18a].

7.6. Decommissioning of other basic nuclear facilities

Apart from nuclear reactors and plants involved in the fuel cycle, there are many other types of nuclear facilities. These include nuclear, operating and waste centers, service facilities, industrial or research irradiators and linear accelerators, among others.

7.6.1. *The centers of nuclear studies*

Most countries with nuclear power generation also have nuclear research centers. These centers are also confronted with the decommissioning of their BNIs. Thus, the NEA details the recommendations of the Canadian regulatory authority for the decommissioning of its laboratories such as Whiteshell Laboratories [NEA 18a]. Only a few examples from France and the United States will be discussed.

7.6.1.1. *Examples of French study centers*

In France, the CEA has a Nuclear Energy Directorate (DEN) in charge of the development of nuclear energy in the civil domain. It has set up a Nuclear Decommissioning and Dismantling Department (DADN). Its aim is to rigorously manage the downstream part of the cycle in order to demonstrate sustainable management of nuclear energy, with strict respect for deadlines and costs. At the CEA, this program is called A/D (for *Assinissement et Démantèlement*, Clean-up and Dismantling). The CEA's objective is to successfully complete its A&D programs at the BNIs, with priority being given to completing the A&D at the nuclear centers in Grenoble and Fontenay-Aux-Roses, and at Marcoule the dismantling of UP1 (military). The Cleanup and Dismantling process is divided into 13 stages.

At the CEA, in 2013, the main civil dismantling sites that had been finished were at Cadarache HARMONIE and Grenoble SILOETTE and MELUSINE. Dismantling was in progress at Saclay LHA and ULYSSE, at Cadarache RAPSODIE, PHEBUS, ATPu, ATUE and STED, at Grenoble SILOE, LAMA and STED, at Fontenay STED, RM2 and building 18 and at Marcoule APM and G1. In the future, the planned dismantling is for OSIRIS at Saclay and PHENIX at Marcoule [DES 13].

In France, several CEA research centers are in the process of denuclearization (military centers in Ile de France, DIF, Fontenay-aux-Roses and Grenoble). In 2015, 21 civilian BNIs were being dismantled at the CEA (out of 43 in total). These activities mobilized nearly 800 employees. They were financed on a multi-year basis through a dedicated fund.

In Grenoble, the PASSAGE project was a unique example of dismantling an entire site for conversion (R&D activities in microelectronics, alternative energies, etc.). The PASSAGE program aimed to decommission six nuclear facilities (Siloette, Mélusine, Siloé, Lama and two STED INB 36 and INB 79) from 2003 to 2012. The decommissioning mobilized a maximum of 250 people, CEA personnel and subcontractors [NEA 18a]. It was a rich experience for all the actors of the nuclear industry, which was completed at the end of 2012 as planned. In 2013, final clean-up operations took place: the "conventional" deconstruction of the Siloé reactor building, as well as the demolition of the basement of Siloé in a nuclear construction site. Today, the Siloé reactor has been decommissioned and the decommissioning of the Active Materials Analysis Laboratory (Lama) was expected in 2015. The cost of the remediation–decommissioning project was just over 300 million euros.

The Fontenay-aux-Roses nuclear center is the "birthplace" of French nuclear power. It is currently being converted to research and innovation activities in life sciences and biotechnology. The program for Fontenay-aux-Roses is called ALADIN (*Assainissement des LAboratoires et Démantèlement des Installations Nucléaires*). The sites are complex because it was on this site that the very first research in high-activity chemistry on the fuel cycle took place. The end of the clean-up and dismantling of the site's former nuclear installations is scheduled for 2034. The main technical difficulty of the site is the dismantling of a set of equipment called "Petrus" comprising armored chains, galleries and underground tanks to store and then treat high-activity liquid effluents. The dismantling will be carried out remotely using remote-controlled machines. This program currently has a budget of 30–40 million euros each year.

Since the beginning of the 1980s, the Marcoule site has been engaged in projects that are among the most important in the world, on all kinds of installations, representative of those that will be dismantled in the coming decades (about one-third concern deconstructions of nuclear power reactors and two-thirds dismantling of fuel cycle installations, experimental reactors, research laboratories and other storage facilities) [CEA 15].

At Cadarache, two BNIs have been decommissioned (HARMONIE and IRCA, IRadiateur de Cadarache), two have been shut down (ÉOLE and MINERVE) and one INB is being cleaned up or dismantled (RAPSODIE) [CLI 19].

7.6.1.2. *Examples of US study centers*

There are many American nuclear research centers. In Chapter 5, the case of the historical centers of the Manhattan Project was detailed because their vocation was military. The example of the Santa Susana Field Laboratory (SSFL) will be detailed below.

The Santa Susanna SSFL nuclear center has a complicated industrial history with frequent changes of study subjects and owners. At the center, 10 reactors have operated, as well as plutonium and uranium fuel fabrication facilities and a "hot laboratory" where highly irradiated fuel from the US nuclear complex has been shipped for unloading and examination. During the life of the laboratory, many reactor accidents occurred. One of the most serious American nuclear accidents occurred in 1959 on the SRE reactor, operating with a graphite moderator and a sodium coolant. A major controversy developed concerning atmospheric releases, especially of radioactive iodine. It was nevertheless restarted and about 10 days later it was finally shut down. A third of the fuel elements had melted [AMI 19a]. This accident was reported to the authorities only after several weeks. A $40 million, multi-year radiological investigation by the Environmental Protection Agency (EPA) revealed that hundreds of sites in Santa Susana were contaminated with radionuclides, including strontium-90, cesium-137 and plutonium-239. The site has also been severely polluted by other discharges over time. It has never been decontaminated, despite a program established in 2010 by the California Environmental Protection Agency (CalEPA) [HIR 19].

A huge fire, known now as the Woolsey Fire, broke out in this area on November 8, 2018, affecting the former nuclear site and destroying more than 400 km^2. The California Department of Toxic Substances Control (DTSC) and the Department of Energy (DoE) stated that no radioactive or chemical pollution was released during the fire. This is clearly false, at odds with the levels of radioactive contamination of volatile radionuclides present at the site [HIR 19].

7.6.2. *The centers of industrial operation*

7.6.2.1. *French industrial operation centers*

The particularity of the Marcoule center is that it contains civil and military installations. This site was detailed in Chapter 5.

At La Hague, a lot of nuclear waste is still on the site. The ASN has also prescribed to AREVA the procedures for taking back and conditioning old radioactive waste (RCD). The ASN and ASND have decided to examine, at the national level, the strategic procedures envisaged by ORANO for the management of all its waste and for the dismantling of its installations. These strategies must be consistent with the national procedures, codes and the different operating strategies of all ORANO group facilities, both civil and military. AREVA's responses were submitted in June 2016. ASN believes that ORANO must strengthen its ability to prioritize operations according to the stakes of the facilities to be dismantled. ORANO's human and technical resources must be strengthened to meet the deadlines for these projects. The ASN and ASND have mobilized significant expertise for the appraisal of this strategy and took a position on this issue in 2019 [ASN 19d].

On the La Hague site, the two vitrification workshops for fission products and minor actinides have two warehouses, R7 for the UP2-800 plant and T7 for the UP3 plant, with more than 10,000 places for packages. In addition, in 2016, 56 t of plutonium (PuO2) was stored on the site [ZER 18].

The effluent treatment plant (STE2) at La Hague (INB 38) has been chemically treating low and medium activity radioactive effluents since 1966, in order to reduce their radioactive load to a minimum before discharge at sea. The co-precipitation sludge produced by these treatments was stored in seven silos or tanks, representing about 9,300 m^3 of sludge [ZER 18].

On January 6, 1981, a fire broke out in silo no. 130 (north-west of the site). This silo includes two pits. The first holds 750 tons of various types of solid waste from the reprocessing of UP2-400, deposited from 1969, and a second pit where 518 tons of waste has been stored since 1973. This silo was flooded and the situation remained as it was 37 years later [ZER 18]. On November 12, 2019, the ASN made decisions regarding the operations that ORANO will have to comply with for silo 130. The following decisions were made: the start of operations to take back the waste stored in silo 130 by December 31, 2019, the end of solid waste disposal by December 31, 2022 and the end of operations to recover and recondition solid waste, sludge and water by December 31, 2025 [ASN 19b].

7.6.2.2. *British industrial operation centers*

The British nuclear industrial centers are Dounreay, specializing in experimental and prototype reactors, and Sellafield-Windscale, known for the reprocessing of nuclear fuel.

In the United Kingdom, the Dounreay nuclear site on the north coast of Scotland was the British center for research and development of fast reactors between 1954 and 1994, including two fast breeder reactors, a material testing reactor, reprocessing and fuel fabrication plants (U, Pu) and many other research and waste management facilities [BEC 12]. This site includes 43 decommissioning projects. It is therefore essential to prioritize the sites. To carry out this operation, the British authorities used a test in the form of a questionnaire. The questions were divided into four main categories: safety and security, environmental factors, costs and public perceptions, each of which was broken down into several items. Depending on the answer, a certain number of points were awarded [BAY 03].

The site is located approximately 15 km west of Thurso on the north coast of Scotland. Dounreay Castle was used by the nuclear industry. Two sources of radioactive contamination have contaminated the site. The first was the result of experiments carried out in the mid-1950s to study the dispersion at sea of radionuclides, mainly fission products. The second was the result of low-level discharges at Dounreay from 1970 to 1980. To clean up the site, the authorities had a choice between four main options. The first two were to do nothing or to carry out a complete clean-up. This option would require the removal of the castle's structure since contamination, through migration, had probably affected the foundations. The third option was targeted remediation. This would have been achieved either by removal of the contaminated material by excavation and replacement with subsequent waste disposal or treatment, or by *in situ* treatment. The last option was to achieve encapsulation. This was done by installing a concrete barrier to reduce the external dose by protecting and preventing airborne contamination. In the end, the chosen option was complete remediation [BAY 03].

Following remediation, an area of 900 m^2 was excavated to a maximum depth of 3 m, yielding a total of 1,540 tons of LLW, of which 1,109 tons was assessed as very low-volume, high-level waste (alpha between 0.4 and 1 $Bq.g^{-1}$). The excavation was partially backfilled with a clean, inert, imported material and excavated material. Site remediation was essentially completed within the defined criteria of 1 $Bq.g^{-1}$ for artificial alphas and 4 $Bq.g^{-1}$ for beta. A new outfall was successfully constructed and commissioned. The radiological safety of personnel and the environment were controlled and remained within the set limits, despite the difficult environmental conditions and the extended duration of the project. The maximum risk of death to a potential future resident is estimated at $2.7.10^{-6}.yr^{-1}$, which is consistent with the current government and consultant guidelines, which indicate that an annual risk of death between 10^{-5} and 10^{-6} is acceptable [BAY 03].

The Nuclear Decommissioning Authority (NDA) was created in 2005 with the specific mission of addressing the UK's nuclear legacy. Sellafield is the largest and most dangerous site in the authority's domain and is home to an extraordinary accumulation of hazardous waste, much of which is stored in obsolete nuclear facilities. The Sellafield site has been operational for 67 years, covers an area of 6 km^2 and has over 1,300 buildings on site. More than 95% of the radioactive materials accumulated in the United Kingdom reside on the site. The number of accidents and incidents have been numerous. From November 1990 to July 2012, there were 2 INES level 3 accidents, 12 level 2 incidents and 249 level 1 incidents. Events prior to the adoption of the INES scale were noted retrospectively, and at Sellafield (previously Windscale), there were two accidents, one level 5 on the plutonium-producing reactor (Oile-1) and the other level 3 in the cesium extraction plant [SLA 13].

The House of Commons noted that clean-up deadlines for Sellafield have not been met, while the total cost of decommissioning the site continues to rise each year and now stands at £67.5 billion. Finally, a huge amount of public money, some £1.6 billion, is spent at Sellafield each year [HOU 13]. This spending still persists and Sellafield accounts for 75% of the NDA's long-term cost estimate. In 2017–2018, Sellafield accounted for 61% of the NDA's total expenditure of £3.3 billion [NAO 18].

Sellafield's most dangerous facilities include four ponds where liquid radioactive waste has been dumped, as well as several canisters that contain large quantities of nuclear materials and the facilities that house most of the UK's plutonium stocks. Sellafield Limited, the Site License Company (SLC) that manages day-to-day work on the site, has five long-term programs in place to address these risks.

In January 2015, the NDA announced the termination of its contract with the private sector consortium that managed Sellafield Limited, a decision approved by the then Department for Energy and Climate Change and the UK's Executive Department for Public Expenditure (HM Treasury). In April 2016, Sellafield Limited became a wholly-owned subsidiary of the NDA.

The UK National Audit Office, following its audit of the NDA, believes that in 2012 the NDA's 14 major projects at Sellafield did not deliver value for money, with significant cost increases and delays, that in 2013 the efficiency savings reported by the NDA at Sellafield were moderate, that in 2015 the costs and delays of major projects at Sellafield had further increased, and that in 2017 the NDA's procurement and management of its contract for decommissioning the 12 Magnox sites were failures [NAO 18].

7.6.3. *Service facilities*

Each nuclear center has a large number of basic nuclear facilities outside of the nuclear reactors that provide input to the overall operations. Taking the example of the Cadarache Nuclear Center, the following observations can be made. Two BNIs were shut down, the *Station de Traitement des Effluents liquides aqueux* STE (BNI 37B) and the *Magasin Central de Matières Fissiles* MCMF (BNI 53). The STE was commissioned in 1963 with the purpose of treating active effluents by evaporation. The activity ceased in 2013 because the seismic resistance of the installation was insufficient. It was replaced by AGATE. The facility is currently intended to be sanitized and then dismantled. Preparatory operations for dismantling have been initiated. The function of the MCMF, put into service in 1962, was the reception, storage and dispatch of non-irradiated fissile materials (enriched uranium and plutonium) intended for use in the fuel cycle or temporarily out of use. Since the seismic resistance of the installation was insufficient, all the material was evacuated to the new MAGENTA installation.

Three BNIs are being remediated and dismantled (A/D), ATPu (INB 32), ATUE (INB 52) and LPC (INB 54). The ATPu and the LPC constituted the Cadarache Fuel Fabrication Complex (CFCa), whose industrial operator was AREVA NC until 2017. The *Atelier de technologie du plutonium* (ATPu), put into service in 1962, manufactured fuel elements based on uranium and plutonium for fast neutron reactors, then from 1988 to 1997 the production of MOX fuel (a mixture of uranium and plutonium oxides) for the reactors of EDF and Swiss and German electricity companies. Production was stopped in 2003. Final shutdown and dismantling were authorized in 2009. Dismantling is to be completed in 2024. The *Atelier de l'uranium enrichi* (ATUE) had been converting uranium hexafluoride into uranium oxide since 1964, intended for the manufacture of fuel for nuclear reactors. It also carried out the chemical reprocessing of waste from the manufacturing of fuel elements with a view to recovering enriched uranium. Production was stopped in 1996. Dismantling was authorized in 2006. Modification of the dismantling plan led to a new public inquiry in 2017. The decree authorizing the dismantling of the ATUE is in preparation. The *Laboratoire de purification chimique* (LPC), put into service in 1965, was the complementary installation of the ATPu for the control of nuclear materials, to treat effluents and radioactive waste, with a view to their evacuation to an appropriate channel, including a unit for cryotreatment of plutonium-laden waste. Final shutdown and dismantling were authorized in 2009. Dismantling is to be completed by 2024 [CLI 19].

Outlets from nuclear centers to the atmosphere often consist of high chimneys equipped with filters to retain part of the gaseous radioactive releases. At the end of a BNI's life, the chimneys must be destroyed. However, over a lifetime that can extend over decades, the chimneys can become contaminated as a result of the deposition of radioactive substances, such as aerosols, on the internal surfaces of the chimney. The dismantling of the chimneys is therefore a difficult operation [IAE 05]. Two solutions are available for their dismantling. The first solution is complete destruction using explosives but the dust created is radioactive and can be widely dispersed. In addition, there may be other industrial facilities nearby, or even homes, and strict precautions must be taken. The second solution is a gradual levelling from top to bottom but this requires a substantial amount of scaffolding, serious dust control and a much longer time for the chimney to disappear. The case of the Windscale chimney is detailed in Chapter 5.

For low-level radioactive liquid discharges, nuclear facilities use outlets to release these discharges into ponds, streams or seas. One of the most impressive of these outlets is the La Hague pipeline, which sends slightly contaminated water into the Blanchard Raz, one of the strongest currents in the world [AMI 17, AMI 19a]. This pipeline has a total length of 5 km and runs from the La Hague site. It runs from Anse des Moulinets, 60 m from the coast, to an outlet 1.5 km off the coast of Nez de Jobourg and sinks to a depth of 60 m [ORA 17]. Its history was marked by a leak due to corrosion in 1979–1980. The principal radionuclides involved were strontium 90 and ruthenium 106 [BAR 00]. It was completely dismantled and rebuilt.

In the centers, there are often specially designed rooms for the treatment of highly radioactive materials or liquids. They are often equipped with armored doors and reinforced concrete walls. They are sometimes specialized glove boxes for high alpha activity. In Harwell, UK, 240 of these glove boxes have been dismantled [BAY 03].

7.6.4. *Interim nuclear waste storage centers*

Waste storage sites can fail and it is then necessary either to correct the defects or to move the site to another storage location. In the United States, estimated costs and safety considerations for decommissioning waste storage sites (LLW) have been evaluated.

Site stabilization is expected to cost between \$0.4 million and \$7.5 million, depending on the site and the stabilization option chosen. The authors estimate that long-term remediation costs approximately \$100,000 per year, with somewhat higher costs in the early years due to increased site maintenance and environmental monitoring requirements.

The relocation of all waste from a disposal site is expected to cost over $1.4 billion and take over 20 years to complete. More than 90% of the costs are associated with packaging, transportation and off-site disposal of the excavated waste [MUR 80].

7.6.5. *Other BNIs in the dismantling stage*

The variety of nuclear facilities is great. Among them, irradiators, linear accelerators and nuclear-powered civilian ships will be decommissioned and dismantled at the end of their lives.

Among the irradiators, some are used for research, others are used to irradiate food to make it sterile. There are two types of irradiators for industrial use, electron accelerators and cobalt-60 irradiators [ZER 94].

Two examples of decommissioning will be reported: that of the Cadarache irradiator (IRCA, *IRradiateur de Cadarache*), a former BNI 121, commissioned in 1960 and decommissioned in 2006. Its function was intended for studies on the behavior of bitumens. Cobalt-based radiation sources were removed in 1996. The definitive cessation of operations has not yet been announced and the work corresponding to this phase has therefore not begun [CLI 19]. The irradiator of the *Société Normande de Conserve et Stérilisation* (SNCS) used for food irradiation was authorized by decree no. 2002-447 of March 27, 2002 to proceed with the final shutdown and dismantling of BNI no. 152, located in the commune of Osmanville [AUB 17].

The CEA had to dismantle two large installations, the 700 MeV Saclay Linear Accelerator (ALS, BNI 43) producing electrons and the 3 GeV Saturne Synchrotron (BNI 48) producing ions (protons). To do this, the CEA created a specific entity, the *Service de Déclassement des Accélérateurs* (Accelerator Decommissioning Service). These two installations represent a mass of $4{,}10^4$ t, a volume of $3{,}10^4$ m^3 and about 20,000 objects. The surface area of the two BNIs is $4{,}10^4$ m^2 [DAM 08].

The ALS was operational from March 1969 to June 1990. The laboratory of nuclear expertise in remediation and design (Lenac) at Dapnia took charge of the dismantling of the ALS (INB 43). For the dismantling of ALS-INB43, the experts at Lenac used the techniques already tested and validated by the dismantling of Saturne. These techniques made it possible to make significant savings and to recover waste that would otherwise have been considered lost. After the authorization for final shutdown and dismantling, signed on January 12, 2004, the decommissioning decree of September 13, 2006 makes the ALS buildings available for use without special surveillance. The ALS is no longer a nuclear installation but a simple tunnel buried under the Orme des Merisiers at Saclay.

In accelerators, there is no radioactive contamination but only activation of the surrounding materials, such as concrete. The main radionuclides likely to be present in the various qualities of concrete (heavy, boron or shattered) or in metallic materials (stainless steel, iron, aluminum, copper, lead, etc.) are ^{3}H, ^{22}Na, ^{45}Ca, ^{54}Mn, ^{57}Co, ^{58}Fe, ^{60}Co, ^{62}Ni, ^{65}Zn, ^{108m}Ag, ^{110m}Ag, ^{133}Ba, ^{134}Cs, ^{137}Cs, ^{152}Eu and ^{154}Eu.

During the dismantling of BNI 43 and 48, the CEA recovered 15,000 tons of concrete. The estimation of cost was varied, such as the repair of a runway at Orly airport, of embankments at the CEA/Saclay heliport or in the underlayers of roadway formwork in several communes in the Paris region. The costing of lead has also been varied with the ballasting of old rigging in Honfleur (Marie-Madeleine, Dehel and Files) or the shielding of irradiators and accelerators at the CEA. The 50 magnets (1200 t) were recovered in Japan. The balance sheet of BNI 48 is 9% (1,548 t) of ultimate waste and 91% (15,318 t) of waste and materials recovered [DAM 08].

Among the linear accelerators at Bures-sur-Yvette, BNI 106 (LURE) is operated by the CNRS first in High Energy Physics and then as a national source for the production and use of synchrotron radiation produced by particles (positrons). This installation includes the linear accelerator (LINAC), the storage rings DCI (*Dispositif de Collisions dans l'Igloo*), ACO (*Anneau de Collisions d'Orsay*), Super ACO and the Free Electron Laser CLIO (*Centre Lasers Infrarouge d'Orsay*) built from a small electron accelerator. The Definitive Cessation of Operation (*Cessation Définitive d'Exploitation,* CDE) phase began on January 1, 2004 and was completed at the end of December 2006. The strategy adopted to define the sectors constituting the CDE and the Final Shutdown (*Mise à l'Arrêt Définitif*, MAD) was to separate the accelerators from the experiment halls and the modulator hall. The LURE's Definitive Cessation of Operation phase was completed three years after the accelerators were shut down. It was entrusted to an Own Service Unit, created specifically for this mission one year before the accelerators were shut down, and which was immediately operational, benefiting from the presence of operating personnel before their redeployment to other units and drawing on the experience of the dismantling of SATURNE and the ALS in Saclay and exchanges with the ASN. CNRS management has made available all the necessary human and financial resources, which has made it possible to take into account all the regulatory requirements, particularly with regard to safety, security and radiation protection (Pauwels and Tadjeddine, in [CCS 07]).

Civilian nuclear-powered vessels are mainly Russian icebreakers. There were also a few cargo ships of various nationalities but all of them have been withdrawn or converted to conventional propulsion. Their dismantling is very similar to that of submarines.

7.7. Conclusions

Research or experimental reactors and prototypes are extremely varied. The same is true for basic nuclear facilities other than power reactors. As a result, their decommissioning and dismantling are also very different from one other.

The number of experimental reactors in the world is very large, with very diverse roles. A large number are shut down and are in the process of being decommissioned. Since these facilities are often small, the cost of decommissioning is lower than for nuclear power reactors, and often also faster.

Many states build prototype reactors before launching mass production of a new reactor. All first-generation prototypes are shut down, and the decommissioning and dismantling of these reactors serve as a test bed to develop the methods and techniques that will later be used for the reactors in this series.

The BNIs of the fuel cycle are numerous and varied. Their decommissioning generates large volumes of radioactive waste. This is particularly true for extraction mines, enrichment plants and spent fuel reprocessing plants.

Many nations have developed centers for nuclear studies. Their roles have evolved considerably over time and many basic nuclear facilities, either service or research, have become obsolete and need to be decommissioned and dismantled.

Because of the large volumes of radioactive waste to be managed, representing several categories, some states are obliged to create emergency and temporary nuclear waste storage centers. At the end of their operation, it may be necessary to move the waste and destroy these temporary centers.

General Conclusions

C.1. The failure of non-proliferation

Nuclear non-proliferation is a failure; it has not prevented the number of states possessing atomic weapons from almost doubling. And this is likely to continue, as more states are expected to acquire the nuclear bomb in the coming years. Only the renunciations of Brazil, and especially South Africa, can be credited to non-proliferation.

International treaties are currently undergoing a serious crisis. For example, only one international treaty has been ratified, the Treaty on the Non-Proliferation of Nuclear Weapons (NPT), although that is on hold. For all the others, the number of signatory countries is too low. What will happen to the latest proposed treaty, the Treaty on the Prohibition of Nuclear Weapons (TPNW), which was ratified by 50 states on October 24, 2020 and was expected to enter into force on January 22, 2021? None of the nine nuclear-weapon states, including the five permanent members of the UN Security Council, nor Japan, a nuclear-weapon martyr, has signed it. Therefore, this new treaty is likely to have limited effect.

There are only five geographical zones free of nuclear weapons, which is not a lot, and this situation is not expected to change, either in the short or medium term.

C.2. The rejection of the concept of nuclear deterrence

For many years, the five nuclear powers, the first to acquire nuclear weapons, have claimed that nuclear deterrence has kept the world from going to war since World War II. The claim was a bit precipitous because it was deliberately ignoring multiple local wars, and so very quickly many NGOs were formed to demand total nuclear disarmament. Some of these NGOs have had enough impact to win the Nobel Peace Prize.

Recently, Pope Francis II has changed the Vatican's view on this issue and has taken a much stronger stance in rejecting the use of atomic weapons and balanced nuclear deterrence. Similarly, a majority of states have come together and allowed the adoption of a new Treaty on the Prohibition of Nuclear Weapons (TPNW). This Treaty was approved by 122 states in July 2017 and, as of September 24, 2020, signed by 84 states and ratified by 45 states. This is a profound change in the attitude of world political opinion. Similarly, some militaries reject nuclear deterrence. This is the case for the Frenchman Francis Lenne, General in the second section (retired), former director of education at the École de Guerre, a specialist in nuclear deterrence, who wrote a book about "the mourning of Hiroshima and Nagasaki". He advocates complete disarmament and believes that nuclear deterrence is in fact a denial or a "psychotic temptation" [LEN 17]. This is also the case for Paul Quilès, former French Minister of Defense (1985–1986) and President of the French Defense Committee (1992–2002) of the Assemblée Nationale. He is now President of the association *Initiatives Désarmement Nucléaire* (IDN) and has written several books, including *Nucléaire, un mensonge français* (Nuclear, a French lie) [QUI 12].

C.3. The failure of nuclear disarmament

One of the main purposes of the UN is to resolve conflicts between member states of this international body. The best way to avoid wars is to deprive the protagonists of weapons, especially weapons of mass destruction, hence the creation of a Disarmament Commission within the UN. Unfortunately, since decisions are made only by general consensus, this Commission has been inoperative for many years. The same is true in the Security Council, where the five members with a possible veto are the same ones who have nuclear weapons.

As soon as the United States acquired the atomic bomb, the Soviet Union (USSR) was constantly developing its own, which led to a vertiginous escalation in the manufacture of nuclear weapons, in both number and power. This cold war resulted in stockpiles of nuclear weapons for the two superpowers capable of destroying the planet several times over. Faced with this absurd situation, the two states decided to begin talks to limit the number of weapons. This resulted in a large number of agreements. Unfortunately, the controls were not rigorous, voluntarily or involuntarily. As a result, nuclear disarmament was not, strictly speaking, a success. Indeed, while the number of active nuclear warheads in the world has decreased significantly since the end of the Cold War, it should be noted that new weapons have been built. These are more powerful, more sophisticated, more precise and with multiple, independent nuclear warheads with different targets. What humanity has gained in quantity, it has lost in quality.

C.4. An impressive dismantling market

Throughout the world, there are several hundred nuclear power reactors supplying electric current. Similarly, more than a thousand experimental reactors, associated with nuclear service facilities, such as water treatment plants, have been built. Thus, at the end of 2018, 176 reactors were shut down (between 1950 and 2018) in 21 countries, 53 reactors were under construction in 19 countries and 447 reactors were still connected to the electricity grid in 31 countries [CEA 19].

Any nuclear installation, like any human construction, has a limited lifespan. At the end of its life, it must be decommissioned and dismantled so that it is no longer a radioactive risk for future generations and the environment. The number of decommissioned BNIs is large and will continue to increase significantly in the short term. Therefore, faced with the impressive number of nuclear installations to be dismantled, a specific industry is being created, with the full arsenal of competition. Many operations are carried out with a multitude of subcontracting companies, which greatly dilutes the responsibility and increases the risk of misconduct.

C.5. International recommendations for decommissioning

Two major organizations, the IAEA and the NEA, have set up groups of experts to analyze the feedback from past decommissioning and to reflect on the best way to carry out these often complex and costly operations. These two organizations have published numerous guides to help states possessing nuclear facilities to decommission them under the best possible conditions.

The first recommendation that emerged is to make provision for the deconstruction of the facility as part of the construction concept. An effort has also been made to estimate more realistically the costs of dismantling the various types of facilities.

C.6. National decommissioning policies

Each state develops a different policy for the decommissioning of BNIs according to its history, economy, geography and philosophy. Thus, France advocates immediate dismantling with the "greenfield" concept, which is not always followed by operators. In the United States, deferred decommissioning is quite frequently chosen after the BNI has been placed in a safe place, or even "put in the grave", where the installation is never definitively decommissioned, remaining under nuclear license and under surveillance. This policy, which is more economical in the short term, does not preserve future generations and freezes surface use for centuries. This is only possible if the state has large areas at its disposal.

C.6.1. *Decommissioning of nuclear power reactors*

Power reactors have several particularities, such as the imposing dimensions of their vessels and the high activity induced by the neutrons they contain.

C.6.2. *Decommissioning of experimental reactors*

The main difficulty in dismantling experimental reactors is their extreme variety and the relevant experience may not be available for their particular case.

C.6.3. *Decommissioning of nuclear fuel cycle facilities*

The difficulty comes from their size. They are often large plants, especially the enrichment and reprocessing plants, combined with a large number of service units. When they are dismantled, the extent and levels of internal contamination are high and the volumes of the various types of waste are enormous. It is important to note that what differentiates reactors from other installations in the fuel cycle is the presence of activated materials in the mass.

C.6.4. *Decommissioning of damaged reactors*

An additional difficulty arises in the event of a relatively serious nuclear accident. Indeed, in this case, the structures are destroyed leaving room for chaos where the radioactivity is much higher than a healthy reactor, preventing humans from approaching the site.

Thus, after the Japanese accident at Fukushima, regardless of the future of nuclear energy, Japan was facing five major political problems: spent fuel management, plutonium stockpile management, radioactive waste disposal, human resource management and the restoration of public trust [SUZ 17].

The Fukushima accident had global repercussions. For example, it marked a turning point in French energy policy, as it prompted the country to propose legislation to reduce the share of nuclear power and accelerate the growth of renewable energy [STU 17]. Similarly, several countries, such as Germany and Switzerland, decided to abandon nuclear energy after the accident.

Similarly, in recent years, China has reduced its nuclear capacity expansion target from a target of 70 gigawatts (GW) by 2020 (published in 2009) to only 58 GW by 2020 (issued in 2016). The Chinese government also adopted a set of policy changes following the March 2011 Fukushima Daiichi nuclear disaster in Japan. Since the Fukushima disaster, the Chinese State Council has suspended plans to build domestic nuclear reactors and has restricted the construction of reactors to modern (third-generation) designs [RAM 17].

However, the Fukushima nuclear disaster in 2011 did not hinder the plans of Southeast Asian countries to develop nuclear power plants, despite safety concerns [CAB 17].

Accidents at Three Mile Island, Chernobyl and, more recently, Fukushima have demonstrated the vulnerability of this technology to human error, design flaws and natural disasters, and these accidents have resulted in enormous health, environmental and economic costs that must be taken into account in any energy policy that includes nuclear energy as an option [MOU 17].

C.7. Problems in dismantling of French power reactors

The dismantling of power reactors in France is experiencing some difficulties. This is the case for the prototype reactor at Brennilis. After 18 years of intermittent operation, the reactor has been in the dismantling stage for 47 years. In 2005, before the Conseil d'État intervened to delay decommissioning, the Cour des Comptes (2005) estimated the cost of dismantling the Brennilis plant at 482 million euros, 20 times the cost estimated in 1985.

The same is true for French first-generation graphite-moderated reactors. EDF planned an "under water" dismantling of these UNGG reactors (except Chinon A1 and A2). However, an unexpected change of strategy was taken unilaterally by EDF and the regulatory authority asked the company to justify itself. This situation is said to be due to a technical difficulty that has not been resolved on an industrial scale. The parliamentary information mission [AUB 17] regrets this long postponement of the dismantling of the graphite-gas reactors especially since it sends a bad signal about the overall feasibility of dismantling. It recommends, following the example of the ASN and in accordance with the regulations, that "the shortest possible deadlines be respected".

The dismantling of the Superphénix reactor is a difficult operation due to the presence of sodium. In the course of the operation, EDF was criticized by the ASN and condemned by the courts. The cost could reach 2 billion euros [AUB 17].

C.7.1. *Sensitive radioactive waste management*

Waste management largely determines the dismantling process. However, in France, even though it is not specific to that state, some landfills are saturated. In addition, certain categories of radioactive waste still have no assured procedure. The parliamentary mission is considering whether a "release threshold" should not be established in France.

C.7.2. *Significant recourse to subcontracting*

Many manufacturers, including those in France, almost systematically resort to subcontracting. This is a specific feature in the nuclear industry, and it is an old industrial choice. It is encouraged by regulations that are not very restrictive in practice. Certain abuses have been observed. In addition, there are burdens that can dilute responsibilities and, moreover, the value added by subcontracting remains modest. Parliamentarians emphasize the need to create an economic sector and therefore subcontracting is not the right solution.

C.7.3. *Undervalued decommissioning costs*

The parliamentary mission believes that EDF is making optimistic assumptions about the cost of dismantling these power reactors. Indeed, it will be difficult to undertake all the dismantling that currently needs to be done. The company thinks that it will make economies of scale by decommissioning several reactors of the same time, which the parliamentary mission contests. Likewise, the pooling of resources for one reactor being dismantled and another under construction on the same site is clearly false information.

Numerous expenses not provisioned by EDF such as site remediation, payment of taxes and insurance and reprocessing of used fuel are not taken into account. The social cost of dismantling is not mentioned at all. Moreover, decommissioning expenses are particularly underestimated, such as a curious discount rate, a particularly low inflation rate, and the scope of the provisions to be covered is minimized. Taking the same delivery time period for the two reactor fleets (UNGG and PWR) seems absurd. Parliamentarians are questioning the quality and liquidity of the hedging assets. They question EDF's strategic choice to extend the life of the reactors as a solution [AUB 17].

Unfortunately for the company, institutional comparisons are convergent, from both national and international evaluators. Thus, the Cour des Comptes underlines the weakness of EDF's provisions. Year after year, this body makes the point in all their reports that the assessments of dismantling costs are always very optimistic. The Cour des Comptes regularly finds increases of several percent and calendar extensions of several years [COU 05, COU 20]. This is not specific to EDF, since AREVA and ORANO, for the final shutdown of the La Hague site, the Court noted in 2014 a 47% increase in the estimate between 2006 and 2013 in constant euros (+628 M€$_{2013}$). Since that report, the estimate has been revised upwards on a regular basis, bringing the overall increase to nearly 19% from 2013 to 2018, excluding inflation.

According to the OECD [OEC 16], EDF is not taking all the costs into account and the company has a variable geometry reading of the dismantling perimeter. This international organization lists 15 items of expenditure in the complete dismantling of a nuclear reactor, but EDF only lists six.

C.7.4. *Negative results from dismantling in France*

In conclusion, the parliamentary mission makes four remarks on decommissioning and dismantling in France. First of all, the technical feasibility that many operators consider to be under control is not entirely assured. The second obstacle to dismantling concerns the duration of the work. Indeed, contrary to initial forecasts, the dismantling of nuclear facilities will probably take longer than expected. The mission formulates a third warning, which concerns the financial feasibility of decommissioning. It considers that the cost of decommissioning is likely to be higher than forecast. The fourth and final remark follows on from the previous one. Given the fact that the cost could be higher than forecast, the sums already provided for and those that will have to be provided for would be insufficient to cover all the costs incurred by the dismantling operations. At the end of the mission's work, the parliamentarians therefore concluded that the overall decommissioning strategy should be seriously reconsidered, both technically and financially [AUB 17].

C.8. Results of dismantling BNIs

A global assessment of power reactor decommissioning was proposed at the end of 2018, as given in Table C.1. Of a total of 159 shutdown reactors, only 19 decommissionings and dismantlings were completed, while 67 were placed in safety (sarcophagus) before dismantling was to be resumed after a period of years, often several decades. Phase 1 includes the post-operational phase, the dismantling of systems that are not required for the decommissioning process. The dismantling of the most contaminated parts of the system also begins during this phase. An indicator of the progress of this stage is the complete removal of spent fuel from the reactor core and pools. Indeed, this is crucial for the companies involved later.

Phase 2 includes dismantling activities in the most contaminated area (the so-called hot zone), i.e. the dismantling of highly contaminated or activated parts, for example the reactor pressure vessel (RPV) and its internal parts (RVI, Reactor Vessel Internals), and the biological shield.

Phase 3 includes the removal of operating systems and the decontamination of buildings. This stage ideally ends with the demolition of the buildings and the release of the reactor site that has become virgin again ("greenfield") for unrestricted use. However, the release of the site as a brownfield site is allowed in some countries, which means that the buildings can also be reused, for nuclear or other purposes [SCH 19].

Country	**Reactor shutdown**	**Steps in the decommissioning process**				
		Phase 1	**Phase 2**	**Phase 3**	**Sarcophagus**	**Complete**
Canada	6	0	0	0	6	0
France	12	3	1	0	8	0
Germany	29	10	4	8	2	5 (17%)
Japan	27	26	0	0	0	1 (4%)
United Kingdom	30	0	0	0	30	0
United States	36	6	0	5	12	13 (36%)
Spain	3	1	0	1	1	0
Italy	4	4	0	0	0	0
Lithuania	2	2	0	0	0	0
Russia	8	0	0	0	8	0
Korea	2	2	0	0	0	0
Total	**159**	**54**	**5**	**14**	**67**	**19**

Table C.1. *Decommissioning status of nuclear reactors in 11 states at the end of 2018 (based on [SCH 19])*

The results of the CEA [CEA 19] are different, with a total of 176 reactors shut down at the end of 2018.

C.9. Specificities of nuclear facility decommissioning

The dismantling of a nuclear installation is not a conventional dismantling process because the access to certain zones requires the implementation of particular precautions. This is in order to protect the parties involved against the ionizing radiation emitted by the elements to be broken up, in order to be extracted from the whole and against the aerosols produced during the removal operations. It is also necessary to find specific outlets that allow the elimination of various radioactive wastes, according to their mass activities.

C.9.1. *A period of latency with monitoring*

The fact that it is only possible to intervene after a significant reduction in irradiation levels often leads to a wait of several years before proceeding with the first dismantling work in the active zone. During this time, the installation must be maintained in good condition with all the technical elements that will be necessary such as electrical and ventilation circuits, water and gas networks, and lifting equipment. These requirements also require the maintenance of certain professional skills in order to keep the collective memory of the installation, as its development process is not always officially recorded.

In order to avoid any contamination of the site and its environment, it is necessary to monitor the aging facility. Follow-ups and maintenance are an additional cost.

C.9.2. *Knowledge of the facility*

A prior assessment of the volumes of waste produced and their mass activities must be carried out to organize their evacuation and to evaluate the cost of its operations.

Dismantling requires close contact with irradiating materials and leads to greater handling activities than would have been experienced during the normal operation of the facility. Therefore, interventions must be studied beforehand, taking into account dose forecasts and the safety of each major intervention. This makes it possible to choose intervention modes that are safe with regard to potential risks and are the least penalizing from a dosimetric point of view, in order to define a rigorous planning of interventions.

C.9.3. *Competent stakeholders*

Stakeholders must have a dual skill set, that of mastering the technical operations necessary to ensure dismantling (mechanical and fusion cutting) and of knowing the radiation protection techniques that reduce the dosimetric balance when unforeseen events occur.

C.9.4. *Innovative techniques and robotization*

The breaking up of PWR and BWR reactor vessels, which are massive, highly irradiating elements, has been a major problem during dismantling. Cutting tests have proved to be conclusive. Technical innovations, such as the use of lasers and remote-controlled robots, currently make it possible to envisage faster, cheaper and less polluting dismantling.

C.9.5. *Companies specialized in dismantling*

The growing number of nuclear facilities being shut down will lead to the development of industrial companies that will specialize in decommissioning. Switching from one activity to another will require staff training if the skills acquired are to be retained. However, attention will have to be paid to the quality of subcontracting companies, a solution frequently used in the nuclear field but which in the past has also often been a source of problems involving contamination of personnel and the environment.

C.9.6. *Various risks*

The radiological risk of dismantling operations must not mask the other risks presented by toxic substances such as asbestos used as insulation in all facilities, nuclear or not, beryllium, cadmium, mercury and the sodium used in fast reactors. Let us recall that at Cadarache, at the end of March 1994, there was one death and four wounded by an explosion during the neutralization in a sodium tank of an experimental reactor (Rapsodie) of a residual puddle of 100 kilos of sodium. This operation had been very insufficiently prepared for because of the absence of truly competent chemists for the delicate and innovative operation of rendering sodium safe by adding alcoholates.

In conclusion, the decommissioning and dismantling of nuclear facilities are complex, costly operations, presenting significant risks of environmental contamination and irradiation of personnel, and even of the public, and which generate enormous volumes of radioactive waste for which storage solutions for certain categories such as long-lived and high-level waste are not yet fully mastered. This subject will be the subject of the next volume in the *Radiactive Risk* set of books.

List of Acronyms

AA	Average Activity
ABM	Anti-Ballistic Missile
AEC	Atomic Energy Corporation
AGR	Advanced Gas-cooled Reactor
AMEC	Arctic Military Environmental Cooperation
AP	Additional Protocol
ARMSCOR	Armaments Corporation of South Africa
ASN	*Autorité de sûreté nucléaire* (French Nuclear Safety Authority)
ATPu	*Atelier de technologie du plutonium* (Plutonium Technology Workshop)
ATUE	*Atelier de l'uranium enrichi* (Enriched Uranium Workshop)
AVM	*Atelier de vitrification et d'entreposage des verres de Marcoule* (Glazing and Glass Storage Workshop in Marcoule, France)
BNFL	British Nuclear Fuels Limited
BNI	Basic Nuclear Installations
BWR	Boiling Water Reactor

CAEA	China Atomic Energy Authority
CANDU	Canada Deuterium Uranium
CEA	*Commissariat à l'énergie atomique et aux énergies renouvelables* (French Alternative Energies and Atomic Energy Commission)
CESTA	*Centre d'études scientifiques et techniques d'Aquitaine* (Center for Scientific and Technical Studies of Aquitaine)
CFR	Code of Federal Regulations
CNSC	Canadian Nuclear Safety Commission
COIL	Chemical Oxygen–Iodine Laser
CTBT	Comprehensive Nuclear-Test-Ban Treaty
CTBTO	Comprehensive Nuclear-Test-Ban Treaty Organization
D&D	Decontamination and Decommissioning
DAM	*Direction des applications militaires* (French Directorate of Military Applications)
DEN	*Direction de l'énergie nucléaire* (French Directorate of Nuclear Energy)
DeSa	International Project on Evaluation and Demonstration of Safety for Decommissioning of Facilities Using Radioactive Material
DGA	*Direction générale de l'armement* (French Directorate General for Armament)
DoD	Department of Defense
DoE	Department of Energy
DP	Decommissioning Plan
EA	Environmental Authority
EBR	Experimental Breeder Reactor

EBRD	European Bank for Reconstruction and Development
EDF	*Electricité de France*
EGFWMD	Expert Group on Fukushima Waste Management and Decommissioning
EIA	Environmental Impact Assessment
ENRESA	*Empresa Nacional de Residuos Radioactivos* (Spanish National Radioactive Waste Company)
EPA	Environmental Protection Agency
ETS	Effluent Treatment Station
EURODIF	European Gaseous Diffusion Uranium Enrichment Consortium
EWN	*Energiewerke Nord GmbH*
FAS	Federation of American Scientists
FBR	Fast Breeder Reactor
FMCT	Fissile Material Cut-off Treaty
FNR	Fast Neutron Reactor
FOST	*Force océanique stratégique* (Strategic Oceanic Force)
FP	Framework Program
GCR	Gas-Cooled Reactor
GGA	General Guarantee Agreements
NUGG	Natural Uranium Graphite-Gas
HA	High Activity
HBR	Hydraulic Breaker
HCoC	Hague Code of Conduct against Ballistic Missile Proliferation

HEU	Highly Enriched Uranium
HLW	High-Level Waste
HOA	High Oxide Activity
HSE	Health and Safety Executive
HTGR	High-Temperature Gas Reactor
HTR	High-Temperature Reactor
HWGCR	Heavy Water Gas-Cooled Reactor
HWRR	Heavy Water Research Reactor
IAEA	International Atomic Energy Agency
ICAN	International Campaign to Abolish Nuclear Weapons
ICBM	InterContinental Ballistic Missile
ICEDA	*Installation de conditionnement et d'entreposage de déchets activés* (Activated Waste Conditioning and Storage Facility in France)
IDN	International Decommissioning Network
IGCAR	Indira Gandhi Centre for Atomic Research
ILW	Intermediate Level Waste
INEEL	Idaho National Engineering and Environmental Laboratory
INF	Intermediate-Range Nuclear Forces Treaty
INL	Idaho National Laboratory
IPEN	*Instituto de Pesquisas Energéticas e Nucleare* (Spanish Institute of Energy Research and Nuclear Energy)
IRID	International Research Institute for Nuclear Decommissioning

IRSN	*Institut de radioprotection et de sûreté nucléaire* (French Institute for Radiological Protection and Nuclear Safety)
ISD	*In Situ* Decommissioning
ISDC	International Structure for Decommissioning Costing
ISFSI	Independent Spent Fuel Storage Installations
JCPoA	Joint Comprehensive Plan of Action
JHR	Jules Horowitz Reactor
LA	Low Activity
LAMA	*Laboratoire d'analyse des matériaux actifs* (French Laboratory for the Analysis of Active Materials)
LANL	Los Alamos National Laboratory
LEU	Low-Enriched Uranium
LL	Long Life
LLW	Low-Level Waste
LPC	*Laboratoire de purification chimique* (French Chemical Purification Laboratory)
LWR	Light-Water Reactor
MARSSIM	Multi-Agency Radiation Survey and Site Investigation Manual
MTCR	Missile Technology Export Control Regime
MTR	Material Test Reactor
NDA	Non-Destructive Analysis
NDA	Nuclear Decommissioning Authority
NDP	Nuclear Decommissioning Projects and Programs

NEA	Nuclear Energy Agency of the OECD
NGO	Non-Governmental Organization
NNSA	National Nuclear Safety Administration
NNWS	Non-Nuclear Weapon States
NPT	Nuclear Non-Proliferation Treaty
NRC	Nuclear Regulatory Commission
NSA	Nuclear Submarine Attack
NSG	Nuclear Suppliers Group
NWS	Nuclear-Weapon States
OKG	*Oskarshamns Kraftgrupp*
PAC	Public Accounts Committee
PFP	Plutonium Finishing Plant
PHWR	Pressurized Heavy Water Reactor
PNGMDR	*Plan national de gestion des matières et des déchets radioactifs* (French National Radioactive Materials and Waste Management Plan)
POCO	Post-Operation Clean Out
PUREX	Plutonium and Uranium Refining by Extraction
PWR	Pressurized Water Reactor
RBMK	*Reaktor Bolshoy Moshchnosti Kanalniy* (High-power Channel-type Reactor)
RCD	Recovery and Conditioning of Old Radioactive Waste
RES	*Réacteur d'Essais* (Land Experimental Reactor for Naval Nuclear Propulsion)

RNR-G	Gas-Cooled Fast Neutron Reactor
ROSATOM	Russian State Atomic Energy Corporation
SALT	Strategic Arms Limitation Talks
SAR	Safety Analysis Report
SER	Safety Evaluation Report
SIPRI	Stockholm International Peace Research Institute
SL	Short Life
SLA	Saclay Linear Accelerator
SLOWPOKE	Safe Low-Power Critical Experiment
SORT	Strategic Offensive Reduction Treaty
SRS	Savannah River Site
SSBN	Sub-Surface Ballistic Nuclear
SSD	Special Session Disarmament
START	Strategic Arms Reduction Treaty
THORP	Thermal Oxide Reprocessing Plant
TPNW	Treaty on the Prohibition of Nuclear Weapons
TRIGA	Training, Research, Isotopes, General Atomic
tSWU	ton Separative Work Units
UKNI	United Kingdom–Norway Initiative
UN	United Nations
US NRC	United States Nuclear Regulation Commission
VHA	Very High Activity

VHTR	Very High-Temperature Reactor
VLLW	Very Low-Level Waste
VSA	Voluntary Submission Agreements
VSL	Very Short Life
WMD	Weapons of Mass Destruction
WPDD	Working Party on Decommissioning and Dismantling
WWER	Water-Water Energy Reactor
ZPPR	Zero Power Physics Reactor

References

[ABB 18] Abbas H., *Pakistan's Nuclear Bomb: A Story of Defiance, Deterrence and Deviance*, Penguin Books, Allen Lane, 2018.

[ABE 14] Abelquist E.W., *Decommissioning Health Physics: A Handbook for MARSSIM Users*, 2nd ed., CRC Press, Boca Raton, 2014.

[ABR 09] Abraham I., *South Asian Cultures of the Bomb: Atomic Publics and the State in India and Pakistan*, Indiana University Press, Bloomington & Indianapolis, 2009.

[ACK 17] Ackerman T., Methods for the radiological characterisation of the FiR 1 TRIGA research reactor decommissioning waste, Thesis, Stellenbosch University, Finland, March 2017.

[AE 16] AE (Autorité environnementale), "Avis délibéré de l'Autorité environnementale sur le démantèlement de l'usine Georges Besse I d'Eurodif production (26)", *AE*, no. 2016-86, 2016.

[AEN 05a] AEN, "Le démantèlement d'installations nucléaires : on peut le faire", *NEA, WPDD*, no. 6086, 2005.

[AEN 05b] AEN, "Vers la réalisation d'un dossier de sûreté de démantèlement : rapport de synthèse préparé au nom du WPDD par son Groupe d'étude sur l'analyse du dossier de sûreté de démantèlement", *NEA*, no. 6073, 2005.

[AEN 06a] AEN, "Choisir des stratégies de démantèlement des installations nucléaires. Rapport de synthèse", *NEA*, no. 6160, 2006.

[AEN 06b] AEN, "Libération des sites des installations nucléaires : rapport de synthèse", *NEA*, no. 6188, 2006.

[AEN 08] AEN, "Dispositions réglementaires pour le déclassement des installations nucléaires. Points importants et pratiques émergentes", *NEA*, no. 6402, 2008.

[AEN 10] AEN, "Démantèlement des installations nucléaires", *NEA*, no. 6830, 2010.

[AEN 19] AEN, 2019 Rapport annuel, available at: https://oecd-nea.org/pub/activities/ar2019/ar2019-fr.pdf, 2019.

[AIE 70] AIEA, "Traité sur la non-prolifération des armes nucléaire (1970)", Nuclear Non-Proliferation Treaty [in French], signed in London, Moscow and Washington on July 1, 1968 and entered into force on March 5, 1970.

[AIE 75] AIEA, "Structure et contenu des accords a conclure entre l'Agence et les Etats dans le cadre du traité sur la non-prolifération des armes nucléaires", INFCIRC/153, March 1975.

[AIE 97] AIEA, "Modèle de protocole additionnel à l'accord (aux accords) entre un état (des états) et l'Agence internationale de l'énergie atomique relatif(s) à l'application de garantie", *INFCIRC/540*, 1997.

[AIE 07] AIEA, *Glossaire de sûreté de l'AIEA. Terminologie employée en sûreté nucléaire et radioprotection*, IAEA, Vienna, 2007.

[AIE 16a] AIEA, "Les garanties de l'AIEA au service de la non-prolifération nucléaire", *IAEA Bulletin*, pp. 4–7, June 2016.

[AIE 16b] AIEA, "L'Iran et l'AIEA : vérification et contrôle au titre du PAGC", *IAEA Bulletin*, pp. 26–27, June 2016.

[AIE 17] AIEA, "Déclassement des installations", *Normes de sûreté de l'AIEA pour la protection des personnes et de l'environnement*, Prescriptions générales de sûreté Partie 6, no. GSR, part 6, Publication P1652_F_web.pdf, Vienna, 2017.

[AIE 20] AIEA, Base de données sur les réacteurs de recherche (RRDB), available at: https://www.iaea.org/fr/ressources/base-de-donnees-sur-les-reacteurs-de-recherche-rrdb, 2020.

[ALB 94] ALBRIGHT D., "South Africa and the affordable bomb", *Bulletin of the Atomic Scientists*, vol. 50, no. 4, pp. 37–47, 1994.

[ALB 16] ALBRIGHT D., *Revisiting South Africa's Nuclear Weapons Program: Its History, Dismantlement, and Lessons for Today*, The Naval Postgraduate School Institutional Archive, Calhoun, 2016.

[ALB 17] ALBRIGHT D., "North Korea's nuclear capabilities: A fresh look", *Institute for Science and International Security*, available at: https://isis-online.org/uploads/isis-reports/documents/North_Korea_Nuclear_Capability_Estimates_Summary_28Apr2017_Final.pdf, April 22, 2017.

[AMA 16] AMANO Y., "Les garanties de l'AIEA, une contribution vitale à la paix et à la sécurité internationales", *IAEA Bulletin*, p. 1, June 2016.

[AMF 19] AMFT M., LEISVIK M., CARROLL S., "Applying and adapting the Swedish regulatory system for decommissioning to nuclear power reactors – The regulator's perspective", *Journal of Environmental Radioactivity*, vol. 196, pp. 181–186, January 2019.

[AMI 13] AMIARD J.C., *Le risque radioactif. Devenir des radionucléides dans l'environnement et impacts sur la santé*, Lavoisier, Tec&Doc, Paris, 2013.

[AMI 17] AMIARD J.C., *Les risques chimiques environnementaux. Méthodes d'évaluation et impacts sur les organismes*, 2nd ed., Lavoisier, Tec&Doc, Paris, 2017.

[AMI 18] AMIARD J.C., *Military Nuclear Accidents: Environmental, Ecological, Health and Socio-economic Consequences*, ISTE Ltd, London and John Wiley & Sons, New York, 2018.

[AMI 19a] AMIARD J.C., *Industrial and Medical Nuclear Accidents: Environmental, Ecological, Health and Socio-economic Consequences*, ISTE Ltd, London and John Wiley & Sons, New York. 2019.

[AMI 19b] AMIARD J.C., *Nuclear Accidents: Prevention and Management of an Accidental Crisis*, ISTE Ltd, John Wiley & Sons, New York, 2019.

[ANC 17] ANCCLI, Quelles conditions pour une participation influente des cli et de l'ANCCLI au suivi territorial et national des chantiers de démantèlement ?, White book, no. 6, January 2017.

[AND 18] ANDREASEN S., WILLIAMS I., ROSE B. *et al.*, *Building a Safe, Secure, and Credible NATO Nuclear Posture*, Nuclear Threat Initiative, January 2018.

[ANO 07] ANONYMOUS, "Hanford Site Tour Script (Public)", October 2007.

[ANO 17a] ANONYMOUS, "Chooz A : EDF entame le démantèlement de la cuve du réacteur", *Lenergeek*, available at: https://lenergeek.com/2017/04/26/centrale-de-chooz-edf-demantelement-reacteur/, 26 April 2017.

[ANO 17b] ANONYMOUS, "Démantèlement de Creys, Superphénix : le chantier de démantèlement franchit une nouvelle étape", *L'EnerGeek,* available at: https://lenergeek.com/2017/06/06/superphenix-creys-malville-demantelement-nucleaire-edf/, 2017.

[ANO 19a] ANONYMOUS, "Une nouvelle ère de la dissuasion nucléaire", *Le Monde*, August 3, 2019.

[ANO 19b] ANONYMOUS, "La course aux missiles balistiques reprend de plus belle entre la Russie et les Etats-Unis", *L'Opinion*, available at: https://www.lopinion.fr/edition/international/course-aux-missiles-balistiques-reprend-plus-belle-entre-russie-etats-195261, August 20, 2019.

[ANO 20] ANONYMOUS, Destination orbite : les lanceurs spatiaux dans le monde, available at: https://lanceurs.destination-orbite.net/teledetection.php, 2020.

[ARN 16] ARNAUD, La Corée du Nord dispose-t-elle d'une aviation de frappe nucléaire ? Avions légendaires, available at: https://www.avionslegendaires.net/2016/01/actu/coree-nord-dispose-t-dune-aviation-de-frappe-nucleaire/, January 7, 2016.

[ASN 09] ASN, Déclassement de l'installation nucléaire de base n 41 (réacteur HARMONIE – CEA Cadarache – Bouches-du-Rhône), Decision no. 2009-DC-0133 of March 31, 2009 by the ASN, 2009.

[ASN 16a] ASN, "Guide n 6 relatif à l'arrêt définitif, au démantèlement et au déclassement des installations nucléaires de base", *ASN*, 2016.

[ASN 16b] ASN, "Guide n 14 relatif à l'assainissement des structures dans les installations nucléaires de base", *ASN*, 2016.

[ASN 16c] ASN, "Guide n 24 relatif à la gestion des sols pollués par les activités d'une installation nucléaire de base", *ASN*, 2016.

[ASN 19a] ASN, Le démantèlement des installations nucléaires, available at: https://www.asn.fr/Informer/Dossiers-pedagogiques/Le-demantelement-des-installations-nucleaires, 2019.

[ASN 19b] ASN, Décision no. 2019-DC-0682 de l'ASN du 12 novembre 2019, available at: https://www.asn.fr/Reglementer/Bulletin-officiel-de-l-ASN/Installations-nucleaires/Decisions-individuelles/Decision-n-2019-DC-0682-de-l-ASN-du-12-novembre-2019, 2019.

[ASN 19c] ASN, Installation de conditionnement et d'entreposage des déchets activés (Iceda), available at: https://www.asn.fr/L-ASN/L-ASN-en-region/Auvergne-Rhone-Alpes/Installations-nucleaires/Installation-de-conditionnement-et-d-entreposage-des-dechets-actives-Iceda, 2019.

[ASN 19d] ASN, Rapport de l'ASN sur l'état de la sûreté nucléaire et de la radioprotection en France en 2018, Annual report, ASN, 2019.

[ASN 20] ASN, Site de Cadarache, available at: https://www.asn.fr/L-ASN/L-ASN-en-region/Provence-Alpes-Cote-d-Azur/Site-de-Cadarache, May 29, 2020.

[ATY 12] ATYEO P.J., HOLDROYD S.D., "Site clearance and licence termination in nuclear decommissioning projects", in LARAIA M. (ed.), *Nuclear Decommissioning. Planning, Execution and International Experience*, Woodhead Publishing, Oxford, 2012.

[AUB 17] AUBERT J., ROMAGNAN B., Faisabilité technique et financière du démantèlement des installations nucléaires de base, Report no. 4428 by the Mission d'information, Commission du développement durable et de l'aménagement du territoire, Assemblée national, February 1, 2017.

[BAE 12] BAECKER A., "Decommissioning of Russian-type watercooled water-moderated nuclear reactors (WWERs)", in LARAIA M. (ed.), *Nuclear Decommissioning. Planning, Execution and International Experience*, Woodhead Publishing, Oxford, 2012.

[BAR 00] BARBEY P., DROMBRY-RINGEARD C., GERMAIN P. *et al.*, Rapport concernant le percement de la conduite de rejet en mer de l'usine de retraitement de La Hague (1979–80) pour le Groupe Radioécologie Nord-Cotentin, GRNC, September 2000.

[BAT 97] BATAILLE C., L'évaluation de la recherche sur la gestion des déchets nucléaires à haute activité. Tome II : les déchets militaires, Report no. 179, Assemblée Nationale, available at: http://www.senat.fr/rap/o97-179/o97-179_mono.html, 1997.

[BAT 18] BAT J.P., "Quand l'Afrique du sud de l'Apartheid voulait la bombe", *Libération*, available at: http://libeafrica4.blogs.liberation.fr/2018/04/08/quant-lafrique-du-sud-de-lapartheid-voulait-la-bombe/, April 8, 2018.

[BAU 17] BAUMGARTNER P., BILINSKY D., KOROLL G. *et al.*, "Environmental impact statement: *In situ* decommissioning of wr-1 at the Whiteshell Laboratories site, Pinawa, Manitoba", Rev. 1, available at: https://iaac-aeic.gc.ca/050/documents/p80124/121192E.pdf, 2017.

[BAY 03] BAYLISS C.R., LANGLEY K.F., *Nuclear Decommissioning, Waste Management, and Environmental Site Remediation*, Elsevier, Amsterdam, 2003.

[BEC 12] BECKITT S., "Decommissioning of legacy nuclear waste sites: Dounreay, UK" in LARAIA, M. (ed.), *Nuclear Decommissioning. Planning, Execution and International Experience*, Woodhead Publishing, Oxford, 2012.

[BEL 11] BELGOPROCESS, Eurochemic decommissioning, available at: http://www.eurochemic.be/eng/ontmanteling.html, 2011.

[BER 15] BERGREN C., LONG T., "Deactivation & decommissioning at SRS", *Savannah River Nuclear Solutions*, available at: https://www.srnl.doe.gov/events/docs/Bergren-Long%20D&D%20at%20SRS%20Rev1.pdf, December 8, 2015.

[BER 19] BERGH N., "Lessons learned from international decommissioning planning", *ICOND 2019, International Conference on Nuclear Decommissioning*, Book of Abstracts, p. 49, 2019.

[BEU 17] BEUVE J.L., "L'interminable démantèlement de la centrale nucléaire de Brennilis", *Le Point*, April 11, 2017.

[BIA 05] BIAD A., "L'Armement nucléaire israélien, un tabou", *Annuaire français de relations internationales*, Bruylant, La Documentation française, pp. 712–725, available at: http://www.diplomatie.gouv.fr/fr/IMG/pdf/49_712-725.pdf, 2005.

[BIS 01] BISCHOFF H., "Decommissioning of nuclear installations in the research framework programmes of the EC", *NUCLEAR DECOM 2001: Ensuring Safe, Secure, and Successful Decommissioning (Proc. Int. IMechE Conf. London, 2001)*, Professional Engineering Publishing Ltd, London, available at: http://www.eu-decom.be/introduction/initintroduction.htm, 2001.

[BLA 17] BLACHIER J.P., GUERNET G., RIOUFOL R., Enquête publique "Demande d'autorisation de mise à l'arrêt définitif et de démantèlement de l'installation nucléaire de base n. 93 (usine Georges Besse), EURODIF du jeudi 12 janvier 2017 au vendredi 10 février 2017", Report, Commission et conclusions motivées, 2017.

[BOD 17] BODEN S., ROGIERS B., VERSTREPEN, G. *et al.*, "WP3. Sampling strategy. Sampling plan for use case 2 BR3 Bioshield", *Improved Nuclear Site Characterization for Waste Minimization in DD Operations under Constrained Environment Research and Innovation Action (INSIDER)*, NFRP-2016-2017-1, 2017.

[BON 04] BOND A., PALERM J., HAIGH P., "Public participation in EIA of nuclear power plant decommissioning projects: A case study analysis", *Environmental Impact Assessment Review*, vol. 24, pp. 617–641, 2004.

[BOU 14] BOUVERET P., "Les ONG, moteur du désarmement", *Revue internationale et stratégique*, no. 96, pp. 123–131, 2014.

[BOU 15] BOUCAU J., "Opérations de découpe d'interne de réacteurs et aspects sûreté", *ASN, "Les enjeux du démantèlement"*, *Contrôle*, no. 199, pp. 29–30, 2015.

[BOU 17] BOUVERET P., COLLIN J.M., Le refus français de faire progresser le désarmement nucléaire va à l'encontre de nos valeurs humanistes, available at: https://www.huffingtonpost.fr/patrice-bouveret/le-refus-francais-de-faire-progresser-le-desarmement-nucleaire-v_a_22013682/, March 27, 2017.

[BOW 18] BOWEN W.Q., ELBAHTIMY H., HOBBS C. *et al.*, *Trust in Nuclear Disarmament Verification*, Palgrave Macmillan, Cham, 2018.

[BRA 96] BRADLEY D.J., "Overview of contamination from U.S. and Russian nuclear Complexes", in LE SAGE L.G., SARKISOV A.A. (eds), *"Nuclear Submarine Decommissioning and Related Problems"*, *Proceedings of the NATO Advanced Research Workshop on Nuclear Submarine Decommissioning and Related Problems*, Moscow, Russia, June 19–22, 1995, Kluwer Academic Publishers, Dordrecht, pp. 299–320, 1996.

[BRA 16] BRAUN C., HECKER S., LAWRENCE C. *et al.*, North Korean nuclear facilities after the agreed framework, available at: https://fsi-live.s3.us-west-1.amazonaws.com/s3fs-public/khucisacfinalreport_compressed.pdf, 2016.

[BRE 16] BRENDEBACH B., "Decommissioning of nuclear facilities: Germany's experience", *IAEA Bulletin*, vol. 57, no. 1, pp. 24–25, April 2016.

[BRI 01] BRIGGS J.D., Historical time line and information about the Hanford site, Richland, Washington, Report, PNNL-13524, Pacific Northwest National Laboratory, 2001.

[BRO 12] BROWN G.A., "Decommissioning of gas-cooled nuclear reactors (GCRs)", in LARAIA M. (ed.), *Nuclear Decommissioning. Planning, Execution and International Experience*, Woodhead Publishing, Oxford, 2012.

[BRU 19] BRUNETTI G., "Nuclear decommissioning in the EU. Strategies and funding", *ICOND 2019, International Conference on Nuclear Decommissioning*, Book of Abstracts, p. 34, 2019.

[BUK 95] BUKHARI O., HANDLER J., "Russian nuclear-powered submarine decommissioning", *Science & Global Security*, vol. 5. pp. 245–271, 1995.

[BYR 01] BYRD DAVIS M., *La France nucléaire. Matières et sites*, Wise, Paris, 2001.

[CAB 17] CABALLERO-ANTHONY M., TRAJANO J.C.I., "Enhancing nuclear energy cooperation in ASEAN: Regional norms and challenges", in VAN NESS P., GURTOV M. (eds), *Learning from Fukushima Nuclear Power in East Asia*, Australian National University Press, 2017.

[CAP 98] CAPETTE I., "Les essais nucléaires indiens et pakistanais : un défi lancé à la non-prolifération nucléaire", *Actualité et Droit International*, available at: http://www.ridi.org/adi/199812a3.html, 1998.

[CAR 18] CARLTON D., "International systemic features inhibiting disarmament and arms control", in FORADORI P., GIACOMELLO G., PASCOLINI A. (eds), *Arms Control and Disarmament. 50 Years of Experience in Nuclear Education*, Palgrave Macmillan, Cham, 2018.

[CCS 07] CCSIN (Conseil supérieur de la sûreté et de l'information nucléaires), Rapport du CSSIN sur le démantèlement des installations nucléaires de base, Report, May 2007.

[CCS 19] CCSN (Commission canadienne de sûreté nucléaire), Centrales canadiennes, available at: https://www.cnsc-ccsn.gc.ca/fra/reactors/power-plants/index.cfm, March 19, 2019.

[CE 05] CE, European website on decommissioning of nuclear installations, available at: https://web.archive.org/web/20070704090027/http:/www.eu-decom.be/siteentrance/, 2005.

[CEA 06] CEA, *Les réacteurs nucléaires à caloporteur gaz*, Editions Le Moniteur, Paris, 2006.

[CEA 10] CEA, "Étude d'impact, Préambule", Dossier de demande d'autorisation de mise à l'arrêt définitif et de démantèlement, et de traitement des objets sodés du CEA, INB 71, Centrale Phénix, Folder 2, Piece 7, available at: http://www.gard.gouv.fr/content/download/10622/69497/file/PHENIX%20Etude%20d%27impact.pdf, 2010.

[CEA 12a] CEA, Les réacteurs à neutrons rapides de 4e génération à caloporteur gaz – le réacteur expérimental Allegro. Les autres filières à neutrons rapides de 4e génération. *CEA, DEN*, vol. 4, available at: http://www.cea.fr/multimedia/Documents/publications/rapports/rapport-gestion-durable-matieres-nucleaires/Tome%204.pdf, December 2012.

[CEA 12b] CEA, Les réacteurs nucléaires expérimentaux, Mongraph, Monograph, DEN, 2012.

[CEA 14] CEA, Les réacteurs nucléaires à caloporteur sodium, Monograph, DEN, Editions Le Moniteur, 2014.

[CEA 15a] CEA, De l'assainissement-démantèlement des anciennes installations nucléaires au pôle de valorisation des sites industriels, Press kit, April 1, 2015.

[CEA 15b] CEA, Démantèlement, Press kit, April 4, 2015.

[CEA 15c] CEA, Avancées des recherches sur la séparation-transmutation et le multi-recyclage du plutonium dans les réacteurs à flux de neutrons rapides, Report, DEN, 2015.

[CEA 17] CEA, L'assainissement-démantèlement des installations nucléaires, Mongraph, DEN, Editions Le Moniteur, 2017.

[CEA 18] CEA, L'instrumentation et la mesure en milieu nucléaire, Monograph, DEN, Editions Le Moniteur, 2018.

[CEA 19] CEA, ELECNUC : les centrales nucléaires dans le monde : Nuclear power plants in the world, available at: https://www.cea.fr/multimedia/Documents/publications/ouvrages/Elecnuc-2019.pdf, 2019.

[CHA 92] CHALINOR S.F., "B204 reprocessing plant decommissioning project at Sellafield", Mechanical Engineering Publications Ltd, London, United Kingdom, *Nuclear Decom '92: Decommissioning of Radioactive Facilities International Conference*, London, United Kingdom, February 17–19, 1992, pp. 55–66, 1992.

[CHA 18] CHAO N., LIU Y.K., XIA H. *et al.*, "A dose assessment method for arbitrary geometries with virtual reality in the nuclear facilities decommissioning", *Radiation Physics and Chemistry*, vol. 144, pp. 238–247, March 2018.

[CHE 10] CHEVENEMENT J.P., Désarmement, non-prolifération nucléaires et sécurité de la France, Information report no, 332, Sénat de la République française, available at: https://www.senat.fr/rap/r09-332/r09-332_mono.html, 2010.

[CHE 18] CHEUTET V., SEKHARI A., CORBEAUX N., APLM approach to support nuclear decommissioning process. *DEM2018: International Conference on Dismantling Challenges: Industrial Reality, Prospects and Feedback Experience*, Avignon, France, October 2018.

[CHI 96] CHILDRESS P.C., ZOTOV A.P., "Liquid waste processing", in LE SAGE L.G., SARKISOV A.A. (eds), *Nuclear Submarine Decommissioning and Related Problems*, Kluwer Academic Publishers, Dordrecht, 1996.

[CHI 03] CHIRON M., DIOP C.M., GAL O. *et al.*, "The use of three dedicated nuclear measurement tools for D&D operations in nuclear facilities", *International Conference on Dismantling Nuclear Facilities. Decommissioning Challenges: An Industrial Reality*? Avignon, France, available at: https://mirion.s3.amazonaws.com/cms4_mirion/files/pdf/technical-papers/3dedinuctools-paper.pdf?1534968952, November 23–28, 2003.

[CHI 19] CHI Y.C., "Current status of nuclear decommissioning in Taiwan", *ICOND 2019, International Conference on Nuclear Decommissioning*, Book of Abstracts, p. 47, 2019.

[CHU 01] CHUEN C., MOLTZ J.C., "Nuclear submarine dismantlement", *NTI*, available at: https://www.nti.org/analysis/articles/nuclear-submarine-dismantlement/, 2001.

[CLI 19] CLI, Les installations nucléaires du centre CEA de Cadarache, CLIC Info no. 66, available at: http://cli-cadarache.org/fileadmin/user_upload/Cadarache/FICHES_DOSSIER/FICHE_DOSSIER_INSTALLATIONS_NUCLEAIRES_DU_CEA_CADARACHE_2019.pdf, 2019.

[COH 98] COHEN A., *Israel and the Bomb*, Columbia University Press, New York, 1998.

[COH 05] COHEN Y., *Whistleblowers and the Bomb. Vanunu, Israel and Nuclear Secrecy*, Pluto Press, London, 2005.

[COH 10] COHEN A., *The Worst-kept Secret: Israel's Bargain with the Bomb*, Columbia University Press, New York, 2010.

[COL 15] COLLIN J.M., DRAIN M., NORLAIN B. *et al.*, Les traités de contrôle et de non-prolifération nucléaire, Fiche d'analyse dossier : le désarmement, la prolifération et la non-prolifération nucléaires, available at: http://www.irenees.net/bdf_fiche-analyse-1045_fr.html, April 2015.

[COL 18a] COLLIN J.M., "Le traité des armes sur l'interdiction nucléaires", *Les rapports du GRIP*, 2018.

[COL 18b] COLLIN J.M., "Les armes nucléaires constituent un danger permanent et bien réel pour la communauté internationale", IRIS (Institut de Relations Internationales et Stratégiques) interview, available at: https://www.iris-france.org/123916-les-armes-nucleaires-constituent-un-danger-permanent-et-bien-reel-pour-la-communaute-internationale/, November 16, 2018.

[CON 15] CONNAISSANCE DES ENERGIES, Lutte contre la prolifération des armes nucléaires, available at: https://www.connaissancedesenergies.org/fiche-pedagogique/lutte-contre-la-proliferation-des-armes-nucleaires, 2015.

[COR 17] CORE (Cumbrians Opposed to a Radioactive Environment), Change to Sellafield skyline as demolition of "seismic risk" stack begins, available at: http://corecumbria.co.uk/briefings/change-to-sellafield-skyline-as-demolition-of-seismic-risk-stack-begins/, November 16, 2017.

[COT 18] COTTIN S., "En Bretagne, le fantôme oublié d'une centrale nucléaire", *Sud-ouest Dimanche*, 28 January 2018.

[COU 05] COUR DES COMPTES, Le démantèlement des installations nucléaires et la gestion des déchets radioactifs, Report to the French President following responses from the administrations and organizations concerned, January 2005.

[COU 12] COUTURE P., "1,5 milliard $ pour démanteler Gentilly-2", *Le Soleil*, September 27, 2012.

[COU 16] COUR DES COMPTES EUROPÉENNES, Programmes d'assistance de l'UE au déclassement d'installations nucléaires en Lituanie, en Bulgarie et en Slovaquie : défis cruciaux en perspective malgré les progrès accomplis depuis 2011, Special report no. 22, 2016.

[COU 20] COUR DES COMPTES, L'arrêt et le démantèlement des installations nucléaires, Communication to the French Senate Finance Committee, February 2020.

[CRO 12] CROSS M.T., GREEN H., ADSLEY I., "Characterisation of radioactive materials in redundant nuclear facilities: Key issues for the decommissioning plan", in LARAIA M. (ed.), *Nuclear Decommissioning. Planning, Execution and International Experience*, Woodhead Publishing, Oxford, 2012.

[CUI 18] CUI X., LIU Y., PENG W. *et al.*, "Energetic particles' fluxes and dose in the radiation gene box measured by space radiation detector onboard SJ-10 satellite", *Radiation Detection Technology and Methods*, vol. 2, no. 43, 2018.

[DAM 08] DAMOY F., "Démantèlement INB 43-48 (accélérateurs)", *SFRP*, Radioprotection et Environnement, February 18–19, 2008.

[DAN 96] DANILYAN V.A., VYSOTSKY V.L., "Safety issues when handling spent nuclear fuel and radioactive waste in the Pacific fleet of the Russian federation", in LE SAGE L.G., SARKISOV A.A. (eds), *Nuclear Submarine Decommissioning and Related Problems, Proceedings of the NATO Advanced Research Workshop on Nuclear Submarine Decommissioning and Related Problems*, Moscow, Russia, June 19–22, 1995, Kluwer Academic Publishers, Dordrecht, 1996.

[DEC 17] DE CHAMPCHESNEL T., "Vers l'interdiction des armes nucléaires ? Autour de l'attribution du prix Nobel de la paix à l'ONG antinucléaire ICAN", *IRSEM, Note de Recherche*, no. 49, p. 13, December 2017.

[DEG 18] DE GEER L.E., WRIGHT C.M., "The 22 September 1979 Vela Incident: Radionuclide and hydroacoustic evidence for a nuclear explosion", *Science & Global Security*, vol. 26, no. 1, pp. 1–35, 2018.

[DEL 85] DELANEY E.G., "Le déclassement des installations nucléaires du Département de l'énergie des Etats-Unis", *AIEA Bulletin*, p. 30, winter 1985.

[DES 13] DESTREZ J.M., NOKHAMZON J.G., "Démantèlement des installations nucléaires", *CEA/DEN/DADN*, Visioatome, May 14, 2013.

[DEV 12] DEVGUN J.S., "Nuclear decommissioning project organization, management and human resources", in LARAIA M. (ed.), *Nuclear Decommissioning. Planning, Execution and International Experience*, Woodhead Publishing, Oxford, 2012.

[DGA 10] DGA, MINISTERE DE LA DEFENSE, Le démantèlement des bâtiments à propulsion nucléaire, available at: http://www.defense.gouv.fr/dga/equipement/dissuasion/le-demantelement-des-batiments-a-propulsion-nucleaire, July 11, 2010.

[DGR 17a] DGRIS, Lutte contre la prolifération des armes de destruction massive et désarmement, available at: https://www.defense.gouv.fr/english/dgris/security-issues/counter-proliferation/lutte-contre-la-proliferation-des-armes-de-destruction-massive-et-desarmement-fr, 2017.

[DGR 17b] DGRIS, Non-prolifération et désarmement nucléaire, available at: https://www.defense.gouv.fr/dgris/enjeux-transverses/lutte-contre-la-proliferation/non-proliferation-et-desarmement-nucleaire, 2017.

[DIX 16] DIXIT A., "Vérification par frottis : collecte et analyse d'échantillons de l'environnement", *IAEA Bulletin*, pp. 22–23, June 2016.

[DOD 10] DoD, Declaration among the department of defense of the United States of America, the Royal Ministry of Defence of the Kingdom of Norway, and the Ministry of Defence of the Russian Federation, on Arctic Military Environmental Cooperation, Report, Department of Defense Environment, Safety and Occupational Health Network and Information Exchange, March 16, 2010.

[DOE 16] DoE, Plan for deactivation and decommissioning nonoperational defense nuclear facilities, Report to Congress, available at: https://www.energy.gov/sites/default/files/2017/01/f34/Deactivation%20and%20Decommissioning%20Report%20to%20Congress.pdf, December 2016.

[DOE 19] DoE, FY 2020 congressional budget request. Environmental management, Report, DOE/CF-0155, vol. 5, March 2019.

[DOE 20] DoE, Dismantlement Y12, available at: https://www.y12.doe.gov/mission/nuclear-deterrence/stockpile/dismantlement, 2020.

[DOR 15] DORISON A., "La valorisation des déchets de très faible activité en dehors de la filière nucléaire est envisageable, mais elle nécessiterait des précautions particulières", *ASN, "Les enjeux du démantèlement", Contrôle*, no. 199, pp. 25–26, 2015.

[DUA 19] DUARTE S., "L'avenir du traité d'interdiction complète des essais nucléaires", *Chronique ONU*, available at: https://www.un.org/fr/chronicle/article/lavenir-du-traite-dinterdiction-complete-des-essais-nucleaires, 2019.

[DUG 15] DUGUEY M., "Stratégie de démantèlement de l'usine Georges Besse d'Eurodif et enjeux associés", *ASN, "Les enjeux du démantèlement", Contrôle*, no. 199, pp. 21–22, 2015.

[DUP 13] DUPRAZ B., Démantèlement des installations Défense du CEA, DSND presentation to the HCTISN, available at: http://www.hctisn.fr/IMG/pdf/Presentation_DSND_Demantelement_INBS_cle4a8e91.pdf, December 12, 2013.

[DUR 18] DURHAM J.M., POULSON D., BACON J. *et al.*, "Verification of spent nuclear fuel in sealed dry storage casks via measurements of cosmic ray muon scattering", arXiv:1710.03098v2 [nucl-ex], March 13, 2018.

[DUT 15] DUTZER M., "Optimiser globalement la gestion des déchets radioactifs de démantèlement", *ASN, "Les enjeux du démantèlement", Contrôle*, no. 199, pp. 27–28, 2015.

[DZI 17] DZINESA G.A., *Disarmament, Demobilization and Reintegration in Southern Africa. Swords into Ploughshares?* Palgrave Macmillan, Cham, 2017.

[EC 00] EC, "Practical use of the concepts of clearance and exemption. Part 1: Guidance on general clearance levels for practices", *Radiation Protection*, no. 122, Part 1, pp. 1–40, EC, Luxemburg, 2000.

[EC 15] EC, Technology readiness levels (TRL), Horizon 2020 work programme 2014–2015 general annexes, Extract, Part 19-commission decision C(2014)4995, available at: http://ec.europa.eu/research/participants/data/ref/h2020/wp/2014_2015/annexes/h2020-wp1415-annex-g-trl_en.pdf, 2015.

[EDF 14] EDF, Déconstruction de la centrale de Brennilis : les chantiers avancent, Press kit, November 2014.

[EDF 17] EDF, Précisions sur la faisabilité technique et financière du démantèlement des centrales nucléaires d'EDF en France, Press release, February 1, 2017.

[EDF 19] EDF, La déconstruction des centrales nucléaires, une responsabilité entièrement assumée par EDF, available at: https://www.edf.fr/groupe-edf/nos-energies/nucleaire/atouts/expertise-nucleaire/deconstruction-des-centrales-nucleaires, 2019.

[EDF 20a] EDF, Actualité de la centrale nucléaire en déconstruction de Brennilis, CLI des Monts d'Arrée, Plenary session, February 10, 2020.

[EDF 20b] EDF, ICEDA, Installation de Conditionnement et d'Entreposage de Déchets Activés, available at: https://www.edf.fr/groupe-edf/producteur-industriel/carte-des-implantations/centrale-nucleaire-du-bugey/iceda, 2020.

[EU 05a] EU, Decommissioning in short, available at: http://www.eu-decom.be/about/initabout.htm, 2005.

[EU 05b] EU, European commission decommissioning projects: Foreword, available at: http://www.eu-decom.be/europe/initeurope.htm, 2005.

[EUR 14] EURATOM, "Directive 2013/59/Euratom du Conseil du 5 décembre 2013 fixant les normes de base relatives à la protection sanitaire contre les dangers résultant de l'exposition aux rayonnements ionisants", *JO L 13*, p. 1, January 17, 2014.

[FAS 20a] FAS, "9. Plutonium Acquisition", *Plutonium: The First 50 Years. United States Plutonium Production, Acquisition, and Utilization from 1944 through 1994*, available at: https://fas.org/sgp/othergov/doe/pu50yc.html#ZZ16, 2020.

[FAS 20b] FAS, Nuclear facilities, available at: https://fas.org/nuke/guide/china/facility/nuclear.htm, 2020.

[FEI 14] FEIVESON H.A., GLASER A., MIAN Z. *et al.*, *Unmaking the Bomb. A Fissile Material Approach to Nuclear Disarmament and Nonproliferation*, The MIT Press, Cambridge, MA, 2014.

[FEL 12] FELLINGHAM L.R., "Environmental remediation and restoration technologies in nuclear decommissioning projects", in LARAIA M. (ed.), *Nuclear Decommissioning. Planning, Execution and International Experience*, Woodhead Publishing, pp. 416–447, 2012.

[FOU 12] FOURIE E., "Decommissioning of nuclear fuel cycle facilities", in LARAIA M. (ed.), *Nuclear Decommissioning. Planning, Execution and International Experience*, Woodhead Publishing, Oxford, 2012.

[FOU 16a] FOURNIER V., "Que trouve-t-on dans la mallette d'un inspecteur ?", *IAEA Bulletin*, pp. 12–15, June 2016.

[FOU 16b] FOURNIER V., "Des matières sous garanties surveillées 24h/24 et 7j/7", *IAEA Bulletin*, pp. 16–19, June 2016.

[FRA 15] FRANCE TNP, Traité interdisant la production de matières fissiles pour des armes nucléaires ou d'autres dispositifs explosifs nucléaires, available at: http://fissilematerials.org/library/france15f.pdf, 2015.

[FRA 19a] FRANCE TNP, Arrêt de la production de matières fissiles pour les armes nucléaires, available at: https://www.francetnp.gouv.fr/arret-de-la-production-de-matieres-fissiles-7, 2019.

[FRA 19b] FRANCE TNP, Le TNP en 8 questions-réponses, available at: https://www.francetnp.gouv.fr/le-tnp-en-8-questions-reponses-623, 2019.

[FRA 19c] FRANCE DIPLOMATIE, La question nucléaire iranienne, available at: https://www.diplomatie.gouv.fr/fr/politique-etrangere-de-la-france/securite-desarmement-et-non-proliferation/crises-et-conflits/la-question-nucleaire-iranienne/, August 2019.

[FRA 19d] FRANCE TNP, L'assistance au désarmement nucléaire, available at: https://www.francetnp.gouv.fr/l-assistance-au-desarmement-nucleaire-59, 2019.

[FRA 19e] FRANCE DIPLOMATIE, Travaux de l'AIEA sur la gestion des déchets nucléaires et le démantèlement, available at: https://onu-vienne.delegfrance.org/Travaux-de-l-AIEA-sur-la-gestion-des-dechets-nucleaires-et-le-demantelement, 2019.

[FRE 18] FRENCH R., CRYER, A., KAPELLMANN-ZAFRA G. *et al.*, "Evaluating the radiation tolerance of a robotic finger", *Towards Autonomous Robotic Systems 2018 (Taros 2018)*, July 25–27, 2018.

[GAI 17] GAIN N., Le désarmement nucléaire ralentit, available at: https://forcesoperations.com/le-desarmement-nucleaire-ralentit/, July 3, 2017.

[GAR 94] GARNER S.E., BOUNDS J.A., ALLANDER K.S. *et al.*, A compendium of results from long-range alpha detector soil surface monitoring: June 1992–May 1994, Report, LA-12861-MS, November 1994.

[GAR 18] GARNIER C., "Naval group commence à déconstruire les premiers sous-marins nucléaires", *L'Usine nouvelle*, September 12, 2018.

[GAZ 10] GAZAL S., AMIARD J.C., *Le tritium. Actualité d'aujourd'hui et de demain.* Lavoisier, Tec&Doc, Paris, 2010.

[GIL 18] GILLIS M., *Le désarmement : un guide de référence*, 4th ed., Bureau des affaires de désarmement, New York, 2018.

[GIR 16] GIRONES P., BACHELOT M., HAMON J.C. *et al.*, Une contribution pour le développement des TICE pour les disciplines de l'Assainissement et Démantèlement des installations nucléaires, Report, CEA-01340921v3, 2016.

[GLA 11] GLASER A., Military fissile material production and stocks in France, available at: https://pdfs.semanticscholar.org/7b93/70191395f81d592550847115e32b7c88160d.pdf, 2011.

[GLO 11] GLOBAL SECUTITY, ADVENA KENTRON CIRCLE, Weapons of Mass Destruction (WMD), available at: https://www.globalsecurity.org/wmd/world/rsa/advena.htm, 2011.

[GOI 18] GOIFFON V., ALLANCHE T., MULLER C. *et al.*, "CAMRAD: Development of a multi-megagray radiation hard CMOS camera for dismantling operations", *DEM 2018, Dismantling Challenges: Industrial Reality, Prospects and Feedback Experience*, Avignon, France, October 22–24, 2018.

[GOL 06] GOLDEMBERG J., "Lessons from the denuclearization of Brazil and Argentina", *Arms Control Today*, vol. 36, no. 3, pp. 41–43, April 2006.

[GOL 07] GOLDBLAT J., How to deter violations of disarmament and non-proliferation agreements, in assessing compliance with arms control treaties, Report, International Group on Global Security, pp. 54–62, September 2007.

[GOL 18] GOLDEMBERG J., ALVIM C.F., MAFRA O.Y., "The denuclearization of Brazil and Argentina", *Journal for Peace and Nuclear Disarmament*, no. 2, pp. 383–403, 2018.

[GON 03] GONZÂLEZ A.J., Opening address. Safe nuclear decommissioning: Need for an international common approach, in IAEA (ed.), *Safe Decommissioning for Nuclear Activities*, Publication 1154, pp. 11–24, 2003.

[GOR 17] GORBATCHEV A., Nucléaire en Russie. État des lieux et perspectives, Report, Ambassade de France en Russie, October 2017.

[GOT 20] GOTSCHAUX, F., Quatre des huit réacteurs des centrales nucléaires de Cruas et Tricastin devraient fermer d'ici 2035, France Bleue, available at: https://www.francebleu.fr/infos/faits-divers-justice/quatre-des-huit-reacteurs-des-centrales-nucleaires-de-cruas-et-tricastin-devraient-fermer-d-ici-2035-1579621930, 2020.

[GOU 18] GOUVERNEMENT DU CANADA, Interdiction de la production des matières fissiles pour les armes nucléaires, available at: https://www.international.gc.ca/world-monde/issues_development-enjeux_developpement/peace_security-paix_securite/banning_materials-interdire_materiaux.aspx?lang=fra, 2018.

[GRA 11] GRALL M., Rapport d'information déposé en application de l'article 145 du Règlement par la Commission de la défense nationale et des forces armées sur la fin de vie des équipements militaires, Assemblée nationale, Report no. 3251, March 16, 2011.

[GRI 12] GRIFFITHS C., "Decommissioning of small nuclear facilities: Industrial, medical and research facilities", in LARAIA M. (ed.), *Nuclear Decommissioning. Planning, Execution and International Experience*, Woodhead Publishing, Oxford, 2012.

[GRO 13] GROIZELEAU V., "Où en est le démantèlement des anciens SNLE français ?" *Mer et Marine*, July 30, 2013.

[GRS 16] GRS (Gesellschaft für Anlagen- und Reaktorsicherheit, gGmbH), Guide to the decommissioning, the safe enclosure and the dismantling of facilities or parts thereof as defined in section 7 of the Atomic Energy Act, Report, GRS, June 23, 2016.

[GUE 06a] GUERZONI R., Il decommissioning della centrale nucleare di Caorso, available at: http://francalombardi.altervista.org/Sito-Caorso/Decommissioning_Caorso.pdf, 2006.

[GUE 06b] GUERZONI R., "Decommissioningdi una centrale nucleare. Esperienze e programmidi", *GEOFLUID*, I siti contaminati Problemi di bonifica, October 4, 2006.

[GUS 96] GUSSGARD K., "Norwegian concerns regarding nuclear submarine decommission activities", in LE SAGE L.G., SARKISOV A.A. (eds), *Nuclear Submarine Decommissioning and Related Problems*, Proceedings of the NATO Advanced Research Workshop on Nuclear Submarine Decommissioning and Related Problems, Moscow, Russia, June 19–22, 1995, Kluwer Academic Publishers, Dordrecht, 1996.

[HAN 00] HANSON L.A., Radioactive waste contamination of soil and groundwater at the Hanford Site, available at: https://www.webpages.uidaho.edu/etox/resources/case_studies/hanford.pdf, 2000.

[HAN 20] HANFORD, Plutonium finishing plant demolition marks the end of an era, available at: https://www.hanford.gov/PFP, 2020.

[HAR 81] HARPER J.R., GARDE R., Decommissioning the Los Alamos molten plutonium reactor experiment (LAMPRE I), Report, LA-9052-MS, UC-70, 1981.

[HAR 19] HARIA K., DE FARAGO M., DAWOOD T. *et al.*, "Integration of regional and asset satellite observations for assessment of infrastructure resilience", *International Conference on Smart Infrastructure and Construction 2019 (ICSIC): Driving Data-informed Decision-making*, ICE, 2019.

[HAU 16] HAUCK D.K., RUSSELL I., "Review of the U.S.-U.K. warhead monitored dismantlement exercise", *57th Annual Meeting of the Institute of Nuclear Materials Management*, 2016-07-24/2016-07-28, Atlanta, GA, 2016.

[HEC 18] HECLA J.J., DANAGOULIAN A., "Nuclear disarmament verification via resonant phenomena", *Nature Communications*, no. 9, p. 1259, 2018.

[HEN 16] HENRIQUES S., "Une journée dans la vie d'un inspecteur des garanties", *IAEA Bulletin*, pp. 8–11, June 2016.

[HER 18] HERTEL O., "Les derniers jours d'un sous-marin nucléaire", *Sciences et Avenir*, no. 860, pp. 70–71, October 2018.

[HIR 19] HIRSCH D., "A failure of governmental candor: The fire at the contaminated Santa Susana field laboratory", *Bulletin of the Atomic Scientists*, available at: https://thebulletin.org/2019/02/a-failure-of-governmental-candor-the-fire-at-the-contaminated-santa-susana-field-laboratory/, February 21, 2019.

[HON 17] HONG Y., NAM J., PARK H., "Framework design for decommissioning information system on nuclear facilities", *Transactions of the Korean Nuclear Society Autumn Meeting Gyeongju*, South Korea, October 26–27, 2017.

[HOU 13] HOUSE OF COMMONS, Nuclear decommissioning authority: Managing risk at Sellafield, Committee of Public Accounts, Report HC 746, House of Commons, January 23, 2013.

[HOU 16] HOU X., "Radiochemical analysis for decommissioning of nuclear facilities", *9th International Conference on Nuclear and Radiochemistry – NRC9*, pp. 89–90, 2016.

[HSE 02] HSE, UKAEA's strategy for the decommissioning of its nuclear licensed sites. A review by HM Nuclear Installations Inspectorate, Report, HSE 2002.

[HSE 04] HSE, A review by HM Nuclear Installations Inspectorate: Urenco (Capenhurst) Ltd's strategy for decommissioning its nuclear licensed site, Report, HSE, 2004.

[IAE 95] IAEA, Safe enclosure of shut down nuclear installations, Technical Reports Series, no. 375, pp. 33–47, IAEA, Vienna, 1995.

[IAE 96] IAEA, Clearance levels for radionuclides in solid materials. Application of exemption principles, Interim report for comment, IAEA-TECDOC-8S5, IAEA, Vienna, 1996.

[IAE 04a] IAEA, Application of the concepts of exclusion, exemption and clearance. Safety guide, Safety Standards Series, N RS-G-1.7, Publication no. 1202, IAEA, Vienna, 2004.

[IAE 04b] IAEA, Format and content of the safety analysis report for nuclear power plants, IAEA Safety Standards Series, Safety Guide no. GS-G-4.1, Publication 1185, IAEA, Vienna, 2004.

[IAE 05] IAEA, Dismantling of contaminated stacks at nuclear facilities, Technical Reports Series No. 440, Publication TRS440, IAEA, Vienna, 2005.

[IAE 06] IAEA, Decommissioning of research reactors: Evolution, state of the art, open issues, Technical Reports Series, no. 446, Publication, IAEA, Vienna, 2006.

[IAE 13a] IAEA, Cost estimation for research reactor decommissioning, IAEA Nuclear Energy Series Technical No. NW-T-2.4, Publication Pub1596, IAEA, Vienna, 2013.

[IAE 13b] IAEA, Safety assessment for decommissioning, Safety Reports Series No. 77, Publication Pub1604, IAEA, Vienna, 2013.

[IAE 16] IAEA, Managing the unexpected in decommissioning, IAEA Nuclear Energy Series No. NW-T-2.8, Publication Pub1702, IAEA, Vienna, 2016.

[IAE 17a] IAEA, Data analysis and collection for costing of research reactor decommissioning, TECDOC-1832, IAEA, Vienna, available at: https://nucleus.iaea.org/sites/connect/IDNpublic/Documents/TE1832_web.pdf, 2017.

[IAE 17b] IAEA, Data analysis and collection for costing of research reactor decommissioning. Report of the DACCORD Collaborative Project, IAEA TECDOC SERIES, IAEA-TECDOC-1832, IAEA, Vienna, 2017.

[IAE 19a] IAEA, Vérification et contrôle en République islamique d'Iran à la lumière de la résolution 2231 (2015) du Conseil de sécurité de l'ONU, GOV/INF/2019/8, July 1, 2019.

[IAE 19b] IAEA, Methodology for assessing the induced activation source term for use in decommissioning applications, Safety Reports Series no. 95, Publication P1823, IAEA, Vienna, 2019.

[IAE 19c] IAEA, Management of project risks in decommissioning, Safety Reports Series no. 97, Publication Pub1839, IAEA, Vienna, 2019.

[IAE 19d] IAEA, Decommissioning after a nuclear accident: Approaches, techniques, practices and implementation considerations, IAEA Nuclear Energy Series No. NW-T-2.10, Publication Pub1811, IAEA, Vienna, 2019.

[IGU 17] IGUCHI Y., YANAGIHARA S., TEZUKA M. *et al.*, "Experiences of development of IT support systems for decommissioning of Fugen NPS", *Workshop on Current and Emerging Methods for Optimising Safety and Efficiency in Nuclear Decommissioning*, Sarpsborg, Norway, February 7–9, 2017.

[INV 17a] INVERNIZZI, D.C., LOCATELLI, G., BROOKES, N.J., "Managing social challenges in the nuclear decommissioning industry: A responsible approach towards better performance", *International Journal of Project Management*, vol. 35, no. 7, pp. 1350–1364, 2017.

[INV 17b] INVERNIZZI D.C., LOCATELLI G., BROOKES N.J., "How benchmarking can support the selection, planning and delivery of nuclear decommissioning projects", *Progress in Nuclear Energy*, vol. 99. pp. 155–164, 2017.

[IOR 18] IORGAA I., DEJUA R., PAVELESCUA A.O. *et al.*, "Technical methodology to evaluate the decommissioning of the contaminated underground structures belonging to the VVR-S nuclear research reactor", *Special Issue of the 7th International Advances in Applied Physics and Materials Science (APMAS 2017), Acta Physica Polonica A*, vol. 134, 2018.

[IRS 10] IRSN, Restitution des travaux du groupe de réflexion "Défense en profondeur tritium", Sources de production et gestion du tritium produit par les installations nucléaires, DSU report no. 217, 2010.

[IRS 16] IRSN, Les réacteurs RBMK, available at: https://www.irsn.fr/FR/connaissances/Installations_nucleaires/Les-accidents-nucleaires/accident-tchernobyl-1986/consequences-industrie-nucleaire/Pages/3-Les_reacteurs_RBMK.aspx#.XowwyHLgrQw, 2016.

[IRS 17] IRSN, Le démantèlement des installations nucléaires, Press kit, October 2017.

[IRS 18a] IRSN, Le démantèlement du réacteur nucléaire EDF de Chooz A, Report, 2018.

[IRS 18b] IRSN, Le démantèlement du réacteur Phénix, Report, 2018.

[IRS 18c] IRSN, La gestion des déchets issus des démantèlements, Report, 2018.

[IRS 19a] IRSN, AIEA, Historique, available at: https://non-proliferation.irsn.fr/Nucleaire/historique/Pages/HistoriqueAIEA.aspx, 2019.

[IRS 19b] IRSN, Les grandes dates de la non-prolifération nucléaire, available at: https://non-proliferation.irsn.fr/Nucleaire/historique/Pages/Lesgrandesdatesdelanonproliferationnucl%C3%A9aire.aspx, 2019.

[IRS 19c] IRSN, Le démantèlement des centrales et des installations nucléaires, Report, 2019.

[IRS 20] IRSN, Le démantèlement de Chooz, available at: https://www.irsn.fr/FR/connaissances/Installations_nucleaires/demantelement/demantelement-France-centrales-installations-nucleaires-EDF-recherche-militaire/demantelement-chooz/Pages/sommaire.aspx#.X3sb4O3grQw, 2020.

[JAE 20] JAEA, History of Fugen, available at: https://www.jaea.go.jp/04/fugen/en/guide/history/, 2020.

[JAS 99] JASANI B., "La télédétection depuis l'espace, facteur de sécurité nationale et internationale", *Forum du désarmement*, pp. 37–46, available at: https://www.peacepalacelibrary.nl/ebooks/files/UNIDIR_pdf-art264.pdf, 1999.

[JAW 16] JAWERTH N., "Vérification nucléaire : établir des faits grâce à la science", *IAEA Bulletin*, pp. 20–21, June 2016.

[JEO 16a] JEONG K.S., CHOI B.S., MOON C.J. *et al.*, "Risk assessment on abnormal accidents from human errors during decommissioning of nuclear facilities", *Annals of Nuclear Energy*, vol. 87, no. 2, pp. 1–6, January 2016.

[JEO 16b] JEONG K.S., CHOI B.S., MOON C.J. *et al.*, "Review establishment of the hazard reduction methodology to be taken into account for safety assessment during decommissioning of nuclear facilities", *Annals of Nuclear Energy*, vol. 87, no. 2, pp. 7–15, January 2016.

[JEO 16c] JEONG K.S., CHOI B.S., MOON C.J. *et al.*, "An evaluation on the scenarios of work trajectory during installation of dismantling equipment for decommissioning of nuclear facilities", *Annals of Nuclear Energy*, vol. 91, pp. 25–35, May 2016.

[JEO 16d] JEONG K.S., CHOI B.S., MOON C.J. *et al.*, "An estimation to measure and to evaluate the work times following the trajectory of workers during decommissioning of nuclear facilities", *Annals of Nuclear Energy*, vol. 94, pp. 10–15, May 2016.

[JEO 16e] JEONG K.S., CHOI B.S., MOON C.J. *et al.*, "The safety assessment system based on virtual networked environment for evaluation on the hazards from human errors during decommissioning of nuclear facilities", *Reliability Engineering and System Safety*, 156, pp. 34–39, 2016.

[JEO 18] JEONG K.S., MOON J.W., CHOI B.S. *et al.*, Apparatus and method for simulation of dismantling operation of nuclear facility, United States Patent Application Publication, No. US 2018/0210979 A1, July 26, 2018.

[JEO 19] JEONG K.S., PARK S.K., HAHM I.H. *et al.*, "Approach to optimization of risk assessment based on an evaluation matrix for decommissioning processes of a nuclear facility", *Annals of Nuclear Energy*, vol. 128, pp. 53–62, 2019.

[JON 01] JONTER T., Sweden and the bomb. The Swedish plans to acquire nuclear weapons, 1945–1972, SKI Report 01:33, September 2001.

[JON 10] JONTER T., "The Swedish plans to acquire nuclear weapons, 1945–1968: An analysis of the technical preparations", *Science & Global Security*, no. 18, pp. 61–86, 2010.

[JOU 18] JOURNEAU C., ROULET D., "Le corium de Fukushima Daiichi: Formation, état actuel et R&D en vue de sa récupération", *16ème congrès francophone de techniques laser pour la mécanique des fluides*, CNRS, CentraleSupélec, Université Paris Saclay, IRSN, Dourdan, France, September 2018.

[KAL 95] KALINOWSKI M.B., COLSCHEN L.C., "International control of tritium to prevent horizontal proliferation and to foster nuclear disarmament", *Science & Global Security*, vol. 5, pp. 131–203, 1995.

[KEM 16] KEMP R.S., DANAGOULIAN A., MACDONALD R.R. *et al.*, "Physical cryptographic verification of nuclear warheads", *Proceedings of the National Academy of Sciences of the USA*, vol. 113, no. 31, pp. 8618–8623, 2016.

[KER 16] KERR P.K., NIKITIN M.B., Pakistan's nuclear weapons, Congressional Research Service, CRS Report, RL34248-1.pdf, 2016.

[KIL 19a] KILE S.N., KRISTENSEN H.M., "British nuclear forces", in SIPRI (ed.), *SIPRI Yearbook 2018. Armaments, Disarmament and International Security*, 2019.

[KIL 19b] KILE S.N., KRISTENSEN H.M., "French nuclear forces", in SIPRI (ed.), *SIPRI Yearbook 2018. Armaments, Disarmament and International Security*, 2019.

[KIL 19c] KILE S.N., KRISTENSEN H.M., "Chinese nuclear forces", in SIPRI (ed.), *SIPRI Yearbook 2018. Armaments, Disarmament and International Security*, 2019.

[KIL 19d] KILE S.N., KRISTENSEN H.M., "Indian nuclear forces", in SIPRI (ed.), *SIPRI Yearbook 2018. Armaments, Disarmament and International Security*, 2019.

[KIL 19e] KILE S.N., KRISTENSEN H.M., "Pakistani nuclear forces", in SIPRI (ed.), *SIPRI Yearbook 2018. Armaments, Disarmament and International Security*, 2019.

[KIL 19f] KILE S.N., KRISTENSEN H.M., "Israeli nuclear forces", in SIPRI (ed.), *SIPRI Yearbook 2018. Armaments, Disarmament and International Security*, 2019.

[KIL 19g] KILE S.N., KRISTENSEN H.M., "North Korea's military nuclear capacity", in SIPRI (ed.), *SIPRI Yearbook 2018. Armaments, Disarmament and International Security*, 2019.

[KIM 16] KIM I., CHOI B.S., HYUN D.J. *et al.*, "Technical note. A framework for a flexible cutting-process simulation of a nuclear facility decommission", *Annals of Nuclear Energy*, vol. 97, pp. 204–207, November 2016.

[KIM 17] KIM I., CHOI B., HYUN D. *et al.*, "Environmental information feedback framework for nuclear facility decommission", *Transactions of the Korean Nuclear Society*, Autumn meeting, Gyeongju, South Korea, October 26–27, 2017.

[KIT 80] KITTINGER W.D., MEYERS G.W., "Decommissioning the sodium reactor experiment: A status report", in OSTERHOUT M.M. (ed.), *Decontamination and Decommissioning of Nuclear Facilities*, Plenum Press, New York, 1980.

[KNO 10] KNOPS R., 212 DSCFC 10 F rév 1. Les armes nucléaires non stratégiques des États-Unis en Europe : un débat fondamental pour l'OTAN, Report, Assemblée parlementaire de l'OTAN, 2010.

[KOC 19] KOCAMAN B., HARMANDALI M., KOPRU M. *et al.*, "Total ionizing dose analysis of a native detector and a satellite on orbit", *AIP Conference Proceedings 2178*, 030006, 2019.

[KOR 19] KORDA M., KRISTENSEN H.M., "US ballistic missile defenses, 2019", *Bulletin of the Atomic Scientists*, vol. 75, no. 6, pp. 295–306, 2019.

[KPM 18] KPMG, Démantèlement d'une centrale nucléaire. Comprendre les obligations de présentation de l'information, available at: https://assets.kpmg/content/dam/kpmg/ca/pdf/2018/06/ca-demantelement-dune-centrale-nucleaire.pdf, June 2018.

[KRA 80] KRATZER W.K., "Decontamination of the Hanford N-Reactor in support of continued operation", in OSTERHOUT M.M. (ed.), *Decontamination and Decommissioning of Nuclear Facilities*, Plenum Press, New York, 1980.

[KRI 11] KRISTENSEN H.M., NORRIS R.S., "British nuclear forces, 2011", *Bulletin of the Atomic Scientists*, vol. 67, no. 5, pp. 89–97, 2011.

[KRI 14] KRISTENSEN H.M., NORRIS R.S., "Worldwide deployments of nuclear weapons, 2014", *Bulletin of the Atomic Scientists*, vol. 70, no. 5, pp. 96–108, 2014.

[KRI 17] KRISTENSEN H.M., NORRIS R.S., "Worldwide deployments of nuclear weapons, 2017", *Bulletin of the Atomic Scientists*, vol. 73, no. 5, pp. 289–297, 2017.

[KRI 18a] KRISTENSEN H.M., KORDA M., "Indian nuclear forces, 2018", *Bulletin of the Atomic Scientists*, vol. 74, no. 6, pp. 361–366, 2018.

[KRI 18b] KRISTENSEN H.M., NORRIS R.S., "North Korean nuclear capacities, 2018", *Bulletin of the Atomic Scientists*, vol. 74, no. 1, pp. 41–51, 2018.

[KRI 18c] KRISTENSEN H.M., NORRIS R.S., DIAMOND J., "Pakistan nuclear forces, 2018", *Bulletin of the Atomic Scientists*, vol. 74, no. 5, pp. 348–358, 2018.

[KRI 19a] KRISTENSEN H.M., "US nuclear forces", in SIPRI (ed.), *SIPRI Yearbook 2018. Armaments, Disarmament and International Security*, Solna, Sweden, 2019.

[KRI 19b] KRISTENSEN H.M., Russia upgrades western nuclear weapons storage sites, available at: https://fas.org/blogs/security/2019/07/russia-upgrades-western-nuclear-weapons-storage-sites/, 2019.

[KRI 19c] KRISTENSEN H.M., "Russian nuclear forces", in SIPRI (ed.), *SIPRI Yearbook 2018. Armaments, Disarmament and International Security*, Solna, Sweden, 2019.

[KRI 19d] KRISTENSEN H.M., KORDA M., "Russian nuclear forces, 2019", *Bulletin of the Atomic Scientists*, vol. 75, no. 2, pp. 73–84, 2019.

[KRI 19e] KRISTENSEN H.M., KORDA M., "China nuclear forces, 2019", *Bulletin of the Atomic Scientists*, vol. 75, no. 4, pp. 171–178, 2019.

[KRI 19f] KRISTENSEN H.M., KORDA M., "French nuclear forces, 2019", *Bulletin of the Atomic Scientists*, vol. 75, no. 1, pp. 51–55, 2019.

[KRI 19g] KRISTENSEN H.M., KORDA M., "Tactical nuclear weapons, 2019", *Bulletin of the Atomic Scientists*, vol. 75, no. 5, pp. 252–261, 2019.

[KRI 20] KRISTENSEN H.M., KORDA M., "United States nuclear forces, 2020", *Bulletin of the Atomic Scientists*, vol. 76, no. 1, pp. 46–60, 2020.

[KUB 94] KUBO S., JOHNSTON E.P., Method of cutting and removing nuclear reactor, United States Patent, vol. 5, no. 329, p. 562, July 12, 1994.

[KÜT 19] KÜTZ M., MIAM Z., PODVIG P., "Global stocks and production of fissile materials, 2017", in SIPRI (ed.), *SIPRI Yearbook 2018. Armaments, Disarmament and International Security*, Solna, Sweden, 2019.

[LAC 80] LACY C.S., MONTFORD B., "Decontamination experience in Ontario hydro", in OSTERHOUT M.M. (ed.), *Decontamination and Decommissioning of Nuclear Facilities*, Plenum Press, New York, 1980.

[LAG 12a] LA GUARDIA T.S., MURPHY K.C., "Financing and economics of nuclear facility decommissioning", in LARAIA M. (ed.), *Nuclear Decommissioning. Planning, Execution and International Experience*, Woodhead Publishing, Oxford, 2012.

[LAG 12b] LA GUARDIA T.S., "Decommissioning of western-type light-water nuclear reactors (LWRs)", in LARAIA M. (ed.), *Nuclear Decommissioning. Planning, Execution and International Experience*, Woodhead Publishing, Oxford, 2012.

[LAG 17] LA GUARDIA T.S., "The real costs of decommissioning", in LARAIA M. (ed.), *Advances and Innovations in Nuclear Decommissioning*, Elsevier, Kidlington, 2017.

[LAG 19] LAGNEAU L., "Les États-Unis soupçonnent la Russie de violer le Traité sur l'interdiction complète des essais nucléaires", *Le Figaro*, available at: https://www.lefigaro.fr/flash-actu/les-etats-unis-soupconnent-la-russie-de-violer-un-moratoire-sur-les-essais-nucleaires-20190529, May 29, 2019.

[LAI 05] LAINETTI P.E.O., "Decommissioning of nuclear fuel cycle facilities – IPEN's experience", *2005 International Nuclear Atlantic Conference – INAC 2005*, Santos, SP, Brazil, August 28 to September 2, 2005.

[LAI 11] LAINETTI P.E.O., "Decommissioning activities at IPEN/CNEN-SP brazil", *Regional Practical Workshop on the Decommissioning of Radioactively Contaminated Facilities*, Necsa, South Africa, August 22–26, 2011.

[LAI 16] LAINETTI P.E.O., "Cutting techniques for facilities dismantling in decommissioning projects", *Journal of Energy and Power Engineering*, no. 10, pp. 513–521, 2016.

[LAR 12a] LARAIA M., "Overview of nuclear decommissioning principles and approaches", in LARAIA M. (ed.), *Nuclear Decommissioning. Planning, Execution and International Experience*, Woodhead Publishing, Oxford, 2012.

[LAR 12b] LARAIA M., "Reuse and redevelopment of decommissioned nuclear sites: Strategies and lessons learned", in LARAIA M. (ed.), *Nuclear Decommissioning. Planning, Execution and International Experience*, Woodhead Publishing, Oxford, 2012.

[LAR 13] LARSSON C.M., "Concluding remarks", in IAEA report on decommissioning and remediation after a nuclear accident", *International Experts Meeting*, IAEA, Vienna, Austria, January 28–February 1, 2013.

[LAR 16] LARSSON A., WEBER I., "Nuclear energy agency task group on radiological characterisation for decommissioning of nuclear installations", *PREDEC 2016: NEA – Radiological Characterisation for Decommissioning*, Lyon, France, February 16–18, 2016.

[LAR 17a] LAROCHE-SIGNORITE V., "États-Unis – Russie : les grandes étapes du désarmement nucléaire", *Le Figaro*, May 24, 2017.

[LAR 17b] LARSSON A., EMPDAGE M., WEBER I., International good practice on practical implementation of characterisation in decommissioning, Abstract intended for HRP/IAEA/NEA Decom. Workshop 2017, Halden, Norway, 2017.

[LAR 18] LARAIA M., "Nuclear decommissioning. Its history, development and current status", *Lecture Notes in Energy*, vol. 66, Springer, Cham, Switzerland, 2018.

[LAS 19] LASYTE D., "Decommissioning of the Ignalina NPP – Planning and execution", *ICOND 2019, International Conference on Nuclear Decommissioning*, Book of Abstracts, p. 46, 2019.

[LAU 17] LAURENT G., "Needs for R&D and innovations to address challenges of nuclear facility decommissioning after its normal shutdown versus advanced approaches required for Fukushima Daiichi NPS decommissioning", *Current and Emerging Methods for Optimising Safety and Efficiency in Nuclear Decommissioning*, Norway, February 7–9, 2017.

[LAZ 06] LAZ K.A., Entre concept et mise en œuvre : le désarmement nucléaire à la croisée des chemins, Thesis, Université du Québec à Montréal, 2006.

[LEA 17] LEAH C.M., *The Consequences of American Nuclear Disarmament. Strategy and Nuclear Weapons*, Palgrave Macmillan, Cham, 2017.

[LEC 12] LE CLAIR A.N., LEMIRE D.S., "Information management for nuclear decommissioning projects", in LARAIA M. (ed.), *Nuclear Decommissioning. Planning, Execution and International Experience*, Woodhead Publishing, Oxford, 2012.

[LEG 11] LE GALL J., "Les forces nucléaires stratégiques sous-marines dans le monde", *Marine & Océans*, no. 230, available at: http://fr.calameo.com/read/000701339ab0cbbc1d550, 2011.

[LEL 19] LE LOIREC C., RIFFARD C., BARRAN F. *et al.*, "Parameters influencing the activation of dismantling nuclear facility structures", *Annals of Nuclear Energy*, vol. 134, pp. 158–169, December 2019.

[LEN 17] LENNE F., "Le deuil de Hiroshima et de Nagasaki. En finir avec la 'dissuasion nucléaire' : diagnostic et thérapie", *Un projet des Nations Unies pour l'humanité*, 2017.

[LES 18] LESSARD L.P., MANCINI A., FURR R.J. *et al.*, Packaging for decommissioned and dismantled nuclear reactors and reactor components, United States Patent Application Publication, No. US 9,984,780 B2, May 29, 2018.

[LI 19] LI R., "The decommissioning planning of 101 heavy water research reactor", *ICOND 2019, International Conference on Nuclear Decommissioning*, Book of Abstracts, p. 45, 2019.

[LIN 02] LINI D.C., RODGERS L.H., Plutonium Finishing Plant Plutonium- Uranium Oxide: Characterization of Items with < 30 Weight Percent Plutonium, Report, HNF-109 19, 2002.

[LOR 14] LOREC A., "Démantèlement : héritage du passé, perspectives d'avenir", *Les Défis du CEA*, no. 189, pp. 12–21, 2014.

[LOV 12] LOVE J., "Public engagement and stakeholder consultation in nuclear decommissioning projects", in LARAIA M. (ed.), *Nuclear Decommissioning. Planning, Execution and International Experience*, Woodhead Publishing, Oxford, 2012.

[LUG 04] LUGARD R., Initiative Nunn-Lugar, available at: https://www.laradioactivite.com/site/pages/Initiative_Nunn_Lugar.htm, 2004.

[MAC 96] MacKinnon III M., Burritt J.G., "Overview of nuclear submarine inactivation and scrapping/recycling in the United States", in Le Sage L.G., Sarkisov A.A. (eds), *Nuclear Submarine Decommissioning and Related Problems*, Proceedings of the NATO Advanced Research Workshop on Nuclear Submarine Decommissioning and Related Problems, Moscow, Russia, June 19–22, 1995, Kluwer Academic Publishers, Dordrecht, 1996.

[MAL 19] Malbrunot G., "Iran : l'Europe hausse le ton sur les missiles balistiques", *Le Figaro*, available at: https://www.lefigaro.fr/international/iran-l-europe-hausse-le-ton-sur-les-missiles-balistiques-20191206, December 6, 2019.

[MAM 16a] Mampaev L., "Le démantèlement des armes et bâtiments nucléaires, Lourd fardeau pour les générations futures", *Reporterre, le quotidien de l'écologie*, available at: https://reporterre.net/Le-demantelement-des-armes-et-batiments-nucleaires-lourd-fardeau-pour-les, July 26, 2016.

[MAM 16b] Mampaey L., "Démantèlement des armes et bâtiments nucléaires. Terrifiant héritage pour les générations futures", *Conference on Nuclear Disarmament*, Université européenne de la Paix, Brest, June 15–16, 2016.

[MAR 98] Marsden B.J., Preston S.D., Wickham A.J. *et al.*, Evaluation of graphite safety issues for the British production piles at Windscale: Graphite sampling in preparation for the dismantling of pile 1 and the further safe storage of pile 2, Technical committee meeting on technologies for gas cooled reactor decommissioning, fuel storage and waste disposal; Juelich (Germany); September 8–10, 1997, IAEA, Vienna, XA9848076, pp. 213–223, 1998.

[MAR 00] MARSSIM (Multi-Agency Radiation Survey and Site Investigation Manual), NUREG-1575, Rev. 1, EPA 402-R-97-016, Rev. 1, DOE/EH-0624, Rev. 1, 2000.

[MAR 02a] Marceau T.E., Harvey D.W., Stapp D.C. *et al.* "History of the plutonium production facilities at the Hanford Site historic district, 1943–1990", *Hanford Cultural and Historic Resources Program*, 2002.

[MAR 02b] Marret J.L., L'action des ONG dans le domaine du désarmement (entre ambitions et limites) et l'exemple britannique, available at: https://inis.iaea.org/collection/NCLCollectionStore/_Public/47/069/47069801.pdf, 2002.

[MAS 96] Masurel A., "French strategy for disposal of nuclear powered ships and spent fuel", in Le Sage L.G., Sarkisov A.A. (eds), *Nuclear Submarine Decommissioning and Related Problems*, Proceedings of the NATO Advanced Research Workshop on Nuclear Submarine Decommissioning and Related Problems, Moscow, Russia, June 19–22, 1995, Kluwer Academic Publishers, Dordrecht, 1996.

[MAS 17] Mascarich F., Wilson T., Dang T. *et al.*, Towards robotically supported decommissioning of nuclear sites, arXiv: 1705.06401v1, May 18, 2017.

[MAT 16] Mattson R.J., "Stealing the atom bomb: How denial and deception armed Israel", *Library of Congress Cataloging*, CreateSpace Independent Publishing, 2016.

[MAV 16] MAVROKEFALOU G., FLOROU H., SYKIOTI O. *et al.*, "Integration of earth observation satellite data and real time ^{137}Cs measurements in the Greek marine environment to gis for advances in radiological remote control", *Proceedings of the 'Living Planet Symposium 2016'*, Prague, Czech Republic, May 9–13, 2016, ESA SP-740, August 2016.

[MCG 97] MCGEHEE E.D., PENDERGRASS A.K., NEPA and NHPA – Successful decommissioning of historic Manhattan project properties at Los Alamos National Laboratory, Los Alamos, New Mexico, Report, CONF-9705100-6, LA-UR-97-1856, 1997.

[MCG 02] MCGEHEE E.D., GARCIA K.L.M., TOWERY K. *et al.*, Decontamination decommissioning of technical area 41. Historic building survey no. 204, Report, Los Alamos National Laboratory, LA-UR-02-2663, 2002.

[MEN 96] MENON S., "Decommissioning of nuclear submarines: Waste minimization by recycling", in LE SAGE L.G., SARKISOV A.A. (eds), *Nuclear Submarine Decommissioning and Related Problems*, Proceedings of the NATO advanced research workshop on nuclear submarine decommissioning and related problems, Moscow, Russia, June 19–22, 1995, Kluwer Academic Publishers, Dordrecht, 1996.

[MER 09] MERCHET D., "Désarmement : quand la France démantèle ses usines de matières fissiles", *Libération*, July 10, 2009.

[MES 80] MESERVEY R.H., "OMRE and hallam decommissioning projects at the Idaho National Engineering Laboratory", in OSTERHOUT M.M. (ed.), *Decontamination and Decommissioning of Nuclear Facilities*, Plenum Press, New York, 1980.

[MES 97] MESERVEY Y.R.H., KENOYER D.J., FRAZEE B.J., "Nuclear decontamination and decommissioning operations at the Idaho National Engineering and Environmental Laboratory (INEEL)", *International Nuclear Conference INC '97*, October 20–21, 1997, Kuala Lumpur, Malaysia, MY9800999, 1997.

[MES 12] MESERVEY R.H., "Decommissioning of legacy nuclear waste sites: Idaho National Laboratory, USA", in LARAIA M. (ed.), *Nuclear Decommissioning. Planning, Execution and International Experience*, Woodhead Publishing, Oxford, 2012.

[MEY 19] MEYER P.A., "Empty promises: The denuclearization of Ukraine and its effects twenty years later'", *Illini Journal of International Security (IJOIS)*, vol. 5, pp. 49–55, Spring 2019.

[MIC 12] MICHAL V., "Remote operation and robotics technologies in nuclear decommissioning projects", in LARAIA M. (ed.), *Nuclear Decommissioning. Planning, Execution and International Experience*, Woodhead Publishing, Oxford, 2012.

[MIN 19] MINISTÈRE DE LA TRANSITION ÉCOLOGIQUE ET SOLIDAIRE, Évaluation environnementale : le guide de lecture de la nomenclature annexée à l'article R.122-2 du code de l'environnement, Theme, available at: http://www.side.developpement-durable.gouv.fr, August 2019.

[MOL 00] MOLTZ J.C., "Russian nuclear submarine dismantlement and the naval fuel cycle", Special report, *The Nonproliferation Review*, pp. 77–87, Spring 2000.

[MON 19] MONIER F., "Un territoire de la dissuasion nucléaire : le plateau d'Albion", *E3S Web of Conferences*, no. 88, 04001, available at: https://doi.org/10.1051/e3sconf/20198804001, 2019.

[MON 91] MONTOYA G.M., Final project report ta-2 water boiler reactor decommissioning project, Report, LA-12049, 1991.

[MOU 17] MOUSSEAU T.A., MØLLER A.P., "Nuclear energy and its ecological byproducts: Lessons from Chernobyl and Fukushima", in VAN NESS P., GURTOV M. (eds), *Learning from Fukushima Nuclear Power in East Asia*, Australian National University Press, 2017.

[MÜL 06] MÜLLER G., Démantèlement des centrales nucléaires. Documentation lacunaire, Report, Conseil national suisse, 06.3162 Interpellation, March 24, 2006.

[MUR 80] MURPHY E.S., HOLTER G.M., "Decommissioning of commercial shallow-land burial sites", in OSTERHOUT M.M. (ed.), *Decontamination and Decommissioning of Nuclear Facilities*, Plenum Press, New York, 1980.

[NAO 15] NAO (National Audit Office), Nuclear decommissioning authority. Progress on the Sellafield site: An update, Report, National Audit Office, March 2015.

[NAO 18] NAO (National Audit Office), The nuclear decommissioning authority: Progress with reducing risk at Sellafield, Report, National Audit Office, June 20, 2018.

[NAU 08] NAUDET G., REUSS P., *Energie, électricité et nucléaire*, EDP Sciences, Les Ulis, 2008.

[NDA 05] NDA (Nuclear Decommissioning Authority), NDA strategy. Draft for consultation, Report, NDA, 2005.

[NDA 19] NDA (Nuclear Decommissioning Authority), Draft business plan 1 April 2020 to 31 March 2023. Cleaning up the UK's earliest nuclear sites, caring for people and the environment, Plan, NDA, 2019.

[NEA 96] NEA, Recycling and reuse of metal scrap, Report, OECD, Paris, 1996.

[NEA 02] NEA, Environmental remediation of uranium production facilities, Joint report, OECD Nuclear Energy Agency and the International Atomic Energy Agency, NEA N 3033, OECD, Paris, 2002.

[NEA 03] NEA, Decommissioning nuclear power plants: Policies, strategies and costs, Report, OECD, no. 3590, Paris, 2003.

[NEA 11] NEA, Decontamination and dismantling of radioactive concrete structures, Report, NEA Co-operative Programme on Decommissioning (CPD), NEA/RWM/R(2011)1, OECD, Paris, 2011.

[NEA 12] NEA, "International structure for decommissioning costing (ISDC) of nuclear installations", *OECD*, Paris, available at: www.oecd-nea.org/rwm/reports/2012/ISDC-nuclear-installations.pdf, 2012.

[NEA 14a] NEA, Nuclear site remediation and restoration during decommissioning of nuclear installations, Report, NEA Co-operative Programme on Decommissioning, NEA no. 7192, Paris, 2014.

[NEA 14b] NEA, "R&D and innovation needs for decommissioning of nuclear facilities", *NEA*, no. 7191, 2014.

[NEA 14c] NEA, "Radioactive waste management and decommissioning in the Russian Federation, National report, 2014.

[NEA 15] NEA, "The practice of cost estimation for decommissioning of nuclear facilities", *NEA*, no. 7237, OECD, Paris, 2015.

[NEA 16a] NEA, "Costs of decommissioning nuclear power plants", *NEA*, no. 7201, OECD, Paris, 2016.

[NEA 16b] NEA, "Management of radioactive waste after a nuclear power plant accident", *NEA*, no. 7305, OECD, Paris, 2016.

[NEA 16c] NEA, "Strategic considerations for the sustainable remediation of nuclear installations", *NEA*, 2016.

[NEA 17a] NEA, "Recycling and reuse of materials arising from the decommissioning of nuclear facilities", *NEA*, no. 7310, Paris, 2017.

[NEA 17b] NEA, "Addressing uncertainties in cost estimates for decommissioning nuclear facilities", *NEA*, no. 7344, Paris, 2017.

[NEA 17c] NEA, "Radiological characterisation from a waste and materials end-state perspective: Practices and experience", *NEA*, no. 7373, Paris, 2017.

[NEA 18a] NEA, "Preparing for decommissioning during operation and after final shutdown", *NEA* no. 7374, Paris, 2018.

[NEA 18b] NEA, The methodology of cost estimation for decommissioning nuclear facilities in the Russian Federation, *International Expert Feedback on the Methodology Developed by ROSATOM*, Final report, Radioactive Waste Management Committee, NEA, Paris, 2018.

[NEA 19] NEA, "Cost benchmarking for nuclear power plant decommissioning", *NEA*, no. 7460, OECD, Paris, 2019.

[NEG 12] NEGIN C.A., SZILAGYI A., "Managing the transition from operation to decommissioning of a nuclear facility", in LARAIA M. (ed.), *Nuclear Decommissioning. Planning, Execution and International Experience*, Woodhead Publishing, Oxford, 2012.

[NEG 17] NEGIN C.A., BOŽIK M., STELMAKH D. *et al.*, "Decommissioning after a severe accident", in LARAIA M. (ed.), *Advances and Innovations in Nuclear Decommissioning*, Woodhead Publishing, Elsevier, Kidlington, 2017.

[NIK 96] NIKISHIN G.D., SHARIKOV V.P., SOKOLOV O.G. *et al.*, "Ensuring of radiation safety when decommissioning, dismantling and recycling nuclear submarines", in LE SAGE L.G., SARKISOV A.A. (eds), *Nuclear Submarine Decommissioning and Related Problems*, Proceedings of the NATO Advanced Research Workshop on Nuclear Submarine Decommissioning and Related Problems, Moscow, Russia, June 19–22, 1995, Kluwer Academic Publishers, Dordrecht, 1996.

[NIK 17] NIKITIN M.B.D., AVERY E.C., MANYIN M.E., "Nuclear negotiations with North Korea: In brief", *Congressional Research Service*, 7-5700, R45033, available at: www.crs.gov, December 4, 2017.

[NOM 00] NOMOTO A., TANAKA H., TAKEDA S., "Chemical oxygen-iodine laser for decommissioning and dismantlement of nuclear facilities", *Journal of Advanced Science*, vol. 12, nos 1 and 2, pp. 146–147, 2000.

[NOR 06] NORRIS R.S., KRISTENSEN H.M., "Where the bombs are, 2006?", *Bulletin of the Atomic Scientists*, vol. 62, no. 6, pp. 57–58, November/December 2006.

[NOR 11] NORRIS R.S., KRISTENSEN H.M., "US tactical nuclear weapons in Europe, 2011", *Bulletin of the Atomic Scientists*, vol. 67, no. 1, pp. 64–73, 2011.

[NOY 12] NOYNAERT L., "Decontamination processes and technologies in nuclear decommissioning projects", in LARAIA M. (ed.), *Nuclear Decommissioning. Planning, Execution and International Experience*, Woodhead Publishing, Oxford, 2012.

[NRC 18] NRC, Backgrounder on decommissioning nuclear power plants, available at: https://www.nrc.gov/reading-rm/doc-collections/fact-sheets/decommissioning.html, 2018.

[NTI 19a] NTI, Nuclear disarmament Ukraine, available at: https://www.nti.org/analysis/articles/ukraine-nuclear-disarmament/, 2019.

[NTI 19b] NTI, Nuclear disarmament Kazakhstan, available at: https://www.nti.org/analysis/articles/kazakhstan-nuclear-disarmament/, 2019.

[OAK 92] OAK RIDGE, Y-12 Plant decontamination and decommissioning program, Surveillance and maintenance plan, FY 1992-2000, Oak Ridge, DoE, January 1992.

[OEC 16] OECD, Cost of decommissioning nuclear power plants, Report, Nuclear Energy Agency, 2016.

[OER 94] OER (Office of Environmental Restoration), "Y-12 plant decontamination and decommissioning technology logic diagram for building 9201-4", vol. 1, Technolgy Evaluation, Y/ER-159/V1, September 1994.

[ØLG 01] ØLGAARD, P.L., NKS-SBA-1 status report: The potential threats from Russian nuclear ships bordering Russia, Report, Nordic Nuclear Safety Research (nks.org), November 2001.

[ONE 12] ONET TECHNOLOGIES, "Études de faisabilité pour le démantèlement du bât 400 et des bâtiments annexes de l'INBS PN Cadarache", *INBS PN Cadarache*, File no. 36, available at: https://fr.groupeonet.com/Nos-metiers/Technologies/Constructions-neuves-et-modifications-d-installations/Ingenierie/INBS-PN-Cadarache, 2012.

[ONU 17a] ONU, Faire avancer les négociations multilatérales sur le désarmement nucléaire, Resolution RES/A/71/258, Assemblée générale, December 23, 2016, 2017.

[ONU 17b] ONU, Traité interdisant la production de matières fissiles pour la fabrication d'armes et autres dispositifs explosifs nucléaires, Resolution RES/A/71/259, Assemblée générale, December 23, 2016, 2017.

[ONU 18] ONU, "Annuaire des nations unies sur le désarmement", *Bureau des affaires de désarmement*, vol. 41, part II, New York, 2018.

[ORA 17] ORANO CYCLE, Rapport d'information du site orano la hague, available at: https://www.orano.group/docs/default-source/orano-doc/groupe/publications-reference/rapport-tsn-la-hague-2017.pdf?sfvrsn=706a36f3_6, 2017.

[ORE 09] OREPA (The Oak Ridge Environmental Peace Alliance), "The future of Y12", *OREPA*, November 2009.

[OSU 09] O'SULLIVAN P.J., "Libération des matériaux et bâtiments radioactifs du contrôle réglementaire", *Revue générale nucléaire*, no. 4, pp. 54–61, July–August 2009.

[PEI 17] PEILLON S., FAUVEL S., CHAGNOT C. *et al.*, "Aerosol characterization and particle scrubbing efficiency of underwater operations during laser cutting of steel components for dismantling of nuclear facilities", *Aerosol and Air Quality Research*, no. 17, pp. 1463–1473, 2017.

[PIC 15] PICAT M., Plateau d'Albion : visite des représentants de la Conférence du désarmement de Genève, available at: https://www.defense.gouv.fr/actualites/articles/plateau-d-albion-visite-des-representants-de-la-conference-du-desarmement-de-geneve, 2015.

[PIK 15] PIKETTY L., "Le démantèlement des installations nucléaires à la direction de l'énergie nucléaire du CEA", *ASN, Les enjeux du démantèlement, Contrôle*, no. 199, pp. 15–17, 2015.

[PIN 17] PINTAT, X., LORGEOUX J., Trillard A. *et al.*, Rapport d'information fait au nom de la commission des affaires étrangères, de la défense et des forces armées par le groupe de travail 'La modernisation de la dissuasion nucléaire', Report no. 560, Sénat, available at: https://www.senat.fr/rap/r16-560/r16-5601.pdf, May 23, 2017.

[POD 05] PODA F., Decommissioning and demolition (D&D), SRS Environmental Report, available at: https://www.srs.gov/general/pubs/ERsum/er05/d_d.pdf, 2005.

[POD 09] PODVIG P., Consolidating fissile materials in Russia's nuclear complex, Research report No. 7, International Panel on Fissile Materials, May 2009.

[POM 00] POMFRET D.G., "Safety and dose management during decommissioning of a fire damaged nuclear reactor", *IRPA-10: 10th International Congress of the International Radiation Protection Association*, T-2-2, P-5-299, Hiroshima, Japan, available at: http://www.irpa.net/irpa10/cdrom/00322.pdf, May 14–19, 2000.

[PRE 09] PRECHTL E., "The decommissioning project multi-purpose research reactor (MZFR). Dismantling techniques applied at MZFR", *Research Reactor Decommissioning Demonstration Project (R^2D^2P)*, IAEA Workshop "Decommissioning Technologies", Forschungszentrum Karlsruhe, July 6–10, 2009.

[QUE 16] QUEVENCO R., "L'imagerie satellite à l'appui des capacités de l'AIEA en matière de garanties", *IAEA Bulletin*, pp. 24–25, June 2016.

[QUI 12] QUILES P., *Nucléaire, un mensonge français : réflexions sur le désarmement nucléaire*, Editions Charles Léopold Mayer, Paris, 2012.

[RAD 18] RADIO-CANADA, "Où en est le processus de déclassement de Gentilly-2 ?", *Radio-Canada*, January 19, 2018.

[RAM 17] RAMANA M.V., KING A., "A new normal? The changing future of nuclear energy in China", in VAN NESS P., GURTOV M. (eds), *Learning from Fukushima Nuclear Power in East Asia*, Australian National University Press, 2017.

[RAN 16] RANNEMALM T., ERIKSSON J., BERGH N., "Decommissioning planning for nuclear units at the Oskarshamn site", *PREDEC 2016: Decommissioning Planning for Nuclear Units at the Oskarshamn Site*, Lyon, France, February 16–18, 2016.

[RAN 17] RANNEMALM T., ELIASSON S., LARSSON A. *et al.*, "Strategic aspects on waste management in decommissioning of nuclear facilities", Abstract intended for HRP/IAEA/NEA Decom. Workshop, Halden, Norway, 2017.

[RAN 18] RANJAN R., BHATTACHARYA S., KARHADKAR C.G., *et al.*, Transition from operation to decommissioning of cirus research reactor, Report, BARC/2018/E/005, 2018.

[RAW 02] RAWOOL-SULLIVAN M., MOSKOWITZ P.D., SHELENKOVA L., "Technical and proliferation-related aspects of the dismantlement of Russian Alfa-class nuclear submarines", *The Nonproliferation Review*, vol. 8, pp. 161–171, Spring 2002.

[REN 18] RENARD J.-D., "La grosse migraine du démantèlement nucléaire", *Sud-Ouest Dimanche*, January 28, 2018.

[RÉP 10] RÉPUBLIQUE FRANÇAISE, Désarmement nucléaire : l'engagement concret de la France – Le démantèlement des usines de production de matières fissiles pour les armes nucléaires, Report, France TNP2010, De l'engagement aux actes, New York, May 3–28, 2010.

[RÉP 19] RÉPUBLIQUE FRANÇAISE, Commission préparatoire de l'Organisation du traité d'interdiction complète des essais nucléaires, available at: https://onu-vienne.delegfrance.org/OTICE-970, 2019.

[RON 15] RONDEAU J.M., "Déconstruction du parc nucléaire à eau pressurisée en exploitation : EDF prépare l'avenir", *ASN, "Les enjeux du démantèlement", Contrôle*, no. 199, pp. 12–13, 2015.

[SAM 16] SAMAH S.F., RODHAIN F., FALLERY B., "Gouvernance de données dans le démantèlement nucléaire : confidentialité, partage et valorisation des données et informations stratégiques entre acteurs", *Journées du Risque : "Nucléaire, Hommes et Sociétés"*, Ecole des Mines, Nantes, France, November 2016.

[SAN 09] SANTOS I., ABRÃO A., CARVALHO F.M.S. *et al.*, "Decommissioning of an uranium hexafluoride pilot plant", *2009 International Nuclear Atlantic Conference*, INAC 2009, Rio de Janeiro, RJ, Brazil, September 27–October 2, 2009.

[SCH 15] SCHILZ F., "Le démantèlement des installations nucléaires de base : le point de vue de l'ASN", *ASN, "Les enjeux du démantèlement", Contrôle*, no. 199, pp. 8–12, 2015.

[SCH 19] SCHNEIDER M., FROGGATT A., "The world nuclear industry. Status report 2019", *A Mycle Schneider Consulting Project*, Paris, Budapest, September 2019.

[SEL 12] SELLING H.A., "Radiological protection in the decommissioning of nuclear facilities: Safety, regulations and licensing", in LARAIA M. (ed.), *Nuclear Decommissioning. Planning, Execution and International Experience*, Woodhead Publishing, Oxford, 2012.

[SHA 14] SHANG J., TSO activities in licensing and supervision procedure during the construction, operation and dismantling of the Vitrification facility Karlsruhe, Paper, IAEA-CN-214-paper 10, 2014.

[SHE 15] SHEA T.E., ROCKWOOD L., IAEA verification of fissile material in support of nuclear disarmament, Report, The Project on Managing the Atom, Belfer Center for Science and International Affairs, John F. Kennedy School of Government, Harvard University, May 2015.

[SIN 11] SINCLAIR K.M., Hanford's public tour program, an excellent educational tool, Session 54, #11439, HNF-47419-FP, Revision 0, 2010, WM 2011 Conference, Phoenix, AZ, February 28–March 3, 2011.

[SIP 18] SIPRI, *SIPRI Yearbook 2018. Armaments, Disarmament and International Security*, Oxford University Press, Oxford, 2018.

[SIP 19] SIPRI, *SIPRI Yearbook 2019. Armaments, Disarmament and International Security*, Oxford University Press, Oxford, 2019.

[SKO 19] SKOROBOGATY P., "Vu de Russie. Deux minutes avant l'apocalypse nucléaire", *Courrier International*, July 2019.

[SLA 13] SLATER S., "Decommissioning and remediation after a nuclear accident, 3 case studies", *International Experts' Meeting on Decommissioning and Remediation after a Nuclear Accident*, Sellafield Ltd, Vienna, Austria, January 28–February 1, 2013, available at: http://www-pub.iaea.org/iaeameetings/IEM4/30Jan/Slater.pdf, January 30, 2013.

[SOL 00] SOLOMON W.C., CARROLL D.L., KING D.M. *et al.*, "Commercial applications for COIL", in NAKAI S., HACKEL L.A., SOLOMON W. (eds), *High-Power Lasers in Civil Engineering and Architecture, Proceedings of SPIE*, vol. 3887, pp. 137–151, 2000.

[SOR 18] SORTIR DU NUCLÉAIRE, De l'interdiction des armes nucléaires à leur élimination, Brochure, Sortir du nucléaire, 2018.

[SRN 13] SRNL, "Innovations from Savannah River National Laboratory", *SRNL*, January 2013.

[SRS 09] SRS, Deactivation and decommissioning, available at: https://www.srs.gov/general/programs/dandd/index.htm, 2009.

[STE 12] STEINER H., "Dismantling and demolition processes and technologies in nuclear decommissioning projects", in LARAIA M. (ed.), *Nuclear Decommissioning. Planning, Execution and International Experience*, Woodhead Publishing, Oxford, 2012.

[STO 95] STOUT D.S., "Decommissioning of surplus facilities at Los Alamos national laboratory", *LA-UR-95-380, WM '95 Conference*, Tucson, Arizona, February 26–March 2, 1995.

[STR 01] STRAHLENSCHUTZVERORDNUNG, "Verordnung über den schutz vor schäden durch ionisierende strahlen (Strahlenschutzverordnung – StrlSchV)", *Bundesgetzhlant*, G 5702, no. 38, p. 1714, July 20, 2001.

[STR 19] STROOBANTS J.P., SEMO M., Armes nucléaires : la fin d'un symbole, available at: https://www.lemonde.fr/international/article/2019/08/02/les-etats-unis-sortent-officiellement-du-traite-sur-les-armes-nucleaires-intermediaires_5495989_3210.html, August 2, 2019.

[STU 17] STUART C., "The French exception: The French nuclear power industry and its influence on political plans to transition to a new energy system", in VAN NESS P., GURTOV M. (eds), *Learning from Fukushima Nuclear Power in East Asia*, Acton, Australian National University Press, 2017.

[SUH 18] SUH Y.A., HORNIBROOK C., YIM M.S. *et al.*, "Decisions on nuclear decommissioning strategies: Historical review", *Progress in Nuclear Energy*, vol. 106, pp. 34–43, July 2018.

[SUZ 17] SUZUKI T., "Nuclear energy policy issues in Japan after the Fukushima nuclear accident", in VAN NESS P., GURTOV M. (eds), *Learning from Fukushima Nuclear Power in East Asia*, Acton, Australian National University Press, 2017.

[SZI 12] SZILAGYI A., US DOE Idaho national laboratory reactor decommissioning, Paper, NEA/RWM/R(2012)8, Annex 6a, pp. 89–97, 2012.

[TAC 00] TACHIBANA M., SHIMADA T., YANAGIHARA S., "Development of remote dismantling systems for decommissioning of nuclear facilities", *WM'00 Conference*, Tucson, AZ, February 27–March 2, 2000.

[TEC 14] TECHNICAL SECRETARIAT, Inauguration of the blast and paint facility for reactor compartment units of the dismantled nuclear submarines in the Russian far east, Technical Secretariat on Cooperation for the Elimination of Nuclear Weapons Reduced in the Former Soviet Union, available at: http://www.tecsec.org/en/?p=1467, July 8, 2014.

[TEC 15a] TECHNICAL SECRETARIAT, Ex-post evaluation of cooperation program for dismantling decommissioned nuclear submarines in the Russian far east, Summary report, Technical Secretariat of the Committee on Cooperation to Assist the Destruction of Nuclear Weapons Reduced in the Russian Federation, 2015.

[TEC 15b] TECHNICAL SECRETARIAT, Construction of a long-term storage facility for the reactor compartment units of dismantled nuclear submarines "Reactor Compartments", Technical Secretariat on Cooperation for the Elimination of Nuclear Weapons Reduced in the Former Soviet Union, available at: http://www.tecsec.org/en/?page_id=279, 2015.

[TEC 15c] TECHNICAL SECRETARIAT, Program for dismantling decommissioned nuclear submarines "Star of Hope", Technical secretariat on cooperation for the elimination of nuclear weapons reduced in the former soviet union, available at: http://www.tecsec.org/en/?page_id=271, 2015.

[TÉN 10] TÉNEZE N., "La doctrine de dissuasion d'une puissance atypique : Israël", *Les Études de L'IRSEM*, no. 3, pp. 167–185, available at: https://www.defense.gouv.fr › content › download › file › Etudes n 3 Israel, May 2010.

[TER 17] TERRANEAR PMC, Lawrence livermore national security, LLC management support for environmental and D&D programs, Livermore, CA, available at: http://www.terranear.com/Projects_DnD.php, 2017.

[THI 12] THIERFELDT S., "Safe enclosure and entombment strategies in nuclear decommissioning projects", in LARAIA M. (ed.), *Nuclear Decommissioning. Planning, Execution and International Experience*, Woodhead Publishing, Oxford, 2012.

[TIS 19] TISDALL S., "Prolifération. En favorisant les saoudiens, Trump relance la course au nucléaire au Moyen-Orient", *Courrier International*, July 2019.

[TOU 13] TOUCHARD L., Quand l'Afrique (du Sud) avait la bombe #3, available at: https://www.jeuneafrique.com/169588/politique/quand-l-afrique-du-sud-avait-la-bombe-3/, July 19, 2013.

[TRI 14] TRI-PARTY AGREEMENT, 2014 Hanford lifecycle scope, schedule and cost report, Report, Tri-Party Agreement, February 2014.

[UKN 10a] UKNI, "The United Kingdom – Norway Initiative: Research into the verification of nuclear warhead dismantlement", *The 2010 Review Conference of the Parties to the Treaty on the Non-Proliferation of Nuclear Weapons*, May 2010.

[UKN 10b] UKNI, Presentation on the UK/Norway Initiative on nuclear warhead dismantlement verification: May 2009, Report, UKNI, 2010.

[ULI 17] ULIN S.E., VLASIK K.F., GRACHEV V.M. *et al.*, "The measuring complex for detection of radioactive waste in near-earth space", *Journal of Physics: Conference Series*, 798, 012122, 2017.

[UN 18] UN, "Securing our common future. An agenda for disarmament", *United Nations*, 2018.

[UNI 17] UNIDIR, "Executive Summary", in BORRIE J., CAUGHLEY T., WAN W. (eds), *Understanding Nuclear Weapon Risks*, available at: http://www.unidir.org/files/publications/pdfs/understanding-nuclear-weapon-risks-en-676.pdf, 2017.

[USD 13] USDOE, "Hanford site. Cleanup completion framework", *USDOE*, DOE/RL-2009-10, Rev. 1, January 2013.

[USN 06a] USNRC, Consolidated decommissioning guidance decommissioning process for materials licensees, Final report, NUREG-1757, vol. 1, Rev. 2, 2006.

[USN 06b] USNRC, Consolidated decommissioning guidance. Characterization, survey, and determination of radiological criteria, Final report, NUREG-1757, vol. 2, Rev. 1, 2006.

[USN 12] USNRC, Consolidated decommissioning guidance. Financial assurance, recordkeeping, and timeliness, Final report, NUREG-1757, vol. 3, Rev. 1, 2012.

[VAL 12] VALENCIA L., "Radioactive waste management in nuclear decommissioning projects", in LARAIA M. (ed.), *Nuclear Decommissioning. Planning, Execution and International Experience*, Woodhead Publishing, Oxford, 2012.

[VAV 18] VAVREK J.R., HENDERSON B.S., DANAGOULIAN A., "Experimental demonstration of an isotope-sensitive warhead verification technique using nuclear resonance fluorescence", *Proceedings of the National Academy of Sciences of the USA*, vol. 115, no. 17, pp. 4363–4368, 2018.

[VEO 20] VEOLIA, Veolia franchit un nouveau cap avec le DoE, available at: https://www.nuclearsolutions.veolia.com/fr/actualites/veolia-franchit-un-nouveau-cap-avec-le-doe, 2020.

[VON 15] VON HIRSCHHAUSEN C., GERBAULET, C., KEMFERT C. *et al.*, "German nuclear phase-out enters the next stage: Electricity supply remains secure – Major challenges and high costs for dismantling and final waste disposal", *DIW Economic Bulletin*, vol. 5, nos 22/23, pp. 293–301, 2015.

[VON 18] VON HIPPEL F., BOUKHARINE O., "US-Russian cooperation on fissile material security and disposition", in FORADORI P., GIACOMELLO G., PASCOLINI A. (eds), *Arms Control and Disarmament. 50 Years of Experience in Nuclear Education*, Palgrave Macmillan, Cham, 2018.

[WAH 80] WAHLEN R.K., "Hanford production reactor decommissioning", in OSTERHOUT M.M. (ed.), *Decontamination and Decommissioning of Nuclear Facilities*, Plenum Press, New York, 1980.

[WAL 17] WALKER, S., Russia begins cleaning up the Soviets' top-secret nuclear waste dump, available at: https://www.theguardian.com/environment/2017/jul/02/russia-begins-cleaning-up-the-soviets-top-secret-nuclear-waste-dump, 2017.

[WEI 11] WEISS L., The 1979 South Atlantic Flash: The case for an Israeli nuclear test, available at: http://www.npolicy.org/article_file/The_1979_South_Atlantic_Flash__The_Case_for_an_Israeli_Nuclear_Test.pdf, 2011.

[WIL 18] WILLIAMS J., Your nuclear security entreprise, available at: https://www.osti.gov/servlets/purl/1594262, October 19, 2018.

[WNN 07] WNN, "US nuclear weapons dismantling speeds up", *World Nuclear News*, October 3, 2007.

[WNN 08a] WNN, "Russia shuts second plutonium-producing reactor at Seversk", *World Nuclear News*, June 6, 2008.

[WNN 08b] WNN, "Dismantling of unfinished Ignalina unit completed", *World Nuclear News*, September 16, 2008.

[WNN 11a] WNN, "Contract to manage Idaho mixed waste", *World Nuclear News*, June 6, 2011.

[WNN 11b] WNN, "Plutonium exposure incident at Idaho site", *World Nuclear News*, November 9, 2011.

[WNN 13] WNN, "Dismantling of Chernobyl stack begins", *World Nuclear News*, November 5, 2013.

[WNN 14a] WNN, "Tenders for Russian submarine fuel removal", *World Nuclear News*, February 7, 2014.

[WNN 14b] WNN, "Dismantling of Beloyarsk units begins", *World Nuclear News*, May 13, 2014.

[WNN 15a] WNN, "USA's experimental breeder reactor-II now permanently entombed", *World Nuclear News*, July 1, 2015.

[WNN 15b] WNN, "Swedish dismantling work for Westinghouse", *World Nuclear News*, November 2, 2015.

[WNN 16a] WNN, "EDF modifies dismantling plans for first generation units", *World Nuclear News*, June 17, 2016.

[WNN 16b] WNN, "Ignalina turbine hall dismantling completed", *World Nuclear News*, August 22, 2016.

[WNN 16c] WNN, "Dismantling work to start at Hamaoka units", *World Nuclear News*, February 3, 2016.

[WNN 17a] WNN, "EnBW all set to decommission Neckarwestheim 1", *World Nuclear News*, February 6, 2017.

[WNN 17b] WNN, "EnBW clear to decommission Philippsburg 1", *World Nuclear News*, April 11, 2017.

[WNN 17c] WNN, "Russia begins fuel unloading of Northern Fleet", *World Nuclear News*, May 18, 2017.

[WNN 17d] WNN, "First used fuel shipment leaves Andreeva Bay", *World Nuclear News*, June 27, 2017.

[WNN 17e] WNN, "Westinghouse signs Bohunice V1 dismantling contract", *World Nuclear News*, September 28, 2017.

[WNN 18a] WNN, "GNS-led consortium wins German decommissioning contract", *World Nuclear News*, January 23, 2018.

[WNN 18b] WNN, "Unterweser gets decommissioning approval", *World Nuclear News*, February 7, 2018.

[WNN 18c] WNN, "Contract awarded for Vandellós 1 engineering support", *World Nuclear News*, February 22, 2018.

[WNN 18d] WNN, "Russia clear to remove used fuel from Lepse cargo ship", *World Nuclear News*, October 16, 2018.

[WNN 18e] WNN, "Decommissioning milestone for space fuel plant", *World Nuclear News*, November 1, 2018.

[WNN 18f] WNN, "Orano prepares Fugen ATR fuel for reprocessing", *World Nuclear News*, November 16, 2018.

[WNN 18g] WNN, "First Taiwanese reactor to enter decommissioning", *World Nuclear News*, December 5, 2018.

[WNN 19a] WNN, "Final oak ridge gaseous diffusion facilities demolished", *World Nuclear News*, May 23, 2019.

[WNN 19b] WNN, "Contract for early Chernobyl dismantling work signed", *World Nuclear News*, August 5, 2019.

[WNN 19c] WNN, "Viewpoint: The positive role of D&D", *World Nuclear News*, September 30, 2019.

[WNN 19d] WNN, "Sogin takes on research reactor for decommissioning", *World Nuclear News*, October 1, 2019.

[WNN 19e] WNN, "IAEA showcases environmental remediation success", *World Nuclear News*, October 1, 2019.

[WNN 19f] WNN, "New techniques used in US legacy reactor decommissioning", *World Nuclear News*, October 4, 2019.

[WNN 19g] WNN, "Orano and Tepco cooperate further on decommissioning", *World Nuclear News*, October 4, 2019.

[WNN 19h] WNN, "Centrus completes oak ridge D&D contract", *World Nuclear News*, October 10, 2019.

[WNN 19i] WNN, "Cyclife to process Swedish decommissioning waste metal", *World Nuclear News*, October 14, 2019.

[WNN 19j] WNN, "EnergySolutions adds TMI-2 to decommissioning projects", *World Nuclear News*, October 16, 2019.

[WNN 19k] WNN, "Demolition of Oak ridge centrifuge facility begins", *World Nuclear News*, October 31, 2019.

[WNN 19l] WNN, "Decommissioning nuclear facilities", *World Nuclear News*, October 31, 2019.

[WNN 19m] WNN, "Dismantling of Zorita containment dome begins", *World Nuclear News*, November 1, 2019.

[WNN 19n] WNN, "EBRD announces breakthrough in clean-up of Andreeva Bay", 2 *World Nuclear News*, November 7, 2019.

[WNN 19o] WNN, "China clear to decommission its first research reactor", *World Nuclear News*, November 20, 2019.

[WNN 19p] WNN, "GEH contracted for pilgrim internals dismantling", *World Nuclear News*, December 4, 2019.

[WNN 19q] WNN, EDF, "Veolia team up to decommission graphite reactors", *World Nuclear News*, December 10, 2019.

[WNN 19r] WNN, "Dismantling of Oskarshamn reactor internals completed", *World Nuclear News*, December 19, 2019.

[WNN 19s] WNN, "Year-end brings trio of reactor retirements", *World Nuclear News*, December 20, 2019.

[WNN 19t] WNN, "Revisions to Fukushima Daiichi decommissioning roadmap", *World Nuclear News*, December 30, 2019.

[WNN 19u] WNN, "Ringhals 2 enters retirement", *World Nuclear News*, December 31, 2019.

[WNN 19v] WNN, "Review: Planning ahead for decommissioning", February 22, 2019.

[WNN 19w] WNN, "Accelerated decommissioning for Crystal River", *World Nuclear News*, May 31, 2019.

[WNN 19x] WNN, "Regulator approves decommissioning plan for Ohi 1 and 2", *World Nuclear News*, December 11, 2019.

[WNN 20a] WNN, "German reactor enters decommissioning", *World Nuclear News*, January 2, 2020.

[WNN 20b] WNN, "Reactor shutdowns outweigh start-ups in 2019", *World Nuclear News*, January 3, 2020.

[YAN 06] YANG Z., Research reactors decommissioning in China, available at: https://nucleus.iaea.org/sites/connect/IDNpublic/R2D2/Workshop%2006/china.pdf, 2006.

[YAN 15] YAN J., GLASER A., "Nuclear warhead verification: A review of attribute and template systems", *Science & Global Security*, vol. 23, pp. 157–170, 2015.

[YTO 16] YTOURNEL B., MACPHERSON I., CLEMENT G. *et al.*, "Applicability of learning from experience to Sellafield post-operation clean out and decommissioning programmes", *PREDEC 2016: Applicability of Learning from Experience to Sellafield Post-operation Clean Out and Decommissioning Programmes*, Lyon, France, February 16–18, 2016.

[YU 01] YU C., ZIELEN A.J., CHENG J.J. *et al.*, User's manual for RESRAD version 6, Manual, ANL/EAD-4, 2001.

[YU 03] YU C., LE POIRE D.J., CHENG J.J. *et al.*, User's manual for RESRAD-BUILD version 3, Manual, ANL/EAD/03-1, 2003.

[YU 17] YU J.H., SOHN W., "Development of historical site assessment (HSA) procedure for decommissioning nuclear power plants", *Transactions of the Korean Nuclear Society*, Autumn meeting, Gyeongju, South Korea, October 26–27, 2017.

[YUN 16] YUN T., JUNG H., KIM Y., "Review of impact factors on decommissioning strategies", *Transactions of the Korean Nuclear Society*, October meeting, Gyeongju, South Korea, October 26–28, 2016.

[ZER 94] ZERBIB J.C., "Accidents sur irradiateurs : les retours d'expériences", *Annales de l'Association Belge de Radioprotection*, vol. 19, no. 4, pp. 645–663, 1994.

[ZER 14] ZERBIB J.C., Retrait des combustibles de la piscine du réacteur no. 4 de Fukushima, available at: http://www.acro.eu.org/retrait_combustibles_piscine4.pdf, March 24, 2014.

[ZER 18] ZERBIB J.C., Usines de retraitement des combustibles irradiés de La Hague et leurs problèmes, Report to the Commission d'enquête of the Assemblée nationale on nuclear safety and security, May 31, 2018.

Index

O, P, R

S, T

U, V, W

Other titles from

ISTE

in

Ecological Science

2021

JOLY Fernand, BOURRIÉ Guilhem
Mankind and Deserts 2: Water and Salts
Mankind and Deserts 3: Wind in Deserts and Civilizations

2020

JOLY Fernand, BOURRIÉ Guilhem
Mankind and Deserts 1: Deserts, Aridity, Exploration and Conquests

BRUSLÉ Jacques, QUIGNARD Jean-Pierre
Fish Behavior 1: Eco-ethology
Fish Behavior 2: Ethophysiology

LE FLOCH Stéphane
Remote Detection and Maritime Pollution: Chemical Spill Studies

MIGON Christophe, NIVAL Paul, SCIANDRA Antoine
The Mediterranean Sea in the Era of Global Change 1: 30 Years of Multidisciplinary Study of the Ligurian Sea
The Mediterranean Sea in the Era of Global Change 2: 30 Years of Multidisciplinary Study of the Ligurian Sea

ROSSIGNOL Jean-Yves
Climatic Impact of Activities: Methodological Guide for Analysis and Action

2019

AMIARD Jean-Claude
Industrial and Medical Nuclear Accidents: Environmental, Ecological, Health and Socio-economic Consequences
(Radioactive Risk SET – Volume 2)
Nuclear Accidents: Prevention and Management of an Accidental Crisis
(Radioactive Risk SET – Volume 3)

BOULEAU Gabrielle
Politicization of Ecological Issues: From Environmental Forms to Environmental Motives

DAVID Valérie
Statistics in Environmental Sciences

GIRAULT Yves
UNESCO Global Geoparks: Tension Between Territorial Development and Heritage Enhancement

KARA Mohamed Hichem, QUIGNARD Jean-Pierre
Fishes in Lagoons and Estuaries in the Mediterranean 2: Sedentary Fish
Fishes in Lagoons and Estuaries in the Mediterranean 3A: Migratory Fish
Fishes in Lagoons and Estuaries in the Mediterranean 3B: Migratory Fish

OUVRARD Benjamin, STENGER Anne
Incentives and Environmental Policies: From Theory to Empirical Novelties

2018

AMIARD Jean-Claude
Military Nuclear Accidents: Environmental, Ecological, Health and Socio-economic Consequences
(Radioactive Risk SET – Volume 1)

FLIPO Fabrice
The Coming Authoritarian Ecology

GUILLOUX Bleuenn
Marine Genetic Resources, R&D and the Law 1: Complex Objects of Use

KARA Mohamed Hichem, QUIGNARD Jean-Pierre
Fishes in Lagoons and Estuaries in the Mediterranean 1: Diversity, Bioecology and Exploitation

2016

BAGNÈRES Anne-Geneviève, HOSSAERT-MCKEY Martine
Chemical Ecology

2014

DE LARMINAT Philippe
Climate Change: Identification and Projections

Printed and bound by CPI Group (UK) Ltd, Croydon, CR0 4YY
15/09/2026
14962363-0002